Modern Methods of Valuation

The new and improved eleventh edition of this essential valuation textbook reflects the changes in the property market since 2009, whilst presenting the tried and tested study of the principles governing the valuation of land, houses and buildings of the previous editions.

The eleventh edition is fully up to date with the latest guidelines, statutes and case law, including the implications of the latest RICS *Red Book* and the Localism Act. Its comprehensive coverage of the legal, economic and technical aspects of valuation make this book a core text for most university and college real-estate programmes, and it provides trainees (APC candidates) and practitioners with current and relevant guidance on the preparation of valuations for statutory purposes.

Over the 28 chapters, the author team of experienced valuation experts presents detailed accounts of the application of these principles to the everyday problems met in practice. This new edition continues to be of excellent value to both students and practitioners alike as it provides the reader with a clear understanding of the methods and techniques of valuation.

Eric Shapiro is Director of Valuations at Chesterton Humberts, Chartered Surveyors. He has had over 45 years' experience in dealing with the management, valuation, sale and letting of residential and commercial property. He is a co-author of *Valuation: Principles into Practice*.

David Mackmin BSc, MSc, FRICS is a chartered surveyor, Emeritus Professor of Real Estate at Sheffield Hallam University, and Visiting Professor in Valuation at the Technische Universität Wien. He is the author and co-author of a number of valuation books, including *Valuation and Sale of Residential Property* and, with Professor Andrew Baum and Nick Nunnington, *The Income Approach to Property Valuation*.

Gary Sams is a consultant surveyor specialising in compulsory purchase and compensation. He is a part-time lecturer at Reading University and visiting lecturer for the College of Estate Management. He is Editor and joint author of *Statutory Valuations*, and Legal Editor for the *Journal of Property Valuation and Investment*.

Modern Methods of Valuation

11th edition

Eric Shapiro, David Mackmin and Gary Sams

Routledge
Taylor & Francis Group

LONDON AND NEW YORK

EG Books

First edition published 1943
by EG Books

This edition published 2013
by Routledge
2 Park Square, Milton Park, Abingdon, Oxon OX14 4RN

Simultaneously published in the USA and Canada
by Routledge
711 Third Avenue, New York, NY 10017

Routledge and EG Books are imprints of the Taylor & Francis Group, an informa business

British Library Cataloguing in Publication Data
A catalogue record for this book is available from the British Library.

Library of Congress Cataloging-in-Publication Data
Shapiro, Eric F.
Modern methods of valuation / Eric Shapiro, David Mackmin and
Gary Sams. – 11th ed.
p. cm.
Includes bibliographical references and index.
Real property–Valuation–Great Britain. 2. Real property–Great
Britain. I. Mackmin, David. II. Sams, Gary, 1951– III. Title.
HD1387.L3 2013
333.33'20941–dc23   2012013456

ISBN: 978-0-415-53801-5 (hbk)
ISBN: 978-0-08-097116-2 (pbk)
ISBN: 978-0-08-097117-9 (ebk)

Typeset in Times
by Cenveo Publisher Services

MIX
Paper from
responsible sources
FSC® C004839

Printed and bound in Great Britain by the MPG Books Group

Contents

Foreword

The preceding edition of this valuable book was published in 2009, only some three years ago. The fact that there is already a demand for a new edition is a measure of the velocity of change, and the consequent increase in the quantity of developments, in the modern world. This is scarcely a revelation, but it is a reminder of the ever-growing need for those who hold themselves out as professional advisers and professional witnesses to keep up-to-date in any field in which they practise. The accelerating speed of change is attributable to many factors, but the root cause is the exponential growth in electronic technology, which has led to an enormous growth in the quality, quantity and penetration of information, ideas and expertise in all areas.

The property world, which is so important to all aspects of modern life – residential, commercial, agricultural and leisure – is especially susceptible to these changes. Accordingly, particularly in a market economy, that means that it is absolutely essential that anyone involved in property valuation is thoroughly up-to-date in the fast-changing fields of the law, the practice and the principles relating to property valuation.

But the past as well as the present has much to contribute when it comes to most subjects, and property valuation is no exception. While it is vital to keep in touch with the most recent advances, most of the fundamental principles of valuation remain unchanged, and so experience is an enormously valuable asset. Eric Shapiro has a very substantial wealth of in-depth expertise acquired over many years as a practical valuer, an expert witness, a lecturer and an author, on which he draws in this book. I therefore warmly welcome and commend this, the latest, edition of *Modern Methods of Valuation*.

Lord Neuberger of Abbotsbury
Master of the Rolls
Royal Courts of Justice, London
June 2012

Preface to the eleventh edition

Although it is only three years since the last edition was published, it was felt necessary to produce a new edition for two reasons. First, we felt the style of the book was becoming somewhat dated and we wanted to bring this more in line with today's students' requirements; second, the legislation and regulatory framework on which much of valuation is based had changed since the previous edition, requiring various deletions and additions.

There has also been a change in authorship, with Keith Davies, to whom we owe much of the law input, deciding to retire from the authorship of this edition having co-authored the seventh to tenth editions. His contribution over many years has been much appreciated. Gary Sams, co-author of our sister publication *Statutory Valuations*, joins the team. Gary specialises in compulsory purchase and compensation.

As with the two previous editions, we have excluded agricultural valuations as we believe this to be too specialist a topic for a general student textbook, and have again excluded the valuation of life interests as the only change since the ninth edition is the change in the mean expectation of life tables, which are regularly updated; the latest tables can be obtained from the Office for National Statistics. For text on this subject see the seventh edition of this book.

In previous editions we have considered whether there is a demand for a change from valuations based on rents being received annually in arrears to quarterly in advance. We considered that whilst students should be aware of this discussion there was no mainstream enthusiasm for a change, and this remains our current view. Those wishing to analyse transactions to establish the equated or equivalent yield achieved from an investment can do so using DCF techniques, which we do consider. Students should also be aware that a head of steam is now building up, especially from retailers, for rents to be paid monthly in advance.

This textbook is designed for the general study of valuations and therefore, for those interested in a more in-depth study or for a study of specialist properties, reference should be made to other textbooks such as *Valuations Principles into Practice*, edited by R.E.H. Hayward, which was designed to follow on from this book.

We would like to thank those mentioned in the acknowledgments to this book who have contributed specialist chapters or made specialist inputs, and also The Right Honourable The Lord Neuberger of Abbotsbury PC MR for generously writing a foreword to this and the previous edition of this book.

Eric Shapiro, David Mackmin, Gary Sams
February 2012

Acknowledgements

The authors sincerely thank all those who have contributed their time and expertise to this, the eleventh edition of *Modern Methods of Valuation*. As with the tenth edition, we are grateful to Patrick Bond BSc (Est. Man.) FRICS Dip Rating IRRV, Deputy Director of Rating at the Valuation Office Agency, for revising Chapter 20 on rating. For this edition we thank Richard E. Smith LLB, Solicitor, Principal Lecturer in Law at Sheffield Hallam University, for revising Chapter 3, 'Property in Land', and Chapter 14, 'Principles of the Law of Town and Country Planning'. As with the tenth edition, we have used Argus Software to illustrate the scope of property development software, and we thank Bee Gan BSc MSc, Senior Lecturer at Sheffield Hallam University, who is Argus Software Certified (ASC), for reworking the examples in Chapter 15; and our thanks go also to John D. Armatys MRICS for his advice on the compensation chapters.

All extracts from the *RICS Valuation – Professional Standards* are from the 2012 edition effective from March 2012 – the 2012 Red Book – which incorporates and is fully compliant with the International Standards; or, where relevant, the seventh edition.

Our thanks to our editor and support staff at Taylor & Francis who have revised the tables of cases, statutes and statutory instruments.

Table of cases

Table of statutes

NIA	Net Internal Area	The area as measured between internal surfaces of external walls but excluding various areas such as toilets, fire corridors, lift shafts, as per the RICS Code of measuring practice
PV	Present Value	
RICS	The Royal Institution of Chartered Surveyors	
SDLT	Stamp Duty Land Tax	
UKVS	UK Valuation Standards	As set out in the RICS Valuation Standards (Red Book)
VS	Valuation Standards	As set out in the RICS Valuation Standards (Red Book)
YP	Years' Purchase	A multiplier used in investment valuation

Principles of valuation

1. The valuer's role

Anyone interested in land and buildings may require the services of a valuer. Valuers may, for example, be instructed to advise:

- owners on the price or rent they should ask for their property;
- buyers on offer prices;
- tenants on the rent they should pay;
- mortgagees on the value of property for loan security; and
- owners dispossessed under compulsory powers on their right to compensation.

In most cases the valuer's role is to estimate the market value (MV) or market rent (MR), that is the capital sum which, at a point in time on specified terms and subject to any legal constraints or restrictions, might be paid for a particular interest in property; or the annual rent at which the property might be let. Market value and market rent are defined in the International Valuation Standards (IVS), effective from 1 January 2012, published by the International Valuation Standards Council (IVSC) and adopted by the Royal Institution of Chartered Surveyors (RICS) in *RICS Valuation – Professional Standards* (the Red Book), the new Global and UK edition, effective from 30 March 2012, and likely to be known as the 2012 Red Book. Extracts in the text are based on the 2012 Red Book. Every effort has been made to ensure that these are correct, but all readers and practitioners must check the current Red Book when using or quoting from the Red Book and need to be aware of proposals for a revised 2013 Red Book.

The services of a professionally qualified valuer with detailed knowledge of the property market is considered to be essential because:

1. the market for property is imperfect;
2. landed property is heterogeneous;
3. the legal interests therein are complex; and
4. the laws relating to landed property are complicated.

The purpose for which a valuation is required is important as the value of property is not necessarily the same for all purposes. Personal details of a client may be necessary as the value of property may vary for different individuals according to, for example, their tax liability.

The special areas of the law relating to landed property are considered in later chapters and the varied interests that can exist in landed property are considered in Chapter 3.

2. The property market

The nature of landed property, the method of conducting transactions in it and the lack of public information relating to transactions all contribute to the imperfections of the market. Apart from possible differences in construction, each piece of landed property is unique by reason of location. The majority of transactions in the property market are conducted privately, and even when the results of the transactions are known they are not helpful in the absence of detailed information on matters such as the size and state of the buildings and the legal title. The degree of imperfection varies across the market and whilst first-class shop investments are relatively homogeneous, because shops in prime locations attract similar tenants, public houses in rural locations will be more heterogeneous due to variations in catchment area and trading potential.

The property market is not a single entity; there are a number of markets, which may be local, national or international. There are sectors and sub-sectors such as residential, agricultural, commercial and industrial; markets for occupation and markets for ownership either for owner occupation or for letting. The market for residential property for occupation is generally local but that for prime property investments is international with buyers, such as insurance companies and pension funds as well as individuals, owning all types of property in all the major cities. The market for development properties can be local, national or international. Small sites for the erection of a few houses will interest small local builders or developers, but large sites for residential or commercial development will be of interest to larger organisations over a wider area and sometimes on an international basis.

3. A definition of value

The RICS, which is the regulatory institution governing the conduct and judgment of chartered surveyors, publishes and regularly revises the Red Book standards. These cover the definitions of value which apply in different circumstances, the process of valuation and the monitoring and regulation of RICS members who prepare valuations. Members carrying out Red Book valuations will also be members of the RICS Valuer Registration Scheme (VRS); it is only in those cases where a valuer provides valuations solely within one or more of the exception areas listed below that registration is not needed. It is mandatory for all members of the RICS to comply with Valuation Standard 1 (VS1), which covers matters relating to conduct and ethics including the key requirements in the RICS Rules of Conduct and/or the IRRV code of Conduct; these cover ethical behaviour, competence and service. If the valuation is required for the purpose of one of the following exceptions (VS1.1), then VS2 to VS6 are not mandatory, but relevant national association standards may apply. The exceptions are when:

- The advice is expressly in preparation for, or during the course of, negotiations or possible litigation.
- The valuer is performing a statutory function or has to comply with prescribed statutory or legal procedures.
- The valuation is provided solely for internal purposes.
- The valuation is provided in connection with certain agency or brokerage work.
- A replacement cost figure is provided for insurance purposes, whether separately or within a valuation report (see VS 1.1.5).

It is not appropriate to comment on all the definitions and procedures here but merely to provide a definition of value which will serve the present purpose; attention will be concentrated on market value. Concepts of social value, aesthetic value or other values are not appropriate to this book but it should be remembered that value can be considered from these points of view.

The market value of a particular interest in landed property is simply the amount of money that can be obtained for it on a particular day from persons able and willing to buy it; it is an estimate of the contract price at that time and under those market conditions. Value is not intrinsic but results from estimates, made subjectively by able and willing buyers, of the benefit or satisfaction to be derived from the ownership of the interest. In order to value an interest in property a valuer must be able to assess the benefits to potential buyers. What is valued is not the physical land or buildings but the legal interest which gives legal rights of use or enjoyment of the land or buildings. The IVS definition of MV adopted by the RICS is:

> The estimated amount for which an asset or liability should exchange on the *valuation date* between a willing buyer and a willing seller in an arm's-length transaction after proper marketing wherein the parties had each acted knowledgeably prudently and without compulsion. (RICS VS 3.2)

VS 3.2 continues with a requirement that:

> In applying *market value*, regard must also be had to the conceptual framework set out in paragraphs 31–35 of the IVS Framework, including the requirement that the valuation amount reflects the actual market state and the circumstances as of the effective valuation date.

The valuation date is "The date on which the opinion of value applies" (RICS VS Glossary).

Buyers are those who propose to tie up capital in land or in land and building; there are three main angles from which they may view the transaction:

1. For occupation having regard to the personal, social or commercial benefits to be derived from that occupation.
2. For long-term investment having regard to the income (annual rent) expressed as a yield (percentage return) on the capital invested.
3. For short-term gain or profit, buying and reselling after improvement or repackaging.

These motives are not mutually exclusive; a transaction may be entered into with more than one motive in mind. The price the buyer is prepared to pay will be influenced by the supply and the demand for that particular type of property at that given point in time. Supply and demand are discussed at length later in this chapter, but it should be noted here that demand must be effective; that is, it must be backed by money so that the desire to own can be translated into the act of buying.

4. Value and valuation

Although the aim of the valuer is to provide an estimate of market value, it should not be assumed that each valuer's opinion will be the same. Different valuers could arrive at different opinions of value because they are making estimates and there is normally room, within limits, for differences of opinion. Under stable market conditions, because market prices result from estimates of value made by buyers and sellers on the basis of prices previously paid for other similar interests, the differences should be relatively small. In times when market conditions are not stable, more serious differences may arise. Here the valuer's estimate of value will be based on prices previously paid, but buyers will adjust their bids to reflect market movements since the previous transaction took place. The correctness of the estimate will, therefore, depend on the valuer's knowledge and understanding of the movements and their skill in quantifying the impact on market prices.

In any market at any time there may be a number of buyers and sellers, all of whom will have their own views, desires and judgments on the value of the commodity in question. A market constitutes an amalgamation of individuals or individual companies who will strike individual bargains, so that a "market price" can only represent an average view of all these factors. Where the commodities are all different, as in the case of property, then the problems of arriving at this average are multiplied. In the case of statutory valuations in hypothetical market conditions these differences of opinion as to value will become even more marked.

Valuation is a process governed in most cases by the requirements of the RICS Standards. The process, provided the valuer is satisfied as to his or her competence to value a specific interest in a specified property for a given purpose (VS 1.6-1.7), will begin with an agreement of the terms of engagement (VS2) and the basis of value (VS3) – market value, market rent, investment value or worth, fair value; this is followed by consideration of requirements for specific applications such as loans and financial statements (VS4); inspections[1] and investigations (VS5), and ends with a written valuation report (VS6).

5. Demand, supply and price

The quantity of property bought will always equal the quantity sold, regardless of price. This does not mean, however, that price is unimportant. Demand and supply are in fact equated at a point in time by some level of price. At any given time, other things being equal, any increase in demand or decrease in supply will cause price to rise; conversely, any decrease in demand or increase in supply will cause price to fall. Therefore, whatever the given demand and supply, in a freely operating market, price will ration the supply and match it to the demand.

Price has, however, a wider role. If, over a long period, price remains persistently high in relation to the cost of producing more of the thing in question, supply will be increased, which will gradually have the effect of pulling down price to a long-run normal or equilibrium level. Price acts as an indicator and as an incentive to increase or decrease supply and demand.

The extent of price changes will depend upon price elasticity, that is the respon-siveness of supply and demand to changes in price. If supply is highly elastic, a change in demand can be matched quickly and smoothly by an expansion of supply and the price change, if any, will be small. Conversely, if supply is highly inelastic, a change in demand cannot be matched quickly and smoothly by an expansion in supply, thus the price change will be more marked. If demand is elastic, a change in supply will quickly generate an appropriate contraction or expansion in demand and the price change, if any, will be small; but if demand is inelastic this means that any change in supply is likely to generate a contraction or expansion in demand less quickly and the price change will be more marked.

This is a simple statement of economic laws which are applicable to all types of commodity or service, but to consider their application to a particular type of commodity or service it is necessary to look at the special characteristics of that commodity or service; a separate discipline of property economics has developed to reflect the unique nature of landed property. Some of these characteristics that affect property elasticity are considered below.

---

1 Valuers carrying out property inspections should adopt appropriate inspection practices as indicated in the RICS Practice Standards guidance note *Surveying safely*. The Red Book and other RICS guidance notes provide property-specific inspection checklists.

6. Demand for and supply of landed property

The term landed property covers such a wide range of types of land and buildings and interests therein that generalisation is difficult, but generally the demand for and the supply of landed property are relatively inelastic with respect to price changes.

An important factor that restricts supply of land and buildings is the need for planning approval for development. There are, of course, also natural limitations on the supply of land; the overall supply is fixed and the supply of land suitable for particular purposes is limited, but these natural limitations are overshadowed by planning limitations. Planning controls must be considered in detail when considering the potential for development and change of use. In this context the following powers are the most important:

1. The power to allocate land for particular uses, for example, agricultural use or residential use. This allocation is shown in the development plans for the area. Thus, it would not normally be possible to erect a factory on a bare site in an area allocated for residential use.
2. The power to restrict changes in the use of buildings. For example, before the use of a house can be changed to offices, planning permission must be obtained and this permission will not usually be forthcoming unless the change is in accordance with the provisions of the planning policy for the area.
3. The power to restrict the intensity of use of land. Thus, the owner of a bare site in an area allocated for residential use is not free to choose between erecting on it, for example, a four-storey or a 15-storey block of flats. Any permission is subject to the further restrictions of building regulations which may negate or limit that which the planners would allow.

Thus, for example, if in a particular area in which all available office space is already taken up, there is an increase in demand for office space, this increased demand might be met by an increased supply in three ways:

1. conversion of existing non-office space to office use;
2. construction of new offices on bare sites; and
3. redevelopment of existing office buildings at a higher density.

In an extreme case, planning permission might be refused for all three. In such a case, in both the short and the long run, the effect of the increase in demand will be to increase the price of the existing office space to the point at which the high price so reduces demand as to equate it once again with the fixed supply.

In a less extreme case, where it would be possible to increase the supply by one or more of the three ways, some, perhaps considerable, time would elapse before the supply could be increased. In the case of conversion of existing buildings, it would be necessary to obtain planning permission, dispossess any tenants and carry out any necessary works of conversion. The time taken would depend on the administrative speed of the planning authority, the security of tenure of the tenants and the extent of the works. With the other methods, planning permission would again have to be obtained, and with the last method, any tenants of the existing buildings would need to be dispossessed and the buildings demolished. The erection of the new buildings could take perhaps one or two years. Thus, in this less extreme case the effect of the increase in demand would, in the short run, be to increase the price of the existing space. In the long run, as the new space becomes available, the price would tend to stabilise or fall. If the increased supply was sufficient to satisfy the whole of the increased demand the price might return to its former level, but if, as is more likely in practice, the increased supply was insufficient, or, if sufficient, was of a higher quality, the new equilibrium price would be higher than the old.

A reduction in the supply of a given type of space is difficult to achieve; the existing stock can only be reduced by demolition or change of use. The loss involved in failing to complete buildings already started will usually be higher than the loss resulting from a fall in price, but their use might be changed during construction.

Although the demand for land and buildings is generally inelastic, the degree of inelasticity will depend to some extent on the purpose for which land and buildings are held and by the economic conditions at the time.

Elasticity of demand for any commodity or service depends on whether the commodity or service is regarded as a necessity and on the existence of satisfactory substitutes. For example, the demand for landed property for residential occupation can generally be regarded as a necessity and no satisfactory substitutes exist. So an increase in the price of living accommodation, within a limited range, would not result in any marked contraction in demand because a particular standard of such accommodation is regarded as a necessity and a caravan or a tent would not, to the majority of people, offer a satisfactory substitute. However, this would not be true if the increase were outside an acceptable range or times were less prosperous, when an increase in price would probably, subject to legal restrictions on subdivision and overcrowding, lead to economies in the use of living space, as occurred during the recession of the early 1990s. The effect of price changes on demand will depend on whether the demand is "need related" or "standards related".

Buyers must be able to translate their desire to own into the act of buying. The ability to buy will usually depend on the availability of loans (mortgages) and on the policies of the lending institutions in respect of the multiples of income and the percentage of capital value (loan to value ratio) on which they will lend. In 2008 the economy was overshadowed by a credit crisis, initially caused by defaulting subprime mortgagees in the USA, which had an international impact as the mortgages were traded world wide. Virtually overnight the lending institutions changed their policies by increasing interest rates where the loan to capital value ratio was high, and they withdrew all lending above a ratio of 80% where before they had been lending up to and in excess of 100%. The effect of this was to push residential prices down and, even though some support has now been offered by the government for first-time buyers, prices and market activity are not likely to recover until mortgage finance becomes more readily available and confidence returns to the market.

The majority of home buyers will normally pay only a small proportion of the price out of their own capital; the remainder is borrowed by way of mortgage from a financial institution such as a building society, an insurance company or a bank. Sometimes this will be done because buyers prefer to invest the remainder of their capital elsewhere but, in most cases, buyers have insufficient capital to pay the whole of the purchase price. However, their income is sufficient to pay interest on the mortgage loan and to repay the capital gradually over a period of years. Therefore, the ability of most buyers of residential property to make their demand effective depends on their ability to borrow money.

Over the years, the financial institutions have experienced extremes in relation to money resources for lending which illustrate well the impact of effective demand on value. In the period 1970–1973, the lending institutions had plentiful sums available for lending, and borrowers were able to obtain mortgage sums for home purchase representing a large proportion of the price and at low rates of interest. This in turn led to a rapid increase in house prices as buyers competed fiercely for the houses available for purchase. At the end of 1973, the situation changed dramatically as lending institutions experienced a rapid fall in the sums they had available to lend.

In the following three years, mortgages became difficult to obtain, certainly for any substantial proportion of purchase price, and then only at high rates of interest, so that house prices actually steadied or even declined before returning to a slow rate of increase. This was particularly marked in the upper price ranges, which suffered most from the reduction in mortgage tax relief and in many cases failed to reach the levels of 1973 until 1976. In 1976 the situation began to ease and the market returned to that seen in the early 1970s.

This rapid increase came to a sudden end following the introduction of high interest rates in 1979, when the market virtually stopped, until the end of 1982, when a combination of low interest rates and aggressive lending policies once again led to a sharp increase in prices which continued, with one or two short pauses for breath, until the end of 1988. During this period prices rose by an average of 125%. This rate of increase was brought to a halt by a combination of a change of government policy in respect of tax deductibility of mortgage interest, changes in government fiscal policies and the economic effects of Iraq's invasion of Kuwait. Multiple income relief for mortgages of up to £30,000 was withdrawn from 1 August 1988 (but announced in the preceding March), after which a combination of high interest rates and increasing unemployment caused a significant fall in demand and an actual fall in prices, so that by the middle of 1993 prices had fallen back to their 1980 levels. Prices continued to fall until 1995 when a recovery began to occur (the market had "bottomed out"), and by the end of 1996 had reached the levels of 1993.

The period from 1996 to 2007 saw almost continuous price rises, due again to low interest rates, an increase in the supply of money and a change in lending criteria. In effect, too much money was chasing too few properties, creating the perfect conditions for price inflation. This was coupled with a surge in demand from non-owner-occupiers, namely buy-to-let investors using borrowed capital, who were desperately seeking higher yielding investments because of the low yields available in the money markets. The credit crunch, which began at the end of 2007, led to a worldwide recession that caused house prices to tumble. Similar affects have occurred in all sectors of the property market. The policy of financial institutions such as building societies and banks has, therefore, an important bearing on the effective demand for landed property.

Legislation often affects, either directly or indirectly, the supply and demand for landed property. The various statutes affecting different types of landed property are considered in later chapters but, as an example, the supply of smaller or poorer types of residential accommodation for letting was directly affected by the Rent Acts which both restricted the rent that owners could charge for their property and limited their powers to obtain possession. The removal of this regulatory system for new lettings in 1989 led to a sharp increase in supply which was matched by a sharp increase in demand due to inward migration from the enlarged EU, as well as from potential owner-occupiers who had been priced out of the market.

The other principal factors affecting the demand for landed property are mainly long term, so that their effects are felt only gradually over a number of years.

Some of the effects of planning in relation to the supply of land and buildings have been noted, but planning may also have important effects on the demand side. The creation of new towns, the extension of existing towns to accommodate overspill from the large conurbations and the redevelopment of central shopping areas are examples of planning schemes which may increase demand in and around the areas concerned, although they may reduce demand in other areas. Thus, a new shopping centre or supermarket may draw demand from adjacent properties and may diminish the demand for those properties.

Changes in the overall size, location and composition of the population will affect the demand for landed property. An overall increase in population, particularly if accompanied by an increase in prosperity, will increase the demand for most types of landed property. The increased population must be housed and its increased demands for necessities and luxuries will have to be met through the medium of shops, factories, offices, hospitals, schools and playing fields. The movement of population from one part of the country to another will have the dual effect of increasing demand in the reception area and reducing it in the area of origin. Changes in the composition of the population also have important effects. For example, a reduction in family size will decrease demand for larger homes, whilst an increase in the number of retired people may increase the demand for accommodation in the favoured retirement areas of the south coast.

Improvements in transport have encouraged people who work in towns to commute, sometimes over long distances, but as the number of commuters increases and transport facilities have become overburdened the process is being reversed, particularly where coupled with increased fuel and fare costs and overcrowded services. The older town centres have again become desirable places in which to live and there has been significant redevelopment of redundant offices, factories and warehouses, particularly in waterfront locations.

Technological developments have their impact on supply and demand. For example, changes in freight handling can render low-height and multi-storey warehouses obsolete. Similarly, offices built in the early post-war years may be unable to accommodate today's electronic requirements. Most recently, the biggest factor affecting demand for all types of property has been the impact of the recession. Currently the issue is that of sovereign debt in Europe and its impact on the euro. Such global factors have to be considered as well as domestic factors such as the possible revisions to the UK planning system. Markets react and, whilst the valuer's role is to interpret the actual impact on property prices, the valuer needs to be aware of the forces at work on a macro level.

7. Landed property as an investment

Every purchase of property can be regarded as an investment, whether it is to secure an income in the form of rent to the owner or for occupation, the benefit in the latter case being the annual value of the occupation. A business may make a conscious choice between investing capital in buying a property for business occupation or renting similar property owned by someone else so that the capital can be invested elsewhere. A further typical case is where a business has invested capital to buy business premises for occupation but now wishes to realise the capital tied up in the property in order to invest it elsewhere in the business. In this case it might enter into a lease-back or a mortgage transaction. In a lease-back it will sell its interest in the property and take the property back on lease, paying an annual rent. In the mortgage transaction it obtains a loan on the security of its title to the property, paying annual interest on the loan.

The MV of a property as an investment will depend upon supply and demand; the price being dependent on the property's net income or rent and on the yield that typical investors would expect to achieve given the general and specific risks associated with owning that property.

Chapter 2

Methods of valuation (valuation approaches)

1. Introduction

Valuations are required of different interests in different types of property for different purposes. Given a range of needs, the approach to the estimation of value in one case may well be inappropriate in another and so, over time, separate approaches or methods of valuation have developed. This book examines each of these methods and explains their application in practice. This chapter looks at the methods in broad terms, providing a framework for later chapters.

2. Comparison

Underlying each method is the need to make comparisons since this is the essential ingredient in arriving at a market view. In any market, sellers compete with other sellers to attract the buyer to them and the buyer is thus faced with a choice. In reaching their decision buyers compare what is available and at what price and buy the good or service which, in their opinion, gives the best return for the price paid; they seek value for money. This will be true in all but the most monopolistic or monopsonistic markets.

The valuer, in arriving at an opinion of value, must try and judge the prices sellers generally seek and obtain, and the choices buyers make. The valuer must therefore assess what is, or has recently been, available in the market and make comparisons between them. In this respect, the valuer acts no differently from any other person making a valuation; for example, a person valuing a car would assess the number and types of cars available and the prices currently being obtained and so arrive at the price which, in their opinion, would attract a buyer for the car in question. The key to valuation accuracy is knowledge of the prices that have been obtained recently for similar goods with which comparison can be made. The availability and the nature of these 'comparables' provides the basis of all the methods of valuation, and the availability and nature of those comparables may determine the choice of the method itself. The importance of comparables is recognised by the RICS in their code of practice (2011) *Comparable Evidence*.

3. Principal methods of valuation

(a) The market approach or comparative method

The most direct valuation approach is to compare the object to be valued with the prices obtained for other similar objects in the same market at the same point in time. The method works best if the comparable objects are identical. For example, if ordinary shares in a company are selling at 150p each, then the value of further shares is likely to be around 150p. Special factors may alter this view; if a large block of shares is on offer, then a buyer might be prepared to offer more than 150p per share. Thus, even in the case of identical goods, some judgment is required in arriving at a value. A property can never be absolutely identical to any other property and so the use of this method is limited. The application of the method is shown in Chapter 4.

(b) The income approach or investment method

A large part of the property market comprises properties where ownership and occupation are separate. The purpose of property is to provide an appropriate environment for different activities: houses for living in, factories for making goods, shops for selling goods, and so on. However, in many cases property is occupied under a contract, with the occupier paying the owner rent in return for the right to occupy; the owner surrenders the occupation rights for rent. The property market is a major source of opportunities for investors looking for a return on their capital.

The valuer is often asked to value an interest in property where the value is clearly dependent on the amount of rent that an occupier would pay for the right to occupy and on the level of return an investor would require on their capital. There are several elements in the valuation, each requiring a different comparable. The basic principle is that investors invest capital to obtain an annual return, in the form of a net income, which represents an acceptable rate of return to compensate for the investment risks involved.

For example, what net income will be required if A wishes to invest £1,000,000 in shop premises and requires a return of 6% on the capital to be invested?

$$6\% \text{ of } £1,000,000 = £1,000,000 \times \frac{6}{100} = £60,000 \text{ net income}$$

For a valuation the valuer will generally assess the property's net income, based on comparable lettings of similar properties, and by using the level of return investors require from similar properties he calculates the value. For example, if shop premises produce a net income of £8,000 p.a. and if the appropriate rate of return is 8%, then the capital sum (value) can be found as follows:

Let the capital sum be C, then:

$$C \times 8\% = C \times \frac{8}{100} = £8,000$$

$$C = £8,000 \times \frac{100}{8}$$

$$C = £8,000 \times 12.5$$

$$C = £100,000$$

Or simply £8,000 ÷ 0.08 = £100,000.

The capital sum or value is found by multiplying the net income by 100 ÷ 8 (the net income is "capitalised"). This reciprocal figure is termed a year's purchase or YP, a traditional expression used for many years by valuers in the UK. Nominally it represents the number of years needed to get back the capital invested, 12.5 years in this case, but this is a misleading concept as it merely expresses the ratio of income to capital and is the same as the present value of £1 per annum in perpetuity (see Chapter 6). The YP at any rate or yield is easy to calculate:

$$\text{YP at } 5\% = \frac{100}{5} = 20\text{YP}$$

$$\text{YP at } 7\% = \frac{100}{7} = 14.286\text{YP}$$

$$\text{YP at } 10\% = \frac{100}{10} = 10\text{YP}$$

The lower the rate of interest, the higher the YP. The assessment of net income is explained in Chapters 5 and 6, the determination of YP and the capitalising of incomes in Chapters 7, 8 and 9.

(c) The residual approach or development method

Property changes over time, bare land is developed with buildings, they become obsolete and are refurbished or redeveloped; this is the life cycle of buildings and the changes are required to meet the changing needs of society. Thus, the valuer often needs to give a valuation of land or buildings which are to be developed or redeveloped. The valuer may be able to arrive at a value by direct comparison with the sale of similar property which is to be developed in a similar manner. This might be done by analysing the comparables at so much per unit of proposed development. However, this may be impracticable due to the unique nature of the property in question and the development proposed.

For example, it may be intended to develop a site with new distribution depots in an area where no such buildings have previously been constructed. It may be possible to predict the rent from similar properties in similar locations, and thus the capital sum which an investor would pay for the completed buildings, using the investment approach. Further, it is possible to assess the cost of building the properties using comparable construction costs, and the profit that someone would require to carry out the development. Given the value of the finished product and the cost of producing it, including profit, then any difference must represent the sum which can be paid for the land. For example, if the value of the finished depots is £80,000,000, the cost of producing them is £50,000,000 and the required profit is £7,000,000, then someone can afford to spend £23,000,000 for the land (£80,000,000 − £50,000,000 − £7,000,000 = £23,000,000). This sum is the residue available – hence the residual method. This is a simple illustration of a method that has developed into a complex calculation which is explained further in Chapter 11.

(d) The profits approach

Many properties depend for their value on various factors which combine to produce a potential level of business profit. In some instances, the factors are so unique that comparison with other similar properties is impossible and the value must therefore be determined by looking at the actual level of business achieved in the property, or achievable from it.

A typical example of this is a petrol filling station. Such premises are relatively similar in design and tend to be located in similar positions on busy roads. However, a comparison of one with another is difficult since each site is susceptible to unique factors which may have a dramatic effect on the sales achieved. Two stations may enjoy almost identical qualities, they could be on either side of the same road, but one may sell considerably more fuel than the other. The level of sales clearly determines the level of potential profits, and the potential profits determine the price someone will pay for the property.

It follows that the value of some properties can be determined from knowledge of their gross profits; for example, if a property produces a profit of £50,000 a year and a buyer will invest capital at 6 times the level of profit, the value of the property is £300,000 (£50,000 × 6). Alternatively, if an occupier is prepared to pay a rent of £20,000 a year in order to earn the £50,000 and a buyer will invest capital at 15 times the rent, then again the value is £300,000 (£20,000 × 15). Even here there is a need to obtain comparable evidence, in this instance to assess the ratio of profits or rent to capital. The profits method is explained in more detail in Chapter 20, where it will be seen that knowledge of the type of business, an ability to interpret accounts and to analyse profits are needed to arrive at the value.

(e) The cost approach or contractor's method

There are some properties that are designed and used for a special purpose to meet specific requirements such as churches, town halls, schools, police stations and other properties which perform non-profitable community functions and are not normally bought and sold.

In nearly all cases, such properties are built by the authority or organisation responsible for the provision of the special service or use, and commonly there is no alternative body that requires the property. In such cases there are no sales in the market and thus no comparables on which to base a valuation. Indeed, such properties are rarely sold and, when they are, they generally need to be replaced by alternative premises which have to be newly built. As a result, the minimum price required by a body owning such properties, if they are being forced to sell and are not willing sellers, is the cost of providing equivalent alternative accommodation. The price for the site will be based on the value of comparable sites, whilst the cost of the building is based on current building costs which are then depreciated to reflect age, condition and aspects of obsolescence.

Thus, the valuation of such properties is derived from the value of alternative sites, plus the cost of the building. This approach is sometimes called the contractor's approach or contractor's test (see also depreciated replacement cost (drc) in Chapter 19). If there is a market for a property then, in the UK, a market-based approach will be used to estimate MV. The cost approach is restricted to special cases as described in Chapters 19, 20 and 26.

(f) Conclusion

It may be possible to approach a valuation by adopting more than one method so as to check one against the other. It is clear that the accuracy of any method depends to a great extent on the comparable evidence that can be obtained by valuers from their detailed knowledge of the market. In the following chapters the examples will make assumptions as to the nature of the evidence; any such assumptions are solely for illustrative purposes and are not intended to indicate either the levels for any specified factor, or the relationship between such factors in the current market. The following definitions of the main valuation approaches are set out in the glossary to the RICS Standards (2012 Red Book).

Cost approach. An approach that provides an indication of value using the economic principle that a buyer will pay no more for an asset than the cost to obtain an asset of equal utility, whether by purchase or construction.

Income approach. An approach that provides an indication of value by converting future cash flows to a single current capital value.

Market approach. An approach that provides an indication of value by comparing the subject asset with identical or similar assets for which price information is available.

Chapter 3

Property in land

I. Land law

The word "property" means the right that a person has in land or goods. So, in valuing land, the valuer is estimating the value of the rights (the property) that a person has in that land. Obviously a valuer has to understand the scope of these rights in order to carry out a competent valuation. If, for example, the valuer is valuing a freehold, it is necessary to understand what legal rights are embodied in that freehold as well as any legal restrictions on those rights. Such rights and restrictions are the subject of land law. Land in this context includes buildings and parts of buildings – section 205(1)(ix), Law of Property Act 1925.

Land law principles are derived from the common law – the law evolved by the judges in thousands of cases over the centuries. This law was modernised, codified and simplified by the reforming legislation of 1925. This consisted of seven Acts of Parliament on various aspects of land law and includes the Law of Property Act 1925 which, as amended, sets out the basic rules governing the ownership, creation and transfer of estates and interests in land.

2. Property rights in general

The two principal rights or interests in land with which the valuer is concerned are known respectively as the freehold and leasehold.

To the valuer the term "freehold" implies a property which the owners hold absolutely (i.e. unconditionally) and in perpetuity, and of which they are either in physical possession or in receipt of rents from leases which have been created out of the freehold interest. The technical legal term for this is the "fee simple absolute in possession". Section 1 of the Law of Property Act 1925 states that this and the "term of years absolute" are the only forms of ownership capable of being a legal estate in land and thus existing at *common law* (see below). "Term of years absolute" is the technical term for the leasehold estate. In practice, the words freehold and leasehold are used instead of the somewhat archaic phrases in the Law of Property Act. However, it should be observed that in law the freehold also includes lesser estates in land than the fee simple (see "Successive interests" below).

Other interests are merely "equitable" (see below). A "legal estate" is "good against the whole world", but an equitable interest is enforceable against some persons and not others. A "fee simple absolute in possession" exists in all land in England and Wales; whereas leaseholds and equitable interests, where they exist, are superimposed on (or "carved out" of) such ownership, and subsequently terminated, depending on circumstances.

In the following sections it will be convenient to consider first the position of freeholders who are in physical possession of their property, and then note the nature and incidence of the various types of leasehold interest which may be created out of the freehold. Then other interests in land will be examined, and finally it will be explained how successive interests in property – such as "entailed interests", "life interests" and "reversions and remainders", which are "equitable interests" only – may sometimes arise under wills or settlements.

At common law, physical ownership of "land" extends indefinitely above and below the surface of the land, within reason, and thus in a normal case includes both air space and subsurface strata of minerals, but not chattels belonging to some other identifiable owner. Boundaries should be unambiguously specified. Ancient hidden gold and silver artefacts belong to the Crown as "treasure trove"; but the Treasure Act 1996 has modernised the rules and the finder and landowner may be granted a reward. The Mines (Working Facilities and Support) Act 1966 strictly regulates all mining operations (not least under another owner's land). Under the Petroleum Act 1998, the Government grants "licences to search and bore for, and get, petroleum", which *in its natural state* is vested in the Crown. The landowner is entitled to compensation for the *access* to his land for the exploitation of the petroleum. This is to be assessed "on the basis of what would be fair and reasonable between a willing grantor and a willing grantee", with an additional 10% for compulsory acquisition. See *BP Petroleum Developments* v *Ryder*.[1]

3. Freeholds

Under the Crown, legal freeholders are inherently the absolute owners of land, and are sometimes said to be able to do what they like with that land, subject only to the general law (notably planning control) and to the lawful rights of others. Legal freeholders may thus in a normal case develop land, transfer it, or create lesser interests in it, without the consent of any other private person. Their ownership of the property is perpetual, even though they are mortal, and so passes to their heirs on death. (If the deceased made no will and has no dependants or relatives entitled to inherit under the intestacy rules, the property reverts to the Crown.)

1 [1987] 2 EGLR 233.

Buildings may become worn out, the use and character of the property may change; but the land remains the permanent property of the freeholders and their successors, whether the latter obtain it from them by gift, sale or disposition on death.

Legal freeholders, as absolute owners of their property, usually hold it without payment in the nature of rent, though in strict legal theory they are always nominally the tenants of the Crown. But in some parts of the country freeholds will be found to be subject to annual payments known as "rentcharges", "fee farm rents" or "chief rents". These usually arise through a vendor of freehold property agreeing to accept an annual sum of money in perpetuity, or for a limited term, in lieu of the whole or part of the purchase price. Such a rentcharge must be distinguished from a normal rent (or "rent service"), which is an incident of a lease. The rentcharge owner has the right, in the event of non-payment, to enter and take the rents and profits of land until arrears are satisfied. The RentCharges Act 1977 prohibits the creation of new rentcharges (they are void) and makes provision for the gradual extinguishment of existing ones by 2037.

A rentcharge or fee farm rent will be treated as an outgoing when arriving at the net income of the freehold for valuation purposes. As for the interest of the rentcharge owner (which is technically a form of legal interest in the land), the rentcharge is usually small in amount compared with the net rental of the buildings and land on which it is secured, and it depreciates in real terms as does any fixed income in periods of inflation.

During a state of national emergency, statute may give the Crown drastic rights of user and possession over the property of its subjects. Even in normal times there are numerous Acts whereby owners can be forced to sell their land to government departments, local authorities or other statutory bodies, and even some private companies, for various purposes. Such compulsory purchases are to be paid for at market value, but this is determined statutorily under the Land Compensation Acts 1961 and 1973.

To develop land, owners must obtain planning permission under the Town and Country Planning Act 1990 (as amended). Legislation governing environmental health, housing and other uses of land, including Local Acts and Building Regulations, restricts the erection or reconstruction of buildings or compels the owners to keep them in a state of good repair and sanitation. Environmental protection legislation governs the emission of pollutants and requires the remediation of contaminated land. Other legislation gives various leaseholders of residential property, including public sector tenants, security of tenure unless the landlord can establish certain grounds for possession. The leasehold reform legislation gives the right to tenants of long residential leases to require their landlords to sell them the freehold reversion or to grant lease extensions.

The owners' freedom to do as they like with their property is also limited by the fact that they must not interfere with the natural rights of others as, for instance, by depriving their neighbours' land of support or polluting or diminishing the flow of a stream, or causing a nuisance. Often the owner's enjoyment is lessened by the fact that some other owner has acquired an easement over their property or is entitled to the benefit of covenants restricting the use of it; these rights are discussed below.

(a) Registered and unregistered land

The title to unregistered land depends on the *title deeds*, a chain of transactions embodied in deeds and other documents leading up to the present owner. This necessitates the storing of old deeds and renders conveyancing a laborious and expensive process. By contrast, if land is registered with the Land Registry details are kept on an easy-to-read electronic database which establishes proof of ownership protected by a state guarantee of the title. Now that the Register is public, anybody can examine the title to a piece of land by searching on the Land Registry website and paying a small fee.

As a "legal estate", a freehold can normally be transferred (by a living owner) only on the execution of a deed, that is to say a document expressly stating that it is "a deed" which is signed and witnessed (Law of Property Act 1925, section 52, and Law of Property (Miscellaneous Provisions) Act 1989, section 1). Under the Land Registration Act 2002, legal freeholds and leaseholds for over seven years must, when transferred or created, be entered in the Land Registry within two months. The register of title consists of a property register, a proprietorship register and a charges register. The property register identifies the property with an Ordnance Survey-based title plan. The proprietorship register identifies the proprietor, the class of title and various other matters including notices and restrictions. The charges register identifies leases, charges (mortgages) and other adverse interests.

Before registration of title came into existence, all equitable interests (see common law equity and Trusts, below) depended for their protection on "notice" in some form (and some still do). In other words, if a purchaser acquired title in good faith without notice of an equitable interest, his ownership was not subject to that interest and it could safely be ignored. This rendered equitable interests somewhat insecure. Now equitable interests must be registered for protection in the Land Register in the case of registered land or in the Land Charges Department (Land Charges Act 1972) for unregistered land.

In registered land there are some interests which, although not registered, are binding on purchasers. Such interests are known as *"overriding interests"* and include persons in actual occupation, unregistrable leases and legal easements and profits (see section 9). The buyer's surveyor will, of course, inspect the property and, in doing so, ascertain whether there are any occupiers and whether there is evidence of easements such as obvious tracks and rights to light. The buyer's solicitor will make preliminary enquiries about any overriding interests and require any occupiers to sign the contract and agree to give vacant possession. (In the case of unregistered land a buyer is deemed to have constructive notice of all the interests of persons in actual occupation if they could be discovered from reasonable inquiries.)

(b) Adverse possession

The common law remedy for an owner wrongly excluded from possession (trespass) is an order of possession obtainable from the court. Damages for financial loss are also obtainable against the trespasser. But if owners are excluded from possession of registered land by trespassers who treat the land as their own (adverse possession), then after not less than 10 years the trespassers may require the Land Registry to grant them a registered title, provided that they have first notified the owners who have then failed to object (Land Registration Act 2002, Schedule 6). (Where the owner does object, the squatter may only be registered as the proprietor if certain conditions set out in the Schedule are met.) This makes the acquisition of title to registered land by adverse possession unlikely, so long as the rightful owner keeps the Registry aware of any change of his address for notification. In unregistered land, however, no notification is required and 12 years' adverse possession may result in the owner's interest being extinguished. This is of concern to large landowners, such as local authorities and universities, whose land is not registered, and provides a strong incentive to register title.

4. Leaseholds

It commonly happens that freeholders, instead of occupying the property themselves, grant or "let" the exclusive possession of the premises to another for a certain period, usually in consideration of the payment of rent, under a lease or tenancy. Grantors of leases are called lessors or landlords and their interest while the lease lasts is the reversion. Grantees are called lessees, leaseholders or tenants. A lease is normally an estate in land, and a contract.

A leasehold estate is in strict legal parlance a "term of years". This expression covers leaseholds of all kinds, no matter how long or short the period for which they are "demised" (i.e. granted) and even if they are periodic. A periodic tenancy – e.g. yearly, quarterly, monthly or weekly – automatically renews itself until terminated by notice to quit from either side, whereas a fixed term tenancy ends on the expiry of the term. In some circumstances, a lease can be cut short by forfeiture (for breach of covenant by the tenant) or by the exercise of a right to break inserted in the lease. Fixed term tenancies may be of any length and a "building lease" of 99 or even 999 years is often met with; occupation leases for terms such as 5, 10, 15, 25 or 50 years are also commonly found in practice.

Like a freehold conveyance, a lease or tenancy taking effect as a legal estate must be granted by deed; but the Law of Property Act 1925, section 54, provides that no formalities whatever are needed for the grant of a tenancy at a full market rent for not more than three years, taking effect immediately. So parties who believe that such an informal lease is unenforceable may be caught out. See, for example, *Hutchison* v *B & DF*.[2]

---

2 [2008] EWHC 2286 (Ch).

Grants of tenancies (but not the obligations contained in them) may be back-dated; or they may be made to begin at a future date but not more than 21 years after the date of the grant (these are known as "reversionary leases"). Leases of more than seven years or leases which take effect in possession more than three months in the future must be registered at the Land Registry.

A lease can be transferred – "assigned" – by gift or sale, or on death, just as a freehold can, although this right can be restricted by the lease. If an assignment takes place in breach of the lease, the assignment as between the assignor and assignee is valid, but is subject to the lessor's right to forfeit the lease for breach of covenant.

A lease can also be granted out of a lease if the lease allows. Such a lease is called a sublease or underlease and must be at least one day shorter than the head lease, otherwise it takes effect as an assignment. The head lessee is the landlord of the sublessee. In theory there can be a multitude of subleases, sub-underleases and sub-sub-underleases carved out of a lease if it is long enough. The term "head rent" is used to distinguish the rent paid by the head tenant to the freeholder from any rent paid in respect of the same property by a subtenant to the head tenant (subrent). Many commercial leases prohibit the head tenant from subletting at less than the head rent.

Freeholders who grant leases or tenancies retain the right at common law to regain physical possession of the property when these leases or tenancies come to an end. For this reason the landlord's interest in the land is known as his "reversion". A leaseholder who has subleased has a leasehold reversion because his right to possession reverts to him at the end of the sublease.

The common law right of the reversioner to regain physical possession of the property at the end of the lease (including, as a normal common law rule, any buildings or other "fixtures" added to the land by the tenant which are not removable by the latter as "tenant's" fixtures) is now much restricted by legislation. The most important statutes are the Housing Act 1988, the Landlord and Tenant Act 1954, the Agricultural Tenancies Act 1995, the Leasehold Reform Act 1967, the Leasehold Reform, Housing and Urban Development Act 1993, and the Housing Act 1996. The provisions contained in these acts are considered in greater detail in Chapters 17 and 18. These statutes usually protect the tenant in actual occupation, so where there is a sublease of the whole of the premises, the subtenant has the protection.

A lease is normally subject to the payment of an annual rent and to the observance of covenants contained in the lease. Commercial leases often include provision for "rent reviews" to take account of general changes (assumed normally to be increases) in property values.

If the rent a lessee has covenanted to pay is less than the true rental value of the premises, the lessee will be in receipt of a net income from the property, a "profit rent", the capitalised value of which, for the balance of the term, will represent the market value of the leasehold interest. Even if a lessee is paying the market or "rack" rent of the property and the lease is subject to regular rent reviews, it may have some value in that it confers a right to occupy the premises for the unexpired term and may enable the lessee to claim a new lease or to continue in possession at the end of the term under the legislation mentioned above. If a lessee is paying a rent above the full value, the lease will have a negative value owing to the liability that it creates.

Lessees are very much more restricted in their dealings with the property than freeholders. In many cases they will be under express covenants to keep the premises in good repair and redecorate internally and externally at stated intervals. Alternatively, they may have to pay the landlord by way of a service charge for repair, maintenance, insurance and the provision of facilities and services. Usually they cannot carry out alterations or structural improvements to the property without first consulting their lessors, and possibly covenanting to reinstate the premises in their original condition at the end of the term. Often they may not sublease or assign their interest or part with possession without first obtaining their landlord's consent, although in most leases that consent cannot be unreasonably withheld. If a tenant wishes to "surrender" his lease to his landlord, this will need the landlord's consent. Whether it is forthcoming will largely depend on market conditions at the time. Where the lease is valuable, the landlord may be happy to purchase it back from the tenant. Where it is not, the tenant may have to pay for release.

If a tenant sublets, even though the sublease contains precisely the same covenants as the headlease, there is the risk of the tenant incurring liability to his landlord if the subtenant fails to keep his part of the bargain. Similarly, if a tenant assigns his lease, he may still be held liable for breaches of the lease covenants by the assignee under the common law or, in leases granted after 1995, under the Landlord and Tenant (Covenants) Act 1995 by virtue of a guarantee given as part of the terms for assignment.

For these and other reasons, leaseholds are less attractive to the investor than freeholds and the income from a leasehold interest in a particular class of property will usually be capitalised at a higher rate per cent than would be used for a freehold interest in the same property.

There follow some types of lease commonly met with in practice.

(a) Building leases

These are leases of land ripe for building, i.e. development, under which the lessee undertakes to pay a yearly ground rent for the land, as such, to erect suitable buildings upon the land, and to keep those buildings in repair and pay all outgoings in connection with them throughout the term. The ground rent represents the rental value of the bare site at the time the lease is granted, and the difference between this and the net rent or premium obtainable from the buildings when erected will constitute the lessee's profit. It is common practice for the lessee to pay a capital sum (a premium) to the lessor at the start of the lease, coupled with a reduction in the ground rent payable. When the building is complete, the builder/lessee will usually assign the lease. The assignee will pay the builder a capital sum for the building but will, of course, be liable to pay the ground rent to the lessor.

Building leases are usually granted for terms of 99 years, although 120-, 125- and 150-year terms have become common in recent years, particularly for large-scale commercial development. Longer terms of 999 years are found occasionally. Lesser terms than 99 years are met with, but it is obvious that no would-be lease-holders will incur the expense of erecting buildings unless they are granted a reasonably long term in which to enjoy the use of them, since they revert to the landlord with the land at the end of the lease. A lessee of residential property exceeding 21 years (but not flats or maisonettes) has the right, under the Lease-hold Reform Act 1967, to a 50-year lease extension at a modern ground rent; alternatively, the lessee may exercise the right to buy the freehold reversion. The freehold ownership of leases of flats and maisonettes exceeding 21 years may be collectively enfranchised by the tenants (through a nominee company in most cases) under the Leasehold Reform, Housing and Urban Development Act 1993, Part I; alternatively, individual tenants may claim a 90-year extension to their leases. When a freehold is enfranchised by tenants it means that the owners of the freehold are compelled to sell the freehold interest to the tenants.

A ground rent should be distinguished from a "peppercorn" rent. Where no rent is payable, for whatever reason (e.g. the rent was capitalised and paid in one lump sum), leases have traditionally reserved the right to some payment, traditionally a peppercorn, in the false belief that without a rent the lease is not valid. (The Law of Property Act 1925 section 205(1) (xxvii) recognises that no rent is required.) Hence the reference to a "peppercorn rent" means that the rent is of no value or a nominal sum (e.g. 1p) which is never collected. As noted above, a ground rent is the market rent for the undeveloped land, not a peppercorn. However, over time the effect of inflation on long building leases without rent reviews may mean that the ground rent becomes hardly worth collecting. So modern building leases usually contain rent reviews.

(b) Occupation leases

This term is applied to leases of land and buildings intended for occupation in their present state by the lessees. In practice the length of the lease varies according to the type of property. Dwelling-houses and older types of commercial and industrial premises are often leased for fairly short terms, such as one, three or five years or on periodic tenancies. In the case of modern commercial property, before 1990 the tendency was to grant leases for 25 years; but since then, because of the slump in the property market, leases for shorter periods, such as three or five years, have become common, or clauses have been included that allow a party to terminate the lease after a short period of years ("break clauses", also known as "options to determine").

In the case of medium- and long-term leases of commercial or industrial premises, it has become standard practice to incorporate a "rent review" clause which enables the rent to be revised, usually upwards, at fixed periods of time. The tendency in recent years, due largely to the effects of inflation, has been for review periods to become shorter, with five years being a fairly standard period. Thus leases are normally granted for multiples of the review period, i.e. 10, 15, 20 or longer periods of years. Before this, multiples of seven years were common.

(c) Leases for life

It was possible until 1925 to grant leases for the duration of the life of the lessee or some other person. By the Law of Property Act 1925, these "leases for life" now take effect as leases for 90 years determinable on the death of the party in question by one month's notice, given either by the lessor or the administrators ("personal representatives") of the deceased lessee's estate, expiring on one of the usual quarter days. This is because a grant for life is an uncertain period and the aim of the 1925 Act is to restrict leaseholds to ascertainable periods.

This provision has little effect on the duration of such a lease because if the premises are held at less than their true rental value, the lessor will take the first opportunity of terminating the lease on the lessee's death, and if the lessees were paying a higher rent than the premises were worth their personal representatives will be equally anxious to determine. From a valuation standpoint, therefore, the period of such a lease may still be safely regarded as that of the lessee's life.

5. Licences

Leaseholds and freeholds must be distinguished from mere permission to enter upon land, even permission by virtue of a contract, e.g. for payment. Such permission is known in law as a "licence", since permission can normally be withdrawn unilaterally on reasonable notice by the licensor (i.e. the landowner) to the licensee. A licence does not create a legal or equitable interest in land; it simply prevents the licensee from being a trespasser so long as it lasts. It therefore has a negligible value in itself; but sometimes an "equitable" right is grafted on to it which does have some value, or at least reduces the value of the licensor's legal title.

An example of this is the decision of the Court of Appeal in *Inwards* v *Baker*[3] in which a father encouraged his son to build a bungalow on the father's land and live there. The son built the bungalow and lived there for many years. When the father died, he left his property by will to a Miss Inwards, who had been living with him for some years and who had borne two children by him. The trustees of the will sought possession of the bungalow on the basis that no title to the bungalow had been transferred to the son – he was a mere licensee. The Court of Appeal held that an equity arises from the expenditure of money by a person in actual occupation of land when he is led to believe that, as the result of that expenditure, he will be

---

3 [1965] 1 All ER 446.

allowed to remain there. The father (or his personal representatives under his will) cannot deny the equity and the son would be allowed to live there as long as he desired to use it as his home.

Licences have been used by landlords to avoid granting tenancies that have protection under various Acts of Parliament – in particular the old Rent Acts which granted a high degree of security of tenure coupled with rent control. (They have been superseded by the Housing Act 1988.) This is acceptable so long as the licence is genuine, i.e. not a lease dressed up as a licence. In the definitive case of *Street* v *Mountford*[4] the House of Lords held that labelling a document a licence does not make it so. The grant of exclusive possession for a term at a rent (the landlord providing neither attendance nor services) is a tenancy.

6. Common law, equity and trusts

The common law is the case law that was developed by judges over the centuries in resolving thousands of disputes. After the invention of the printing press and the publication of law reports, the common law developed into a body of binding precedent law. This means that the legal principles set out in the previous decisions (precedents) of higher courts (appeal courts) are binding on lower courts. Thus, these decisions become part of the law. Land law disputes are civil cases, as distinct from criminal, and are decided in local county courts and the High Court (which sits in London, or sometimes in provincial cities). Appeals go to the Court of Appeal, from which a further appeal may go to the Supreme Court which recently replaced the Judicial Committee of the House of Lords as the highest appeal court. Various specialist tribunals, e.g. the Lands Tribunal (now the Upper Tribunal (Lands Chamber)) from which appeals go direct to the Court of Appeal, exist to deal with particular types of valuation case. Valuers exist to provide professional advice for litigants plus expert evidence in litigation in these various courts and tribunals. But litigation is always expensive and is best not undertaken lightly, wantonly or unadvisedly (e.g. without insurance).

In order to make a claim at common law, the claimant had to use the appropriate form – a "writ" – for each type of claim. As society developed, the limited number of writs available meant that an aggrieved party would not be able to get justice if his particular claim did not fall within an existing writ. So litigants petitioned the King for justice directly. The King passed these petitions on to his secretary, the Chancellor, who resolved them on the principles of fair dealing or "equity". In time the "Chancery" – the Chancellor's department – became a judicial body, the Court of Chancery, and "equity" became a body of rules of precedent alongside the common law. In 1615 it was resolved that where there was conflict between the common law and equity, equity would prevail.

4 [1985] AC 809.

Once equity became a body of law, there was no need for a separate set of courts, and in 1873 the Judicature Act made every court a court of both the common law and equity and ended the different procedures for seeking equitable and common law remedies. Thus the common law remedy of *damages* (financial compensation) is the normal award, but if equity has to be applied, the equitable remedies of *injunction, specific performance, rescission* and *rectification* are available. This can be illustrated by the situation where the parties have entered into a contract for the sale of land. If the vendor or the purchaser now refuses to proceed and "complete" the transaction, the courts will normally grant not damages but an equitable remedy, namely specific performance, in order to enforce the transfer of the title. As a result of this there is a transitional period between contract and completion during which the vendor still has the legal title (he is still the owner) but the purchaser has an equitable right to become the owner, known as an estate contract, which prevails thanks to this equitable remedy of specific performance.

This temporary relationship under an "estate contract" between the legal and equitable ownership is one example of a "trust", because the vendor now holds the legal title "in trust" for the purchaser's right in equity. It is an "implied" trust. Apart from such trusts there are countless "express" trusts in which title is "conveyed", i.e. transferred by deed of grant, to one or more "trustees" to hold expressly on trust for one or more "beneficiaries" for definite or indefinite periods of time. A trustee, as such, has the burden; a beneficiary, as such, has the benefit.

A valuer has to take all this legal background into account if asked to value landed property, which will in some cases be subject to the equitable rights of many beneficiaries under express or implied trusts, or to the legal titles of many separate legal owners such as freeholders, leaseholders, sub-leaseholders or "incumbrancers" (mortgagors and mortgagees). If the valuer is lucky, there may only be one sole unencumbered legal title holder and no equitable owners; but even that sole owner may in fact be a corporate body such as a company or public authority.

Statute Law

Common law and equity should be distinguished from *Statute Law*, which is legislation made by the Crown in Parliament – the House of Commons and the House of Lords with the Royal Assent. Legislation consists of Acts of Parliament ("primary legislation") or rules, regulations and orders made in accordance with them ("secondary legislation"). Such legislation overrides any inconsistent case law.

7. Equitable interests including options

A great many different equitable interests in landed property now exist, but they all rest on the principles mentioned and are enforceable by equitable remedies of one sort or another. All involve the complete or partial separation of the benefit under the equitable interest from the "legal estate", which is the title that exists at common law.

Some of these equitable interests appear to be similar to legal estates and interests (e.g. equitable leases, equitable mortgages etc.) but exist in equity either because they have been created informally (without a deed) or have not been registered as required by the Land Registration Act 2002. (Where a transfer of the freehold is not registered, the Land Registration Act provides that the transferor holds the property on trust for the transferee. Where a registrable lease is not registered, it takes effect as a contract to grant the lease.) Other equitable interests are created by "covenants", i.e. wording *included* in deeds of grant but expressed in terms of a contract, such as restrictive covenants. Estate contracts arising under written contracts for sale, as mentioned above, are implied trusts. There is an important class of estate contracts known as "options"; these are in effect delayed-action contracts under which a prospective purchaser can take up the vendor's offer to sell (or lease) the property within a prescribed period of time: they are conditional contracts in effect. The owners or purchasers of any of these equitable rights will always be well advised to have them recorded against the legal title in the Land Registry; if not, a subsequent purchaser may be able to disregard them, and often they will be lost, for lack of "notice", and be of no value.

8. Co-ownership

In general, any property right can, under English law, be co-owned by two or more people. But in practice this often creates difficulties should they fall out or die, or if they become bankrupt. Common law and equity both recognised a distinction between two main forms of co-ownership: joint ownership and ownership in common. (In legal jargon these two types of co-ownership are called the "joint tenancy" and the "tenancy in common", even though the land which is jointly owned may be freehold or leasehold.) The distinction was and remains this: joint ownership "imports survivorship", so that on the death of one joint owner his or her share goes to the survivor(s), and the last man (or woman) standing takes all and becomes sole owner; whereas under "ownership in common" each owner's share can be disposed of separately on death by will or on intestacy, or else by gift or sale before death ("*inter vivos*" – between the living) and can even be subdivided. Joint ownership gives simplicity, at the cost of inhibiting choice.

The Law of Property Act 1925 simplified matters by enacting that, if co-owned, legal title now can only exist as joint ownership, not ownership in common, and also that co-ownership of land imports a trust of that title. The purpose of this is to keep the trusts off the title and make conveyancing simpler. Since the Trusts of Land and Appointment of Trustees Act 1996 these trusts have been called "trusts of land". The trustees of co-owned land therefore hold the legal title in trust for whoever has the benefit in equity. Often the trustees and beneficiaries are the same people, especially married couples or co-habitees. So a husband and wife who jointly own their home will be trustees of it for themselves as beneficiaries. These trusts can be set up either by the living (especially of a newly purchased matrimonial home) or by will. They can also arise automatically "by operation of law", especially as intestacy means that the deceased's property, not being left by a valid will, has to be held on trust for the next-of-kin. So if (say) a married couple both die intestate in a simultaneous accident, leaving a co-owned matrimonial home, there will in any case have to be a trust, under which the legal title will thereupon devolve on their personal representatives.

So in co-ownership of land the legal title must be held by trustees and the law says it must be a "joint tenancy". What of the co-owners' beneficial equitable ownership? This can be either "joint tenancy" or "tenancy in common". Registered title may omit to say which. If so, the immediate implication is that the benefit is held jointly, as is the registered legal title. But for various reasons it is normally more advantageous for the beneficiaries not to be subject to the survivorship rule. Even if the registered title makes no mention of this, the beneficiaries can, at any time, make a formal written declaration "severing" the joint beneficial interest in equity into ownership in common and so give themselves freedom of disposing separately of each share of the beneficial ownership by will or "*inter vivos*" (with an eye to potential tax advantages as well). In any case, why should a survivor take all? Whatever the position at common law, equity "leans against joint tenancies". This means that any ambiguous wording is to be construed so as to avoid joint ownership of the benefit of the beneficial interest under the trust. Not only that: although equity allows "severance" of joint ownership into ownership in common, it does not allow the reverse: the traffic is one-way. To put it baldly: joint ownership is ideal for trustees; ownership in common is ideal for beneficiaries.

9. Easements and related rights

Two other common law property rights frequently met with are legal easements and legal *profits à prendre* (though they can also sometimes exist as equitable rights). Each is a right belonging to a "dominant" property but existing over an adjacent "servient" property: the former has the benefit, the latter has the burden. Easements are of great importance because the rights of neighbouring property owners are frequently interdependent. Thus, an easement of way entitles owner X as a legal right (i) to walk or drive on a track over the adjoining land of owner Y "as of right" across the land surface, or (ii) to climb by stairs or a lift between floors, in Y's building. An easement of support entitles X as a legal right to enjoy the support of a building from buildings or retaining walls on Y's land. An easement of light entitles X as a legal right to receive natural light to particular windows from across Y's adjacent land. These are the main kinds of easement usually met with in practice; often they are reciprocal between two or more owners, as in shared driveways, blocks of flats, etc. *Profits à prendre* entitle X as a legal right to *take* substances (animal, vegetable or mineral) from Y's land, and include grazing, fishing and shooting rights. But whereas all easements must have a servient and a dominant property, it is possible in some cases of profits *à prendre* for there to be only a servient property (e.g. fishing and shooting rights) and this is more likely today where X may be an individual or syndicate owning no adjoining property. Grazing rights are often enjoyed in common by several dominant owners; hence the expression "common land". Interference with any of these rights is a "private nuisance", to be redressed by damages or injunctions.

Legal as opposed to equitable easements and profits must, strictly speaking, be created by deed, termed an "express grant", but creation by "implied" or "presumed" deed of grant is also possible at common law. Equitable easements or profits will normally exist under written contracts and are enforceable by the equitable remedy of specific performance. In an "implied" grant there is an actual conveyance or lease which, however, does not expressly mention the easement or profit, yet in the circumstances contains a clear implication (i) that both parties have intended to create such a right, or (ii) that its existence is a practical necessity.

"Presumed grants" of easements are *prescriptive* rights, based on long use. "Presumed" is a legal fiction. Common law *presumes* that the grant must have been made long ago, provided that the use has been exercised continuously for more than 20 years "as of right", i.e. not forcibly, nor secretly, nor by permission of the servient owner (*"nec vi, nec clam, nec precario"*, in lawyers' Latin). If the period of long use is 40 years, then *oral* permission will not prejudice the dominant owner if that was given before the prescription period began. Easements of light ("ancient lights") in every case only need a 20-year period. Profits *à prendre* require 30 and 60 years instead of 20 and 40 years. Rights of common, by the Commons Act 2006, will no longer be capable of being created by prescription, but a Town and Village Green (TVG) can be created where a significant number of the inhabitants of a locality have indulged as of right in lawful sports and pastimes on an area of land for a period of at least 20 years. An application to register land as a TVG by the local community has become a popular way of blocking unwanted development where there is sufficient evidence of longstanding use for sports and pastimes.

10. Restrictive covenants

A "covenant" is a contractual obligation in a deed. Normally, at common law, contracts are only enforceable by and against the contracting parties – *"privity of contract"* – and so successors in title cannot sue or be sued on the contract. (There is an exception to this in the case of assignment of a lease or reversion: the assignee tenant or assignee landlord stands in the same position as the assignor with respect to the leasehold covenants and can sue or be sued on those covenants.) Assume, for example, that X has covenanted to maintain a wall on his land which benefits the adjacent land owned by Y. In a strict application of the privity rule, the benefit of the covenant would not run to Y's successors in title, nor would the burden of the covenant run to X's successors in title. However, the common law allows the *benefit* of a covenant to be transferred ("assigned"). This will be an intentional assignment, unless the covenant "has reference to" the land, i.e. "touches and concerns" it. In that case, the benefit of that covenant will "run with the land" automatically, and thus be enforceable by a subsequent freeholder (Law of Property Act 1925, section 78). This applies regardless of whether or not the original covenantor owns any *burdened* land. The leading case is the decision of the Court of Appeal in *Smith* v *River Douglas Catchment Board* (1949).[5]

Although the *benefit* of a covenant may "run" with the land at common law, in freehold land the *burden* does not run. But the rules of equity enable a covenant which "touches and concerns" land to run automatically with that land to burden all subsequent owners who have "notice" of it, provided that it is negative ("restrictive") in nature, and for the benefit of specified neighbouring landowners. So the burden of a positive covenant, such as to maintain a wall, does not normally run as it requires positive acts on the part of the covenantor. But if the covenant is restrictive in nature, for example to prevent land being used for business, or to prohibit the erection of houses, then the burden will run with the land.

Restrictions on the user of land are sometimes imposed solely for the benefit of neighbouring land which the vendor is retaining. In other cases they are intended to safeguard the general development of property in many towns (e.g. Eastbourne) where building estates were sold in freehold plots to various purchasers, who would enter into identical covenants as regards the use of their land. Normally such covenants are mutually enforceable between the various purchasers and their successors in title, so that the owner of any one plot may compel the observance of the restrictions by the owner of any other plot. Yet although restrictive covenants may help to maintain the character of an estate and thus the value of the houses, in some cases – particularly when imposed many years ago – they may restrict desirable development of a property and detract from its value.

A restrictive covenant is an equitable interest in the burdened land. Unless it is an obligation in a lease, it must appear in the register in registered land (Land Registration Act 2002) or be registered as a "land charge" in unregistered land (Land Charges Act 1972). Failure to register such a covenant renders it void against a purchaser. But the burden of a restrictive covenant cannot "run with the land" unless the benefit of it is annexed to nearby land capable of benefiting (often in the form of a scheme of development) or else is assigned together with such land. Enforcement is principally by means of an injunction against infringement, though a court may award damages that are usually based on what the parties would have freely negotiated for the release of the covenant.

The right to enforce restrictive covenants may be lost. This may happen if the party entitled to enforce them has for many years acquiesced in open breaches of the covenant or has consented to a breach of covenant on a particular occasion in such a way as to suggest that other breaches will be disregarded, or if the character

5 [1949] 2 KB 500.

of the neighbourhood has so entirely changed that it would be inequitable to enforce the covenant.

There are also statutory means by which a modification or removal of restrictive covenants may be obtained. Thus the Housing Act 1985, section 610, enables any interested persons or the local authority to apply to the County Court for permission to disregard leasehold or freehold covenants that would prevent a single dwelling house being converted into two or more dwellings if it is not lettable as a single dwelling or if planning permission has been granted for the conversion. More generally, under the Law of Property Act 1925,[6] the Lands Chamber of the Upper Tribunal may wholly or partially modify or discharge restrictive covenants on freeholds if satisfied that by reason of changes in the character of the property or the neighbourhood, or other material circumstances, the restrictions ought to be deemed obsolete or that their continuance would obstruct the reasonable use of the land for public or private purposes without securing practical benefits to other persons. Current planning policies of local planning authorities should be taken into account. These powers also extend to restrictive covenants on leaseholds where the original term was for over 40 years, of which at least 25 years have elapsed.[7]

The Upper Tribunal may make such order as it deems proper for the payment of compensation to any person entitled to the benefit of the restrictions who is likely to suffer loss in consequence of the removal or modification of the restrictions. In practice, these payments are of small amounts (usually less than £1,000) as any larger award would tend to indicate that the covenant is of substantial value and so should not be discharged.

The 1925 Act also gives power to the court to declare whether land is, or is not, affected by restrictions and whether the restrictions are enforceable and, if so, by whom. This is a valuable provision which enables many difficulties to be removed where the existence of restrictions or the right to enforce them is uncertain.

11. Commonhold

In strict legal theory, separate flats or floors in a building can be held in either freehold or leasehold ownership. (A *flying freehold* is a freehold that overhangs another freehold.) But, until recently, legal technicalities have made it difficult in practice to have freehold ownership of such horizontal divisions of landed property. This is because, in general, and as noted above, *positive* obligations do not run with freehold land – they only bind the original covenantor. So although the benefit of a positive covenant would run to a purchaser of the freehold, the burden of such a covenant (e.g. to repair) would not. In leaseholds, however, the

---

6 Section 84 as amended by the Lands Tribunal Act 1949, section 1(14), and the Law of Property Act 1969, section 28.
7 Law of Property Act s. 84(12).

benefits and burdens of positive leasehold covenants run with the land and so are enforceable by and against successor landlords and tenants. So where buildings are multi-occupied, the use of leaseholds avoids the problem of the enforcement of positive covenants (such as repairing obligations) against successors in title.

Long leaseholders of flats campaigned for the creation of a "commonhold" system under which they could become freeholders of their flats and manage the property themselves. So in 2002 the Commonhold and Leasehold Reform Act was passed. Part 1 governs the creation of commonholds and requires the making of a "commonhold community statement" which sets out the rights and duties of the holders of units within the commonhold. (The other parts of the Act give long leaseholders the right to club together to buy out the landlord or to take over the management of the property.) To become "commonhold land", property *must* be held thereafter on registered freehold title and owned by a "commonhold association". The owner of a "commonhold unit" is entitled to be registered as the owner of a freehold estate in the unit. Each unit is freely transferable without the consent of the commonhold association.

Most commonhold properties will in practice be new developments, though existing leaseholds are convertible to commonhold. There must be an accompanying statement listing all the prospective unit holders and the property must be registered with "absolute title" under the Land Registration Acts. The commonhold association in each case is to be a company limited by guarantee, run by directors who have (a) a general duty to manage and (b) a direct responsibility for the common parts of the building. Regulations made by the Lord Chancellor deal with the various rights and duties in the association's memorandum and articles and the contents of each commonhold community statement. These include financial provisions for (i) creating repair and maintenance funds and (ii) setting a levy to produce these funds, comprising contributions from unit holders.

12. Successive interests in property

Provisions are frequently made in settlements by deed or by will whereby, so far as the law permits, the future succession to the enjoyment of the property is controlled.

A once common example (familiar to Jane Austen, but rare in consequence of taxation) is the creation of an "entailed interest" in freehold or long leasehold property whereby the property is "settled" on certain specified persons for life and, after their death, on their eldest son and the "heirs of his body". In this way, the property would descend from generation to generation, unless the entail is barred or unless heirs fail, in which latter case the land will revert to the original grantors or their successors or else pass to any other person entitled under the settlement on termination of the entail.

Under this and other forms of settlement, certain persons may be entitled, either alone or jointly with others, to successive life interests or entails in a property, the ownership finally passing to some other person in fee simple. The first owner enjoys an interest "in possession" as "tenant for life"; but each of the others have, at the outset, an interest which is only a future right, and hold it "in remainder". Such a person is called a remainderman. Sometimes they may be entitled to succeed to a property only subject to certain conditions, for example, that they reach a prescribed age. In such a case, the interest is spoken of as "contingent" and if the contingency is too remote in the future, the interest will infringe what in law is called the "perpetuity rule" and be void (Perpetuities and Accumulations Act 2009). Life interests and future interests in property rank as equitable interests, not legal estates. But from a valuation point of view the distinction is of little consequence. To a valuer the problem they present is that of finding the capital value of an income receivable for the duration of a life or lives, or the capital value of an income for a term of years or in perpetuity receivable after the lapse of a previous life interest. Readers interested in the valuation of such interests should refer to earlier editions of this book.

All persons enjoying the interest of "tenant for life" have vested in them, and may sell, the whole freehold or leasehold interest (the legal estate) in the settled property, the capital money being paid to trustees and held for the beneficial interests, as was the land itself. A valuer advising on a sale of this description is concerned not with the value of the party's life interest, but with the market value of the freehold or leasehold property which is the subject of the settlement; and in such a case the equitable (beneficial) interests in the settlement can be ignored for valuation purposes because the land will be sold free of them for its full unencumbered market value. They are said to be "overreached", that is to say, preserved not against the land but against the capital of the settlement in its new form in money, investments, etc. There have been two main kinds of trust of land: strict settlements and trusts for sale. The Trusts of Land and Appointment of Trustees Act 1996 replaces these by a single "trust of land".

13. Transfers of interests in land and security in land

The rules of the law of property in land have evolved chiefly in connection with transfers of ownership. A life interest is a right that cannot be transferred at the life owner's death because that is the time when it comes to an end, but it can be transferred for so long as the owner remains alive (e.g. for mortgage purposes). The legal freehold or "fee simple", on the other hand, is perpetual in its duration, and so can be transferred by current owners during their lifetime (*inter vivos*) or at death. The same is true in principle of leaseholds, except, of course, that the term of a leasehold may end while its holder is still alive, in which case no property right remains to the holder in respect of it (though, of course, it may be possible to secure a renewal).

Valuers need to take proper account of these various characteristics because what they are valuing is not "land" in the purely physical sense, but property rights in land. Death should perhaps be considered first. Transfer on death is governed by the deceased's will, if any. Failing that, it goes by operation of law, under the Administration of Estates Act 1925 (as amended), to the deceased's next-of-kin, in accordance with various elaborate rules governing "intestacy". Frequently, in practice, this simply involves a sale by the deceased's personal representatives; in which case what the next-of-kin actually receive is a distribution of the money proceeds of the sale. But, where necessary, the Inheritance (Provision for Family and Dependants) Act 1975 empowers the courts to order payments to be made out of a deceased person's estate to any surviving spouses, co-habitees or other dependants who would otherwise in effect be left destitute because no proper provision has been made for them.

Bankruptcy (or liquidation in the case of a company) is treated in a similar way under the Insolvency Act 1986 (as amended), the property in question being sold not by personal representatives but by a "trustee in bankruptcy" or a liquidator and the proceeds of sale (normally inadequate) going not to next-of-kin but to unsecured creditors. This is, in a sense, a sort of financial death, except that a bankrupt individual will probably qualify for return to life in due course by receiving a discharge from bankruptcy. An insolvent owner's land is thus dealt with in much the same way as an intestate owner's land.

Transfer *inter vivos* occurs either by gift or by sale, that is to say, either non-commercially or commercially. The latter can, for this purpose, be taken to cover transfers in return for a financial consideration in the form of a purchase price or a series of regular payments, or both, or in the form of an exchange. Grants of new leases, assignments of existing leases and transfers of freeholds are all varieties of transfer *inter vivos*, whether as gifts or as commercial transactions. A lump sum is termed a "premium".

Transfers of land may also operate to create security for loans and similar financial obligations, because the creditor is granted (expressly or implicitly) a "legal charge" by way of "mortgage" pending repayment in full. This is dealt with in a later chapter.

A gift of land, on death or *inter vivos*, requires the carrying out of the appropriate legal conveyancing formality, which is the execution of a deed of "conveyance" (or a written "assent" by personal representatives) of the legal freehold or leasehold as the case may be. The same is true of commercial transfers. An equitable property right, however, need only be in writing as a rule. Legal leasehold tenancies, if at a market rent and taking immediate effect, may not, however, need to be granted by any formality at all, provided that they are created for a period not exceeding three years (thus oral short-term tenancies of this kind are still "legal estates" in land).[8]

But commercial transfers, unlike gifts, involve another stage in addition to (and prior to) the "conveyance", namely the contract or bargain constituting the sale, lease or mortgage. Section 2 of the Law of Property (Miscellaneous Provisions)

8 Law of Property Act 1925, section 54(2).

Act 1989 provides that a land contract "can only be made in writing". This must be done "by incorporating all the terms which the parties have expressly agreed in one document or, where contracts are exchanged, in each". But these terms can "be incorporated in a document either by being set out in it or by reference to some other document". The contract "must be signed by or on behalf of each party to the contract".

If in correspondence the phrase "subject to contract" is used, there is no contract at all at that stage, merely negotiation,[9] from which either party can back out. But once the contract itself has been made, in accordance with the 1989 Act, it is binding and enforceable by an order of "specific performance" whereby the court compels a defaulting party to proceed to the completion stage. Damages, if appropriate, may be awarded, and a defaulting party will normally have to return (or forfeit) a deposit.

If a definite date is prescribed for completion it is binding at common law, but not in equity unless, on reasonable notice, one of the parties expresses an intention to make a specified date "of the essence of the contract". The appropriate legal remedies for breach of the contract, whether specific performance (or the converse, i.e. *rescission* – setting aside of the contract) or damages, will be available accordingly.[10] The availability as a general rule of specific performance to vendor and purchaser alike (or lessor and lessee) means that *in equity* the property, as a capital asset, passes to the purchaser at the contract stage and is termed an "estate contract" (which should be protected by registration) even though the legal title stays with the vendor until completion. The estate contract amounts to an implied trust (mentioned above) in the intervening period. Options (section 7 above) are similarly protected. Their duration can be extended by agreement.[11]

---

9 *Tiverton Estates Ltd* v *Wearwell* [1974] 1 All ER 209.
10 See the House of Lords decision in *Stickney* v *Keeble* [1915] AC 386.
11 *Rennie* v *Westbury Homes* (2008) 2 P&CR 12.

Chapter 4

The market approach or comparative method

1. Introduction

The simplest and most direct method of valuation is direct comparison. The method is based on comparing the property to be valued with similar properties, and the prices achieved for them, and allowing for differences between them to determine the market value. The principle is that of market substitution, hence it is the market approach.

The method is based on a comparison of like with like. Properties may be similar, but each property is unique; they can never be totally alike. As the comparables move away from the ideal of absolute similarity, they become less reliable. The reasons for dissimilarities are:

(a) Location

The location of every property is unique. The actual location is an important factor in the value of a property and for some, such as shops, a major factor. For example, on an estate built to the same design at the same time, some houses may be nearer to a main road than others; some may be on higher ground with good views whilst others may be surrounded by other houses; some may back south, others back north; some may have larger plots. In a parade of shops some may be nearer a street crossing than others or be next to a pedestrian exit from a car park or nearer to a popular large store.

(b) Physical state

The physical condition of a property depends on the amount of attention which has been given to its maintenance, repair and decoration. Two otherwise identical properties can be in sharp contrast if one has been well maintained and the other has been allowed to fall into disrepair. Repair apart, occupiers of properties typically make alterations that can vary from a minor change to a major improvement, such as the conversion of an integral garage into a study. This is perhaps most apparent with residential properties, where occupiers may install or change the form of central heating, renovate kitchens and bathrooms, change windows, remove fireplaces, add extensions, solar panels and so on, to the extent that houses which from the front elevation are identical are totally different.

(c) Tenure

Freehold, leasehold and commonhold interests can be found in property. Apart from a freehold interest with no subsidiary leasehold interests, the likelihood of interests being similar is remote in view of the almost endless variations that may exist between them. Even on an estate with a standard lease the terms are likely to vary as to expiry date, rent payable or permitted use. In the market at large, the differences are greater and this method is strained to compare, say, the freehold or leasehold interest in a shop where the occupying tenant holds an unexpired lease of 45 years at a rent of £10,000 a year, is responsible for all outgoings apart from external repairs, is subject to no rent reviews and no limitations as to use, with similar interests in a physically identical shop let with eight years to run at a rent of £40,000 a year with a rent review in three years time, the tenant being responsible for all outgoings and the use limited to the sale of children's clothes. It is impossible to assess the values of either type of interest in the latter by directly comparing the price achieved on the sale of either interest in the former. This contrast between the types and nature of interests is the principal barrier to the use of the comparative method for the valuation of interests other than freeholds in possession.

(d) Purpose

The purpose of a valuation is an important element in deciding the approach to use. Where the valuation is, for example, for investment purposes, the market approach might be inappropriate and the income approach might be better.

(e) Time

In normal circumstances, transactions take place in the market regularly. The market is not static, so that the price obtained for a property at one time will not necessarily be achieved if the same property is sold again at a later date. Thus, the reliability of the evidence of prices achieved diminishes as time passes since the transaction took place. In a volatile market only a short time need pass before the evidence becomes unreliable.

It is clear that, given the special characteristics of property and the drawbacks these create, the market approach is of limited application. But it does have wider application to establishing constituent parts, such as the market rent needed for the other approaches. In practice, only three types of property lend themselves readily to the use of the market approach for capital value estimates, and normally only when the freehold interest in the property is free from any leasehold estate. Two types are considered below; the third is agricultural property.

2. Residential property

The majority of all property transactions are of houses and flats with vacant possession; the market approach is used to value all such properties. It is true that flats and some houses are held as leasehold interests, but these tend to be on long leases at relatively small rents and often have similar characteristics to a freehold, so allowing for direct comparison.

Example 4–1
A owns the freehold interest in a 1964 semi-detached house which is to be sold with vacant possession. It has three bedrooms, bathroom and separate WC on the first floor and a living room and kitchen on the ground floor. There is a garage at the rear of the house with an access shared by the adjoining houses. No improvements have been carried out apart from taking down the wall between the former front and rear rooms to form a through living room, and the installation of central heating. The bathroom and kitchen need to be modernised. The property is in a fair state of repair. Several nearby similar properties have sold recently for prices in the range of £265,000 to £305,000, the most recent being in the next pair to A's house, built at the same time with identical accommodation but requiring a new boiler. The bathroom and kitchen were recently refitted. It is in good repair. The house sold for £290,000. Another house in the same street, also built at the same time and with the same basic accommodation but with a newly fitted kitchen and bathroom, a new boiler and with a ground floor cloakroom, and having an independent drive to the garage, all in good repair, sold recently for £305,000.

Valuation
It appears that a house such as A's with the basic amenities and modernised and in good repair is worth around £290,000 if the only issue is the age and serviceability of the central heating boiler.

Assume that with a new boiler the value would be £295,000. This appears to be realistic since a similar house but with ground floor cloakroom and independent drive to the garage sold for £305,000 (an allowance of £10,000 for these further factors seems reasonable).

Allowing £14,000 for the cost and trouble of modernising the bathroom and kitchen, it appears that the value of A's house is in the region of £282,000 in good repair.

Allowing for the lower standard of repair, the value of A's house might be considered to be £278,000.

It is impossible to be precise in a valuation of this kind because different buyers will make different allowances for the cost of modernisation and for the costs of repairs. Some buyers will replace the bathroom suite, even if it is almost new, if they dislike its colour, and so would pay more than £278,000 for this house.

It is clear that various allowances for differences in quality have to be made. The level of allowance is subjective and requires the expertise of an experienced and knowledgeable valuer. The method is simple in its general approach but is dependent on considerable valuation judgment for its application. The valuation of residential property is considered in detail in Chapter 16.

3. Development land

Landed property sold for development or redevelopment is known as development land or building land.

Residential building land can be valued using the comparative method because residential sites share similar characteristics. The valuer, in analysing the comparables, will need to allow for such factors as the type of development contemplated, ranging, for example, from luxury houses to high-density low-cost accommodation; the site conditions, such as whether level or sloping, well drained or wet; and location in relation to general facilities such as transport, schools, hospitals. The comparable sites will vary in size and the valuer therefore reduces the evidence to a common yardstick. This may be value per hectare (or acre), value per plot or per unit, or value per habitable room. For example, if a site of four hectares (10 acres) sold for £5,000,000 with permission to build either 90 four-room houses or 120 three-room flats, the price may be expressed as:

£1,250,000 per hectare (£500,000 per acre); £55,555 per plot; £41,667 per unit; £13,888 per habitable room. The valuer will use the yardstick which is most appropriate to the type of site being valued.

Example 4–2
Value the freehold interest in two hectares (five acres) of residential building land with consent to build 42 houses. Recent comparable land sales include:

(a) three hectares (7.5 acres) with consent to build 60 houses, sold for £3,600,000;
(b) five hectares (12.5 acres) with consent to build 125 houses, sold for £6,250,000; and
(c) 0.4 hectares (one acre) with consent to build six houses, sold for £600,000.

The analysis below indicates that building land is worth between £1,200,000 and £1,500,000 per hectare (£480,000 and £600,000 per acre) or between £50,000 and £100,000 per house. Where more houses are to be built on a given area (an increase in the "density"), the houses are likely to be smaller and less valuable, so that the value per house and the value of each plot both fall, although conversely the value per acre rises since the builder will have higher output and presumably higher profits overall.

Site (c) has values outside the general range since it is a small development which would attract a different type of buyer, and it also has a low density so that large and high-value houses are to be built. The higher value of (c) supports the general value evidence of (a) or (b).

Analysis of comparables

Site	£/hectare	(£/acre)	£/house	Density (houses/hectare)
(a)	1,200,000	(480,000)	60,000	20
(b)	1,250,000	(500,000)	50,000	25
(c)	1,500,000	(600,000)	100,000	15

Hence, the site to be valued, which has a density of 21 houses per hectare (8.4 houses per acre), appears to lie within the general band of values.

Valuation

	2 hectares at say £1,225,000 per hectare	=	£2,450,000
or	42 houses at say, £57,500 per house plot	=	£2,415.000
	Value, say		£2,425,000

Although residential building land may be valued by direct comparison, it is often valued using the residual method, which is the typical method for valuing development land (see Chapters 11 and 15) and is used to cross-check the valuation based on direct comparison.

Chapter 5

Market rent

1. Introduction

The income approach or investment method assesses market value by capitalising the future income (cash flows) obtainable from the property. There are two parts to the valuation, namely the net receivable income and the yield. This chapter and Chapter 6 are concerned with the factors determining the net income, while Chapter 7 deals with the factors determining the yield. Various terms are used to describe rent, such as rack rent, fair rent, full rent; for most valuation purposes the valuer is interested in market rent. This is defined in the Red Book in VS 3.3 as:

> The estimated amount for which a property would be leased on the *valuation date* between a willing lessor and a willing lessee on appropriate lease terms in an arm's-length transaction, after proper marketing where the parties had each acted knowledgeably, prudently and without compulsion.

2. Rent and net income

Property can be let on the basis that the tenant will bear all of the costs and outgoings associated with the property including repairs, insurance, rates, etc. These lettings are normally known as full repairing and insuring lettings or lettings on FR and I terms or FRI leases. An alternative form of this kind of letting is what is known as a "clear lease", i.e. one where the tenant undertakes to keep in full repair the interior of the demise and also to pay a service charge to reimburse the landlord for a proportionate share of the cost of insuring, repairing, maintaining, servicing, decorating, etc. the exterior and common parts.

Alternatively, property can be let on the basis that the landlord bears some or all of the outgoings, in which case the net income is arrived at by deducting outgoings from the rent payable. It is only the net rental income that is capitalised to arrive at the capital value; where reference is made to "rent" it is to be understood that "net rental income" is intended. The outgoings connected with property are outlined in Chapter 6 and in the chapters dealing with particular types of property.

In simple terms, rent may be defined as "an annual or periodic payment for the use of land or of land and buildings". In assessing the market rent a valuer is largely influenced by the evidence that can be found of rents being paid for comparable properties in the same location. But it is essential that the valuer should appreciate the economic factors that govern those rents. Knowledge of economic factors may not be necessary for the immediate purpose of the valuation, but it is vital in understanding and forecasting fluctuations in rental value and in advising on the reasonableness or otherwise of existing rents.

3. Economic factors affecting rent

(a) Supply and demand

In Chapter 1 it was noted that the general laws of supply and demand govern capital values and that value is not an intrinsic characteristic. Variations in capital value may be due either to changes in rental value or to changes in the return that may be expected from a particular type of investment. Either or both of these factors may operate at any time.

Until recently the general trend was for rents to rise over time, but there is no law of economics which dictates that this must happen. Indeed, technological changes can lead to a reduction in demand for certain types of property, e.g. for offices where computerisation is increasingly seeing a shift towards home or remote working. Over-supply and an economic recession led to a sharp fall in market rents in London and the South East from 1989 to 1993 and again in 2008. Increasing obsolescence of design could have the same result even in normal economic conditions.

Movements in property yields follow general movements in interest rates but not necessarily to the same degree. When rental values rise and returns fall, they combine to produce a sharp rise in capital values; if the opposite occurs capital values will fall. Indeed, there is considerable logic in there being this double movement. The fact that rents are seen to be rising due to increased demand is a spur to investors to seek such property, thus pushing down yields, and the reverse is equally true. However, these may be opposing forces.

For example, the rent of shops may increase over a period of years but at the same time there may be changes in the investment market which cause that particular type of security to be less well regarded; a higher yield will be expected and consequently a lower multiplier (YP) will be applied to the net rent to arrive at the MV. The result may be that, whilst the rent has increased, the capital value remains unchanged (see Example 5-1).

Example 5-1

Rental value today	£15,000	Rental value 5 years hence	£18,750
YP to show 6%		Y P to show 7.5% yield in 5 yrs	
Yield now	16.667		13.333
Market value	£250,000	Market value	£250,000

Changes in the rent and changes in the yield are normally caused by changes in supply and demand. In both cases any change can affect the market value of the property. In this chapter the concern is with the influence of supply and demand on rent.

When stating that rent depends on supply and demand, it must be remembered that supply means the effective supply at a given price, not the total amount of a particular type of property in existence. In a particular shopping centre there may be perhaps 50 shops of a very similar type, but 45 may not be available to let at a given time because they are already let or owner occupiers are trading from them and they are not on the market. The effective supply consists of the other five shops whose owners wish to let.

Similarly, demand means the effective demand at a given price, i.e. a desire backed up by money and ability to trade. In the shopping centre example, demand is derived from potential tenants who have weighed up the advantages and disadvantages of the shops offered against other shops elsewhere.

The prevailing level of rents will be determined by the interaction of supply and demand, which is sometimes referred to as "the haggling of the market". If there is a large or increasing demand for a certain type of property, rents are likely to increase; if the demand is a falling one, rents will decrease.

As pointed out in Chapter 1, landed property can be distinguished in two important respects from other types of commodity in that the supply of land is to a large extent fixed and that, in the case of buildings, supply can only respond slowly to changes in demand.

(b) Demand factors

The prime factor in fixing rent is demand. However, before looking at demand factors in detail, it is important to distinguish between residential property and other types of property. The majority of families in the UK own their own homes (i.e. between 65% and 70%) and the capital value with vacant possession is derived from demand and supply for owner occupation and not from capitalising the rent. There is a significant rental market for housing, but this is divided into low cost (usually subsidised) housing and private housing. In the market for private housing to rent the underlying factors may have similarities to those for other types of property; but this does not mean that rent and capital value are identified in the same way as they are for commercial property, and indeed they are not.

The demand for land and buildings is a basic one in human society, as it derives from three essential needs of the community:

(i) the need for living accommodation;
(ii) the need for land, or land and buildings, for industrial and commercial enterprise – including the agricultural industry; and
(iii) the need for land, or land and buildings, to satisfy social demands such as schools and hospitals and recreation.

The demand for any particular property, or for a particular class of property, will be influenced by a number of factors, which it will be convenient to consider under appropriate headings.

General level of prosperity

This is perhaps the most important of the factors governing demand.

When times are good and business is thriving, rents will tend to increase. There will be a demand for additional accommodation for new and expanding enterprises. There will be a corresponding increase in the money available for new and improved housing accommodation and for additional associated amenities.

Changes in market rents will not necessarily compare exactly to changes in the general standard of living. Changes may not apply to all types of property, and the extent may vary across the country or in different parts of the same area due to the influence of other factors such as changing tastes. Experience in this and other countries has shown, however, that market rents generally show a steady upward tendency in times of increasing prosperity.

Population changes

Times of increasing prosperity have in the past been associated with a growing population.

The supply of land being limited, it is obvious that an increase in demand due to increased prosperity or increase of population will tend to increase values. On the other hand, when these causes cease to operate or diminish in intensity, values will tend to fall.

Increase in population will, in the first place, influence the demand for housing, but indirectly will affect many other types of property the need for which is dependent upon the population in the locality – as with, for instance, retail shops. Increases in population within a locality may occur in various ways. The birthrate may exceed the number of deaths so that there is a natural increase in the size of the existing population. Alternatively, population may rise locally due to migration of people from other areas or countries, and in this connection the UK membership of the EU is very relevant. A further change in population may arise for a temporary period; this is typical of holiday and tourist centres, where the numbers rise significantly for certain periods of the year.

Changes in character of demand

Demand, in addition to being variable in quantity, may vary in quality. An increase in the standard of living may cause a change in the character of the demand for many types of property.

For example, in houses, many amenities regarded today as basic requirements would have been regarded as refinements only to be expected in the highest class of property two decades ago. Similar factors operate in relation to industrial and commercial properties. The increased standards imposed by legislation (e.g., the Disability Discrimination Act) or technical improvements in the layout, design and equipment of buildings cause many buildings to become obsolete and, as a result, market rents fall.

Rent as a proportion of personal income

In relation to residential property, the rent paid may be expressed as a proportion of the income of the family occupying the property. It is obvious that, in a free market, the general level of rents for a particular type of living accommodation will be related to the general level of income of the type of person likely to occupy that accommodation. There will, of course, be substantial variations in individual cases owing to different views on the relative importance of living accommodation, motor cars, television sets, holidays and other goods and services. In the UK, until comparatively recently, there was no free market in living accommodation to rent as landlords were faced with their tenants having security of tenure and fixed rents. Thus, apart from those who were able to secure subsidised local authority or housing association property, most people sought to buy and were rewarded with substantial capital gains due to many years of rising values and few years of falling values.

However, confidence in the never-ending rise in values was severely dented between 1989 and 1994/5 when the housing market collapsed and many owners were faced with negative equity and many houses and flats were repossessed by mortgagees. This slump in the housing market, coupled with changes in legislation whereby houses and flats could be let at market rents without security of tenure, led to a revival in the free market with lettings at market rents. The cause was a reluctance to buy and take on mortgage commitments against a background of job insecurity. The UK housing market revived post 1994/5 and once again saw huge increases in capital values caused by increased prosperity, low interest rates and the easy availability of mortgages, frequently without the need for any deposit for first-time buyers and at multiples of income that were unheard of up until then. However, 2008 saw a dramatic change in the availability of mortgages, with all 100% loans being withdrawn and with some interest rates, especially for those who previously had a low fixed rate mortgage, being increased not withstanding the intervention of the UK Government and the Bank of England to reduce Base Rate then, and to this date, to an historic low of 0.5%. Once again, repossessions have increased and there could be a swing towards renting rather than buying as confidence in ownership is once again dented.

There is a free market in residential properties for sale and a useful parallel today in relation to capital, not rental, values is that lenders, in determining an appropriate mortgage loan, are applying lower income multipliers. This may again prevent individuals from gaining a foothold on the property owning ladder thus increasing the demand for rental property.

Rent as a proportion of the profit margin

Both commercial and industrial properties and land used for farming are occupied, as a rule, by tenants who expect to make a profit out of their occupation, and that expectation of profit will determine the rent that such tenants are prepared to pay.

Out of gross earnings tenants have first to meet running expenses. From the profit that is left tenants will require remuneration for their own labours, interest on the capital that has been provided, and a profit as a reward for enterprise. The balance that is left is the margin that tenants are prepared to pay in rent. Obviously, no hard and fast rule can be laid down as to the proportion of profit that rent will form, but the expectation of profit is the primary driver of tenant's demand and the total amount of rent to be paid will be influenced by this estimate of the future trend of receipts and expenses and the profit required.

Rent is, therefore, a function of profit and, as profit is a function of gross earnings, so rent must also be a function of gross earnings. This relationship has led some landlords to move rents to turnover rents, i.e. a percentage of gross takings, or indexed to the Retail Price Index.

Competitive demand

Certain types of premises may be adaptable for use by more than one trade or for more than one purpose and there will be potential demand from a larger number of possible tenants. For instance, a block of property in a convenient position in a large town may be suitable for use either for warehouse or for light industrial purposes and consequently the rent that can be expected will depend on the relative demands; the use having the greater demand produces the greater rent.

In a free market it would be expected that the enterprise yielding the largest margin out of profits for the payment of rent would obtain the use of the premises because, where there are competing demands for various purposes, properties will tend to be put to the most profitable use. It is for this reason that low-margin or low-turnover shops are frequently put out of business by rents that are too high for them to pay because competing users can pay more.

In practice, the operation of this "most profitable" rule may be affected by the need to obtain planning permission for a change of use.

(c) Supply factors

It has already been stated that land differs from other commodities in being to a large extent fixed in amount. This limitation in the supply of land in the UK is exacerbated by planning legislation which has created green belts around cities and has limited development by facilitating zoning policies. This limit on the amount of land available is the most important of those factors that govern supply, and thus the supply of land and building is regarded as being relatively static.

Limitation of supply

At any one time there will be in the country a certain quantity of land suitable for agricultural purposes and a certain quantity of accommodation made available for industrial, commercial and residential purposes.

If these quantities were entirely static, the rents would be affected only by changes in demand. In fact, they are not static, but respond, although slowly, to changes in demand. There is, ultimately, only a certain quantity of land available for all purposes, but increase in demand will cause changes in the use to which land is put.

In agriculture, conditions of increasing prosperity, leading to higher prices, will bring into cultivation additional land which previously it was not profitable to work. Conversely, in bad times land will go out of cultivation and revert to waste land. Similarly, increasing demand will render it profitable for additional accommodation to be built for commercial and residential uses and the supply of accommodation will be increased by new buildings until the demand is satisfied.

These statements assume that the market is working freely. In practice, there is considerable intervention by government, either to support and stimulate some uses or to dampen them down. For example, agriculture generally enjoyed a strong measure of support for most of the post-World War II years from all governments, but this led within the European Community to substantial over-production until 2007. To reduce production to the level of demand within the European Community, a number of restrictive measures, such as milk quotas, were introduced and a substantial amount of land was deliberately taken out of agricultural production. In 2007 there was an abrupt change due to awareness of global warming and the perceived need to switch away from the use of fossil fuels. The effect has been a switch of a large amount of agricultural land from food crops to cash crops usable for biofuels. Falling crop yields in many third-world countries saw food riots in Haiti, India and Indonesia in 2008, caused by rapidly rising prices for basic foods such as rice and wheat. As a result, set-aside land is now being brought back into cultivation. It may be that with the development of genetically modified crops this policy will be reversed again at some time in the future.

Office development has often been curbed by government action in parts of the country, despite a strong demand for new accommodation, in the mistaken belief that this will encourage similar development in other parts of the country.

The increase in rents that would occur if the supply remained unchanged will be modified by additions to the supply. But, as was pointed out in Chapter 1, adjustments in the supply of land and buildings occur slowly in comparison with other commodities.

Where there is a tendency for the demand to fall, the supply will not adjust itself very quickly; building operations may well continue beyond the peak of the demand period and some time may elapse before expansion in building work is checked by a fall in demand.

The principal causes of inelasticity in supply are probably:

- the fact that building development is a long-term enterprise, i.e. one where the period between the initiation of a scheme and completion of the product is comparatively lengthy; and
- the difficulty involved in endeavouring to forecast demand in any locality.

The above considerations introduce the subject of what economists call "marginal land". In the instance given of land being brought into cultivation for agricultural purposes, there may come a point where land is only just worth the trouble of cultivating it. Its situation, fertility or other factors will render it just possible for the profit margin available as a result of cultivation to recompense tenants for their enterprise and for the use of their capital. There will, however, be no margin for rent. This will be "no-rent" land or marginal land.

It is obvious that land better adapted for its purpose will be used first and will yield a rent, but that also, with increasing demand, the margin will be pushed farther out and land previously not worth cultivating will come into use.

Similar considerations will apply to other types of property. For instance, there will be marginal land in connection with building development. A piece of land may be so situated that, if a factory is built on it, the rent likely to be obtained will just cover interest on the cost of construction, but leave no balance for the land. With growth of demand, the margin will spread outwards, the rent likely to be obtained for the factory will increase, and the land will again have a value.

Land may have reached its optimum value for one purpose and be marginal land for another purpose. For example, land in the vicinity of a town may have a high rental value for market-garden purposes and have reached its maximum utility for agricultural purposes. For building purposes it may be unripe for development – in other words, "sub-marginal" – or it may be capable of development but incapable of yielding a building rent – in other words, "marginal".

In practice it is unlikely that land will be developed immediately it appears to have improved a little beyond its marginal point. Some margin of error in the forecast of demand will have to be allowed for, and no prudent developer would be likely to embark on development unless there is a reasonable prospect of profit. In many actual cases planning control will prevent development where land has improved well beyond its marginal point. As an example, the pressure to develop open land around cities is constrained by green belt planning policies.

Relation of cost to supply

Another factor that may influence supply is the question of cost.

If, at any time, values as determined by market conditions are less than cost, the provision of new buildings will be checked, or may even cease, and building will not recommence until the disparity is removed. The disparity that may exist at any time between cost and value may be subsequently removed either by a reduction in cost, or by an increase in demand or a reduction in supply due to demolition of marginal property. As an example, offices and flats built in the 1950s, 1960s and 1970s are being demolished. Some are being rebuilt to a higher specification or at a higher density, but many offices are being rebuilt as (or in many cases converted to) hotels or blocks of flats. Governments may intervene by providing grants or other incentives so as to reduce the effective cost or to stimulate demand. The situation might be further modified by the non-replacement of obsolete buildings, which might in time bring about an excess of demand over supply and a consequent increase in value.

It is important to remember always that cost does not equal value. A developer must not assume that, because a given sum of money has been paid for the land and expended on converting the building, the total cost can be passed on in the form of rent or capital payment. Market rent or market value must be established from the market and not dictated to the market.

4. Estimation of market rent

(a) Generally

In estimating the market value of a property from an investment point of view, the valuer must first determine its market rent.

In doing so, the valuer will have regard to the trend of values in the locality and to those general factors affecting rent discussed above. These will form, as it were, the general background of the valuation, although the valuer may not be concerned to investigate in detail all the points referred to when dealing with an individual property. The two factors most likely to influence the valuer's judgment are:

• the rents paid for other properties; and
• the rent, if any, at which the property itself is let.

When preparing an investment valuation it is necessary to make an assumption as to whether the market rent at the time when the valuation is made will continue unchanged in the future. This assumption, that rent would continue either unchanged or would rise in the future, used to be the accepted rule as normal changes, either upwards or downwards, were reflected in the property yield. The rent payable at the date of the valuation would have been compared with the then current market rents, and regard would be had to the length of the lease, the incidence of rent reviews and the terms of the lease, regarding matters such as repair and user. This assumption may no longer be valid as obsolescence must now also be considered. Where the property is over-rented, i.e. the rent payable exceeds the market value, this must be allowed for in the calculation by reverting on review to a lower rent, or on reversion to either a lower rent with an allowance for a possible void, or to a development value for the actual or an alternative use. Where obsolescence is considered to be relevant to the future rent, this would normally be reflected in the yield. This is further considered in greater detail in Chapter 9.

In general, if rent is likely to increase a lower yield (higher YP) will be applied, but if it looks as if it is going to decrease then a higher yield (lower YP) will be used – low risk low yield, high risk high yield.

Future variations in market rent may sometimes arise from the fact that the premises producing the income are old, so that it may be anticipated that their useful life is limited. Together with the likelihood of a fall in rent there may be a possibility that, in the future, considerable expense may be incurred either in rebuilding or in modernisation to maintain the rent. In such cases, the return which the investment will be expected to yield will be increased and the YP to be applied to the net income correspondingly reduced.

Where a future change in rent is reasonably certain it should, of course, be taken into account by a variation in the net rent. For example, a valuation may be required of a shop which is offered for sale freehold subject to a lease being granted to an intending lessee on terms that have been agreed. These terms provide for a specified increase in rent every five years. It has been ascertained that many similar shops in the immediate vicinity are let on leases for 10, 15 and 20 years with a provision for similar increases in the rents at the end of each five-year period. Assuming that the position is an improving one and that the valuers are satisfied that such increasing rents are justified, the valuers will be correct in assuming that the shop with which the valuers are dealing can be let on similar terms, and the valuers will take future increases in rent into account in preparing their valuation. The method of dealing with such varying incomes from property will be considered in detail in Chapter 9.

(b) Basis of the rent actually paid

Where premises are let at a rent and the letting is a recent one, the rent actually paid is usually the best possible evidence of market rent.

But the valuer should always check this rent with the prevailing rents for similar properties in the area or, where comparison with neighbouring properties is for some reason difficult, with the rents paid for similar properties in comparable locations elsewhere, i.e. where the socio/economic conditions are the same.

There are many reasons why the rent paid for a property may be less than the market rent. For instance, a premium may have been paid for the lease, or the lease may have been granted in consideration of the surrender of the unexpired term of a previous lease, or the lessee may have covenanted to carry out improvements to the premises or to forego compensation due to the lessee from the landlord. In many cases the market rent may have increased since the existing rent was agreed. Differences between rent paid and market rent may also be accounted for by the relationship between lessor and lessee, e.g. father to son, parent company to a subsidiary.

In some instances a rent from a recent letting may be above market rent. For example, an owner may sell an asset to realise capital and take back a lease from the buyer (a sale and leaseback). If the owner is a strong covenant, the buyer may agree to pay above market value and accept a higher rent, relying on the strength of the covenant for security of their income.

If it is clear that the actual rent paid is a fair reflection of MR, then it will be adopted as the basis of the valuation. The valuer will ascertain what outgoings, if any, are borne by the landlord and by making an appropriate deduction will arrive at the net rent.

If, in the valuer's opinion, the market rent exceeds the rent paid under the existing lease, the valuation will be made in two stages. The first stage will be the capitalisation of the present net income for the remainder of the term. The second stage will be the capitalisation of the market rent after the end of the term; the term may be the period to the next rent review rather than the unexpired term of the lease if the rent review clause is not upward only.

If, on the other hand, the valuer considers that the rent fixed by the present lease or tenancy is in excess of the market rent, the valuer will have to allow for the fact that the tenant, at the first opportunity, may refuse to continue the tenancy at the present rent, and also for the possible risk, in the case of business premises, that the tenant may fail and that the premises will remain vacant until a new tenant can be found.

As a general rule, any excess of actual rent over market rent may be regarded as indicating a lack of security in the income from the property. But it must be borne in mind that rent is secured not only by the value of the premises but also by the tenant's covenant to pay, and a valuer may sometimes be justified in regarding the tenant's covenant as adequate security for a rent in excess of the market rent, particularly where there is an upward only rent review.

An example of such cases is a sale and leaseback deal of the type referred to above. Another is where a shop is let on long lease to some substantial concern such as one of the large retail companies with many outlets – a "multiple". Here, the value of the goodwill the tenants have created may make them reluctant to terminate the tenancy, even if they have the right to do so by a break clause in the lease; they may, however, threaten to do so unless the rent is reduced. Where there is no such break, they will in any case be bound by their covenant to pay.

Since rent at the moment is a first charge, ranking even before debenture interest, and since such a concern will have ample financial resources behind it, the security of the income is reasonably assured, but the valuer should be aware of the true financial state of the tenant company before making this assumption. Sometimes the lease is held by a subsidiary of the major company; in the absence of a guarantee from the parent, the covenant is only as good as the accounts of the subsidiary justify.

(c) Comparison of rents

It has been suggested that the actual rent paid should be checked by comparison with the general level of values in the district. Not only is this desirable where the rent is considered to be *prima facie* a fair one, but it is obviously essential where premises are vacant or let on old leases at rents considerably below market rent.

In such circumstances, the valuer has to rely upon the evidence provided by the actual letting of other similar properties. The valuer's skill and judgment come into play in estimating the market rent of the premises under consideration in the light of such evidence. Regard must be had not only to the rents of other properties that are let, but also to the dates when the rents were fixed, to the age, condition and location of the buildings as compared with the subject property, to the terms of the lease, and to the amount of similar accommodation in the vicinity to be let or sold.

In many cases, as for instance similar shops in a parade, it may be a fairly simple matter to compare an unlet property with several that are let and to assume, e.g., that since the latter command a rent of £20,000 a year on lease it is reasonable to assume the same rental value for the shop under consideration which is similar to them in all respects.

But where the vacant premises are more extensive and there are differences of planning and accommodation to be taken into account in comparing them with other similar properties, it is necessary to reduce rents to the basis of some convenient "unit of comparison" which will vary according to the type of property under consideration.

For example, it may be desired to ascertain the market rent of a factory having a total floor space of 1,000 m^2. Analysis of the rents of other similar factories in the neighbourhood reveals that factories with areas of 500 m^2 or thereabouts are let at rents equivalent to £50 per m^2 of floor space, whereas other factories with areas of 4,000 m^2 or thereabouts are let at rents of £40 per m^2 of floor space. This indicates that size is a factor in determining the unit rent. After giving consideration to the situation, the building, and other relevant factors, it may be reasonable to assume that the market rent of the factory in question should be calculated on a basis of £47.50 per m^2 of floor space.

The unit chosen for comparison depends on the practice adopted by valuers in that area or for the type of property. Notwithstanding the transition to metric units, over the past years valuers have shown a marked reluctance to abandon imperial units. Thus, the most commonly met units of measure are still values per square foot or per acre and most, if not quite all, comparative statistics are based on imperial measures. This may change, although it didn't change over the life of the last two editions. Metric units are commonly adopted for agricultural purposes and in valuations for rating purposes. The valuer must be prepared to accept and work with either method. The practitioner will follow normal practice but needs to be aware that the RICS advocates the use of metric units, either with or without the imperial equivalent.

The method of measuring also varies according to the type of property concerned. In some cases the areas occupied by lavatories, corridors, etc. are excluded and in other cases the valuer works on the gross internal area. In an attempt to bring some uniformity, the professional societies have produced recommended approaches. The RICS publishes a *Code of Measuring Practice: A Guide for Property Professionals* (6th edn, 2007). This Code both identifies various measuring practices, such as gross internal area and net internal area, and also recommends the occasions when such practices would be appropriate, for example gross internal area for industrial valuation and net internal area for office valuation. Though the rules of conduct of the professional bodies do not bound valuers to follow the Code, such codes are regarded as "best practice" and comparable evidence and calculation of market rent are usually on the same basis when obtained from other surveyors. Failure to follow the Code could in some circumstances be regarded as negligent practice.

When quoting rental figures, the amount per annum should always be qualified by reference to the length and terms of the lease such as the liability for rates, repairs and insurances. The date and the purpose of the valuation should also be stated.

(d) Rent and lease terms

The amount of rent that a tenant will be prepared to offer will be influenced by other terms of the lease. Rent is but one factor in the overall contract and if additional burdens are placed on the tenant in one respect the tenant will require relief in another to keep the balance. For example, if landlords wish to make a tenant responsible for all the outgoings they cannot expect the rent offered to be the same as where the landlords bear responsibility for some of them.

Similarly, if a landlord offers a lease for a short term with regular and frequent upward revisions of the rent, he or she cannot expect the same initial rent as would be offered if a longer term were offered with less frequent reviews, particularly if his/her terms are untypical of the general market arrangements; if a landlord seeks to limit severely the way in which premises may be used, the rent will be less than where a range of uses is allowed.

It is clear that the market rent must reflect the other terms of the lease and cannot be considered in isolation; indeed, a rent cannot be determined until the other terms of the lease are known.

In making comparisons, it must be remembered that the terms of the lettings of different classes of property vary considerably. It is convenient, therefore, to compare similar premises on the basis of rents on the terms usually applicable to that type of property. Good shop property and industrial and warehouse premises are usually let on terms where the tenant is liable for all outgoings including rates, repairs and insurance. Thus, net rents are compared on this type of property, assuming a full repairing and insuring lease. With blocks of offices let in suites, the tenant is usually liable only for internal repairs to the suite and rates and the landlord for external repairs, repairs to common parts and insurance. The landlord provides services such as lifts, central heating and porters, but the cost of these are dealt with separately by means of a service charge on the tenants (and which may also include the landlord's liability to repair and insure). Thus, rents used for comparing different suites of offices would be inclusive of external repairs, repairs to common parts and insurance but exclusive of the service charge. Having arrived at the rental value of the office block on this basis, the net income would be arrived at by deducting the outgoings which are included in the rents and for which the landlord is liable. In cases where landlords include within the service charge their liability to repair and insure and also the supervisory management fees, the rent will be the net income. This latter form of lease is known as a "clear lease" as the landlord recovers all of his/her costs from the tenants and thus the rent is usually comparable with the rent on an FRI lease.

Example 5–2

A prospective tenant of a suite of offices on the third floor of a modern building with a total floor space of 250 m^2 requires advice on the rent which should be paid on a five-year lease. The tenant will be responsible for rates and internal repairs to the suite. The landlord will be responsible for external repairs, repairs to common parts and insurance. Adequate services are provided, for which a reasonable additional charge will be made. The following information is available of other recent lettings on the same lease terms in similar buildings, all of suites of rooms:

Floor	Area in m^2	Rent, £	Rent, £ per m^2
Ground	200	36,000	180
	300	48,000	160
1st	150	18,000	120
2nd	160	20,000	125
4th	400	52,000	130
5th	500	67,500	135
6th	200	26,000	130

Estimate of rent payable

As these other lettings are on the same terms as that proposed, it is unnecessary for purposes of comparison to consider the outgoings borne by the landlord or to arrive at net rent.

The broad picture that emerges from this evidence is that the highest rent is paid on the ground floor but thereafter the rents are approximately the same. This is what would be expected in modern buildings with adequate high-speed lifts. In older buildings with inadequate lifts or no lifts, the higher the floor the lower will be the rent. The inconsistencies in the rents might be due to market imperfections or to differences in natural light or noise.

Subject to an inspection of the offices under consideration, so that due weight can be given to any differences in quality and amenities, a reasonable rent would appear to be, say, £130 per m² or £32,500 per annum.

The different methods used in practice for assessing rent by comparison are referred to in more detail in later chapters dealing with the various classes of property.

5. Effect of capital improvements on market rent

Since, in most cases, owners would be unlikely to spend capital on their property unless they anticipated a fair return by way of increased rent, it is reasonable to assume that money spent by a landlord on improvements or additions to a property will increase the rent by an amount approximating to simple interest at a reasonable rate on the sum expended. The rate at which the increase is calculated will naturally depend upon the type of property.

Since, however, the real test is not solely what owners expect to obtain for their premises, but more importantly what tenants are prepared to pay for the accommodation offered, then in practice any estimate of increased rent should be carefully checked by comparison with the rents obtained for other premises to which similar additions or improvements have been made or already exist.

Capital expenditure is likely to increase rent or capital value only where it is judicious and suited to the type of property in question. For example, house owners may make additions or alterations to their house which are quite out of keeping with the general character of the neighbourhood and are designed solely to satisfy some personal whim or hobby. So long as they continue to occupy the premises they may consider they are getting an adequate return on their money in the shape of personal enjoyment, but there will be no increase in the rent or capital value of the premises as the alteration is unlikely to appeal to the needs or tastes of prospective tenants or buyers.

Again, expenditure may be made with the sole object of benefiting the occupier's trade, as, for instance, where a manufacturer spends a considerable sum in adapting premises to the needs of its particular business. In this case, the occupier may expect to see an adequate return on its capital in the form of increased profits, but it is unlikely that the expenditure will affect the market rent unless, of course, it is of a type which would appeal to any tenant of the class likely to occupy the premises such as the installation of air conditioning in an office building which significantly improves the working conditions.

On the other hand, the increase in rent due to capital expenditure may considerably exceed the normal rate of simple interest on the sum spent. This will be so in those cases where the site has not been fully developed or where it is to some extent encumbered by obsolescent or unsuitable buildings. For example, an area in the business quarter of a flourishing town may be covered by old, ill-planned and inconvenient premises. If the owner can improve these premises so as to make them worthy of their location, then the increase in rent will not only include a reasonable return on the capital sum spent, but will also include a certain amount of rent which had been latent in the site and which will have been released by the improvements.

6. Hierarchy of market evidence

Evidence of market rent can come from many sources. The best evidence is that from recent market lettings where negotiations have been conducted on a market rent basis as defined by the RICS. Other evidence may be from arbitrations, settlements by independent experts, and agents asking rents. All of these provide added information for the valuer, but the need for interpretation is great given the range of lease terms and definitions of rent that find there way into leases. The use of lease incentives will mean that not all rents agreed can be taken at their face value (see RICS *Valuation standards* UKGN 6 *Analysis of commercial transactions* and Chapter 17); issues relating to the analysis of rents are considered further in property-specific chapters. Chapter 6 now looks at some of the outgoings that have to be assessed in order to arrive at the net rent.

Chapter 6

Outgoings

1. Introduction

An important difference between property and other forms of investment is that regular expenditure is needed to maintain, insure, manage and, when let, to secure tenants; all such expenses are referred to as outgoings. Where a property is owner-occupied, the owner is responsible for meeting all these costs. Where the property is let, the lease specifies who is liable for each outgoing, subject only to any statutory requirements.

In the UK the landlord normally seeks to pass all of the outgoings on to the tenant by means of a full repairing and insuring (FRI) lease so that the rent, as investment income, is free from any deduction. This is particularly the case for non-residential property let to a single tenant. In the case of property in multi-occupation, where services and other activities, such as the cost of maintaining a lift serving different floors, can only be met through a central or common approach, landlords tend to recover the expenditure by means of a charge, additional to rent, known as a service charge.

Landlords tend to be concerned about outgoings because rents are normally fixed for a number of years and most outgoings are subject to unpredictable and substantial changes due to their nature and to inflationary increases. Where a rent is fixed for a period and the landlord has to pay for outgoings from the rent, then, if their cost rises, the net return to the landlord will fall. A variable income is unsatisfactory to an investor at all times and during a time of inflation it could result in a negative income.

Tenants are concerned because the cost of outgoings is an additional payment to the rent required and so they must consider the aggregate cost when deciding whether or not to take a lease of a property, not just the rent. If the burden and cost of outgoings increases, then there will be a smaller sum available out of the total budgeted figure to offer as rent. This is known as the "total cost concept".

For example, if two factories of equal quality are available, but in one the tenant must meet the £6,000 a year cost of the outgoings whilst in the other the landlord will bear half, then in the second case the tenant could offer to pay around £3,000 p.a. extra in rent (see Chapter 5).

In the case of similar properties but with different burdens, a tenant will pay a higher rent for the building with the lowest outgoings. For example, a prospective tenant may be offered two similar offices which in other respects are of equal value; one of them the tenant finds to be of indifferent construction, requiring considerable external painting with a high repair cost. Logically, the prospective tenant will be willing to offer a higher rent for the offices where the repair costs are likely to be lower. It follows that if the level of outgoings has an effect on rent negotiations then it must be important for the valuer to be able to determine those costs.

2. Repairs

(a) Generally

The state of repair of a property is an important factor to be taken into account when arriving at the market value. Likewise, the burden of keeping a property in repair is an important factor in estimating market rent. Hence, regard must be had to the need for immediate repairs, if any; to their probable annual cost; and to the possibility of extensive works of repair, rebuilding or refurbishment in the future.

The age, nature and construction of the buildings will affect the annual cost of repairs. Thus, a modern well-designed structure with minimum external paint-work will cost less to keep in repair than a structure of some age, of indifferent construction and with extensive paintwork.

(b) Immediate repairs

If immediate repairs or renewals are necessary, then the usual practice is to calculate the capital value of the premises in good condition and to deduct the estimated cost of putting them into that condition. This is known as making an "end allowance".

(c) Annual repairs

The cost of repairs will vary from year to year, but to assess the net income or net operating income of a property it is necessary to reduce the periodic and variable costs to an average annual equivalent. This may be done by reference to past records, if available; by an estimate based on experience, possibly expressed as a percentage of the market rent; by an estimate based on records of costs incurred on similar buildings, expressed in terms of pounds per m^2; or by examining the cost of the various items of expenditure and their periods of recurrence.

Where reference is made to past records, it is essential to check the average cost thus shown by an independent estimate of what the same item of work would cost at the present day.

One of the dangers of using a percentage of market rent to estimate repair costs is that if two shops are very similar physically but one is in a high rent area and the other in a low rent area, then using percentages could produce very different estimates even though the cost of repair is likely to be similar. For example, one shop may have a market rent of £50,000 p.a. and the other may be in a secondary position with a market rent of £15,000 p.a. If the annual costs of repair for each are £2,000 p.a., in the first case this is 4.0% of the rent, whilst in the second it is 13.3%. This variation will remain even if the standard of repair of the better placed property is higher than the other. It is clear that the repair costs of the better shop would need to be £6,670 p.a. if 13.3% were to be the general percentage of rent spent on repairs. The percentage basis is only reliable where the properties are similar physically and attract a similar rent. While the percentage basis may be of some use, it is better to judge each case on its merits and to make an estimate of the cost of the work required to put and keep the premises in good condition.

An estimate may be based on costs incurred on similar buildings. If the costs of keeping in repair several office buildings of similar quality and age are known, an analysis may show that they tend to cost a similar amount per m^2. For example, if the analysis showed that around £20 per m^2 was being spent, it would be reasonable to adopt this sum when valuing another office building in the area of similar quality and age[1].

The fourth method of estimating the average annual cost of repairs is illustrated below.

Example 6–1

A valuer is required to estimate the average annual cost of repairs of a small and well-built house of recent construction, comprising three bedrooms, two reception rooms, kitchen and bathroom, let on an assured shorthold tenancy at a rent of £12,000 p.a. The landlord is responsible for all repairs. The house is brick-built, in good condition, with a minimum of external paintwork.

(a) Experience might suggest an allowance of 15% of the rent, say, £1,800.
(b) This estimate might be checked by considering the various items of expenditure and how often they are likely to be incurred, as follows:

(1)	External decorations every 4th year, £2,000	£500	p.a.
(2)	Internal decorations every 7th year, £2,100	£300	p.a.
(3)	Pointing, every 25th year, £5,000	£200	p.a.
(4)	Sundry repairs, gutters, boiler service say	£400	p.a.
		£1,400	p.a.

The method used in the example can be applied to other types of property, regard being had in each case to the mode of construction, the age of the premises and all other relevant factors that will affect the annual cost. Here, a simple average cost has been taken, but techniques such as life cycle costing can be used to reflect the time value of money and inflation.

---

1 A useful guide is the Building Cost Information Service (BCIS) *Building Running Costs Online*.

(d) Conditions of tenancy

Regard must be had to the conditions of the tenancy in estimating the annual cost. Where premises are let on lease, the tenant is usually liable for all repairs and no deduction has to be made (FRI lease). In other cases, the liability may be limited to internal repairs (IR lease), particularly where a building is in multi-occupation; in valuing the landlord's interest, a deduction for external repairs must be made. This will be so even if the landlord has not covenanted to carry out external repairs and is therefore not contractually liable for such repairs (unless statutory provisions override the contractual position) since the rent paid is assumed to exclude any liability on the tenant for this element. In practice, most landlords would regard it as prudent property management to keep the property in repair. On the other hand, a service charge might be levied on tenants which includes the recovery of the cost of external repairs; in such cases, no deduction from the rent will be made.

Before the valuation is carried out, the solicitor acting for the client should be contacted, or the lease(s) should be examined, to verify the exact repairing liability of the parties. In the case of short lettings it must be remembered that at the end of the tenancy the premises will need redecoration before re-letting or sale; even if in a fair state of repair at the date of valuation. The same may be true where a longer lease is coming to an end, depending on the extent of the tenant's repairing liability. A tenant under a general covenant to repair must leave the premises in good or tenantable repair on termination of the lease and any works required at that time to bring the premises into proper condition are termed "dilapidations". Nonetheless, a valuer would need to decide if some allowance would be appropriate in valuing the landlord's interest, particularly if the tenant is of a poor covenant and there is the risk of repairs not being done.

If the cost of repairs is expressed as a percentage of the market rent, it must not be forgotten that the landlord may also have agreed to pay for other outgoings besides repairs; the percentage to be used in such cases, where applied to the inclusive rent, will be somewhat more than where only repairs are included within the gross rent.

(e) Future repairs

In considering the value of property, it may be necessary to make allowance for the possibility of extensive works of repair, improvement, or even ultimate rebuilding in the future. Where the date for such works is long deferred it is not usual in practice to make a specific allowance, but where such works are possible in the near future (say up to 10 years or so from the date of valuation) the present cost of the works deferred for the estimated period should be deducted. In the case of older properties, the works will tend to reflect a considerable element of improvement to allow for the changes in standards that have occurred since the property was erected. In fact, the rate of improvement has tended to increase over recent years as changes in standards accelerate, and a generous allowance for repairs as an annual outgoing may be prudent to reflect the cost of these works, which are often cosmetic rather than actual repair. Alternatively, a provision for a capital investment in the property on the termination of the lease might be more appropriate. This would be so even where a tenant will contribute to dilapidations since these only cover putting into repair what was there, whereas the landlord might wish to carry out more extensive works of improvement or refurbishment.

3. Sinking funds

Sometimes, in addition to the cost of repairs, provision is made for a sinking fund to cover the reinstatement of the buildings over the period of their useful life. In some countries it is standard practice to do so.

In the case of properties where the buildings are estimated to be nearing the end of their useful life, this provision may have to be made, but in most valuations the risk of diminution in the value of buildings in the future is reflected in the yield used to capitalize the net income. This is considered further in Chapters 9 and 18.

4. Business rates, water services and council taxes

Rates and council taxes are normally the liability of the tenant, but in some cases, such as lettings on weekly tenancies, the landlord may undertake to pay rates or council taxes.

Business rates are levied on the rateable value at a poundage that is consistent across the country (the Uniform Business Rate or UBR). Council taxes vary from one local authority to another. Water services may be metered, calculated on the number of employees in occupation or as a percentage of the historic rateable value (see Chapter 20).

To arrive at a fair amount to be deducted, regard must be had to the actual amount currently paid and to possible changes in the rate poundage and future assessment. This is particularly important if the rent cannot be increased on account of the increase in charges. However, in most cases, increases in charges will be borne by the tenant, either because the tenant bears them directly or because the landlord has reserved the right to recover such increases.

As regards changes of assessment, the present rateable value of a property must not necessarily be taken to be correct. All assessments are capable of revision to take account of changes to the property or its use, or under a general revaluation. If an assessment appears to be too high, regard should be had to the possibility of its amendment at the earliest possible moment. If too low, the possibility of an upward revision should be anticipated.

On first view, it would seem incorrect, where no provision exists, to adjust the rent paid in respect of alterations in rates, or to base a valuation on the assumption that rates will continue at their present level, when there is every possibility of an increase in assessment. However, as it is not possible to predict the future changes in the rates, it may be more appropriate to adopt a higher yield or capitalisation rate than would otherwise be the case, rather than to guess what the future change might be.

With regard to possible variation in rate poundage, unless there is some clear indication of a higher than average change in the future, it is usual to assume continuance at the present amount if the rental income can be varied in the short term. Where the rent is fixed for the medium or long term, then regard must be had to future changes in these outgoings. This can be done by changing the net income from year to year or by deducting a sum above the present level of payment or by adopting a higher yield to reflect the prospect of a falling net income.

Any alterations in rates due to changes in assessment and in rate poundage are passed on to the tenant if the property is let on inclusive terms where there is an "excess rates" clause in the lease. If there is such a provision, then no account need be taken of possible changes in the future for the unexpired term of the lease or agreement.

5. Income tax

The income from landed property, like all income, is subject to income or corporation tax, but tax as an outgoing is normally disregarded as most types of income are equally liable (see Chapter 9).

Allowance may have to be made for tax, however, when valuing leaseholds and other assets of a wasting nature, such as sand and gravel pits. The reason for this is that tax is levied on the full net annual income of the property without regard to the necessity, in such cases, of a substantial proportion of the income having to be set aside to cover capital replacement or reinstatement of land, which might be by means of a sinking fund (see Chapters 9 and 12).

6. Insurances

(a) Fire insurance

This outgoing is sometimes borne by the landlord and will be deducted in finding the net income, but the normal practice, when premises are let on lease, is for the lessee to pay the premium or to reimburse the premium paid by the landlord. Where the tenant's occupation is for some hazardous purpose, such as a woodworking factory or chemicals store, the tenant may, in addition, have to pay the whole or a portion of the premium payable in respect of other premises owned by the same landlord where the insurance cost is increased by virtue of the adjoining hazardous use.

The cost of fire insurance is small in relation to the market rent, particularly for modern, high-value premises. In the case of older premises of substantial construction, where the cost of remedying damage will be expensive, the premium may be substantial. The use of a percentage of market rent can be misleading and it is advisable to establish the current rebuilding cost and premium payable. The premium is calculated on the replacement cost of the building and not on its value; cost assessments for fire insurance purposes are considered in Chapter 19.

(b) Other insurances

In many cases of inclusive lettings, e.g. where blocks of flats or offices are let in suites to several tenants, allowance must be made for special insurance, including insurance of lifts, employers' liability, third-party insurance and national insurance contributions. But these may be recoverable through the service charge. The cost of plate glass insurance for shop fronts is normally carried by the tenant.

7. Voids

In some areas and for some properties, it is customary to make an allowance for those periods when a property will be unlet and will not produce revenue. For example, a poor-quality block of offices in multiple occupation may have some part of the accommodation empty all the time. These periods, termed voids, may be allowed for by deducting an appropriate proportion of the annual rent as an outgoing. It might be that the property is only ever 80% let.

In addition, a landlord will be liable for rates, insurance and service charges during such periods, so there may be a need to allow for these outgoings in respect of void parts in a property where such costs would otherwise be recoverable from the tenants.

Where a property is unlet but it is anticipated that an early letting can be achieved, an allowance for voids may still be required if the normal market terms include a long initial rent-free period. This might be achieved by deferring the income flow for six months or deducting six months' rent from the capital value. However, deductions for outgoings may not be needed.

8. Service charges

Where buildings are in multiple occupation, the responsibility for maintenance of external and common parts is usually excluded from the individual leases; for example, the maintenance of the roof benefits not only the tenant immediately below it but all the tenants of the building, and so it would be inequitable for one tenant to be responsible for the repair of the roof.

The responsibility may be retained by the landlord, although, in some cases, particularly blocks of flats which are sold on long lease, the responsibility may be passed to a management company under the control and ownership of the tenants.

Where a landlord retains the responsibility, he/she will normally seek to recover the costs incurred from all of the tenants by means of a service charge. The goal is to recover the full costs so that the rent payable under the various leases is net of all outgoings. Such a situation arises with office blocks let in suites, but is also found in shopping centres, industrial estates and other cases of multiple tenancies.

Typical items covered by a service charge are repairs to the structure; repair and maintenance of common parts, including halls, staircases, lifts and shared toilets; cleaning, lighting and heating of common parts; employment of staff such as a receptionist, caretaker, maintenance worker and security staff; insurance and management costs of operating the services. In addition to these types of expenditure, which relate to the day-to-day functioning of the building, service charges are increasingly extended to provide for the replacement of plant and machinery such as lifts and heating equipment.

It is desirable for costs to be spread as evenly as possible over time so that an annual charge is normally levied, with surpluses representing funds which build up to meet those costs that arise on an irregular basis, such as major repairs or replacements of plant. These latter funds are known as "sinking funds" or "reserve funds".

Various methods of allocating the costs to individual tenants are adopted, including proportion of total floor area occupied by each tenant, and proportion of rateable value of each part to the aggregate rateable values. In shopping centres a weighted floor area apportionment may be adopted to reflect the fact that not all service costs increase directly with floor area. Where some tenants do not benefit from services as much as others, then special adjustments are needed. For example, where there is an office building with lock-up shops on the ground floor, it would be unfair to seek to recover costs relating to the entrance hall, staircases and lifts from the tenants of the shops who make no use of and receive no benefit from these facilities.

The existence of service charges should allow valuers to assume that the lease rents received are net of further outgoings, provided they are satisfied that the charges are set at a realistic level; the lease terms permit the landlord to levy such a charge; all the items are included; and that statutory requirements which apply, e.g., to blocks of flats are met. Some service charges now include indexation to the Retail Price Index but the majority provide for annual adjustment linked to annual budgeting. However, if there are any voids or costs not recoverable, the landlord will become responsible for the charges relating thereto. If the valuer feels that voids are likely to occur fairly frequently, and past evidence may support this view, consideration needs to be given to making some deduction on this account. For example, if the space is extensive it may be reasonable to assume that 5% to 10% will be empty at any one time and so to allow, as an outgoing for the landlord, 5 to 10% of the predicted service charges. (For further reading on service charges and value added tax (VAT) on outgoings see Freedman, Shapiro and Slater, *Service Charges: Law & Practice* (2012, Jordans) and the RICS *Code of Practice for*

Service Charges in Commercial Property, 2nd edition, RICS Practice Standards, UK (GN 24/2011).

9. Management

Agency charges on lettings and management must be allowed for as a separate outgoing in certain cases. An allowance should be made even where the investor manages the property, as even here there is an opportunity cost which should be reflected. In properties where there is a service charge, this charge often includes the cost of management, but the valuer should check the service charge provisions in order to decide whether or not an allowance needs to be made. The existence of a service charge does not automatically mean that all costs incurred by the landlord are recoverable.

Where the amount of management is minimal, as with ground rents and property let on full repairing terms, the allowance is normally ignored. In other cases, such as agricultural lettings and houses let on yearly tenancies or short agreements, the item is often included in a general percentage allowance for "repairs, insurance and management".

Where a separate deduction for "management" is to be made, it can usually be estimated as a percentage of the gross rents or market rents. Advice should be sought from management agents as these percentages will vary considerably between one or two per cent and 15%. The use of percentages here is as difficult as with repair allowances, as two similar properties could require the same management but, due to location, command dissimilar rents. Management fees on service costs are higher and can be as much as 20% of the cost of the services.

The allowance to be made for this type of outgoing, and for voids and losses of rent referred to below, cannot be dissociated from the yield to be used in the valuation. In making an analysis of sales of properties let at inclusive rents, the outgoings have to be deducted before a comparison can be made between net income and sale price to arrive at a yield. If management is allowed for in the analysis, the net income is reduced and a lower yield is shown to have been achieved than would have been the case if management had not been allowed for. Accordingly when, in making a valuation, a yield is used derived from analysis of previous sales, it is important that similar allowances are made for outgoings as were made in the sales analysed.

10. Value added tax

In the case of some commercial property, value added tax (VAT) is chargeable on rents and service charges and all VAT expenditure is recoverable by the landlord. However, in many cases the property will not be "vatable". In such cases, any outgoings on which VAT is charged are deducted gross of VAT. This is considered in further detail in Chapter 21.

II. Outgoings and rent

The impact that outgoings may have on a tenant's bid offer is illustrated in the following chart, where it is assumed that a prospective tenant is offered a property on three different bases, and the costs of meeting the various liabilities are as indicated.

Terms of lease	External repairs	Internal repairs	Insurance	Management	Rent	Budget figure
FRI lease	13,000	12,000	1,600	1,400	120,000	148,000
Internal repairing and insuring lease	–	12,000	1,600	1,400	133,000	148,000
Fully inclusive terms	–	–	–	–	148,000	148,000

In practice the effects on rent will not be quite so clear cut, but the chart does demonstrate how rents are affected by lease liabilities. It also shows that rents are, in relative terms, at their lowest when a lease is on FRI terms. The valuer, when comparing rents on different lease terms, will need to adjust by adding back or subtracting allowances for outgoings as appropriate.

12. Current issues

In the current market landlords are experiencing changeable market conditions. Lease lengths are shortening[2] and there is some reluctance from tenants to commit to FRI leases or to meeting the landlord's costs of external repairs. There have also been increases in the volume of vacant properties and the size of lease incentives in some areas. The valuer in every valuation has to check carefully, by reading the lease or leases, to see precisely what the gross income is and what deductions or allowances need to be made to estimate net income. Where leases are on gross terms the valuer needs to check for escalation or inflation clauses which may allow for annual increases in the event of, say, increases in utility costs. Lease structures are becoming more complex and the net income assumptions of the FRI lease may no longer be the norm.

2 Investment Property Databank and British Property Federation Annual lease review report 2010 gives the 2009/10 average commercial lease length as 5.0 years.

Chapter 7

Yield

1. Introduction

To use the income approach the valuer must determine the yield. This represents the rate of return that buyers, at the valuation date, are seeking in relation to the particular interest in that type of property, of that investment quality, in that location. The yield[1] is normally based on the analysis of comparable transactions. Valuers must do more than simply analyse the yield – they must also have a clear idea of what the market is doing, why the market is doing it and, if they are to advise adequately on the quality of the investment, what the market is likely to do in the future.

The valuer needs to have an understanding of the current returns from most types of investment and of the principal factors which influence them. This is because property is only one form of investment and it must compete for funds with all other forms. In some cases, the characteristics of a property investment will be similar to stock market quoted investments and thus the yields will be related to each other. This chapter considers the principles governing interest rates generally and the yields on the main types of landed property.

However, it is first necessary to dispose of two preliminary points that can cause confusion.

(a) Nominal and actual or effective rates of interest

The nominal rate of interest, or dividend, from savings or an investment in stocks or shares is the annual return to the investor in respect of every £100 saved or of every £100 face value of the stock. Where stock is selling at face value, that is at par, the nominal rate of interest and the actual rate of interest, or yield, are the same. For example in the case of Government Stock such as 2.5% Consolidated Stock ("Consols"), the nominal rate of interest is fixed at 2.5%; that is, £2.50 interest will

---

1 Yield in the normal context of the income approach is called the 'all risks yield' (ARY) as it is the single yield most used to analyse comparable sales evidence of investment property sales. The term is considered to cover the initial yield, equivalent yield and reversionary yield (see Chapter 9).

be received each year for each £100 face value of the stock held. If the stock is selling at £100 for each £100 face value, then investors will receive £2.50 interest each year for every £100 invested, which gives a yield of 2.5%. But if that £100 face value of 2.5% Consols is selling at £75, or £25 below par, then each £75 of capital invested will be earning £2.50 interest annually:

$$\therefore \text{Yield} = \frac{2.5}{75} \times \frac{100}{1} = 3.333\%$$

Thus, the actual rate of interest is 3.33% whilst the nominal rate of interest remains 2.50%.

If an industrial concern declares a dividend of 25% on its ordinary shares, then the company will pay 25% of the nominal value of each share as the dividend. Hence, if the shares are £1 shares, the dividend per share will be 25% of £1 = 25p per £1 share. But if the price of each £1 share on the market is £4, then:

$$\text{Yield} = \frac{25p}{400p} \times \frac{100}{1} = 6.25\%$$

If the share price rises to £4.50 and the same dividend of 25% is paid, then:

$$\text{Yield} = \frac{25p}{450p} \times \frac{100}{1} = 5.56\%$$

From these examples, two important points can be noted:

(i) A comparison of income receivable from various types of investment can only be made on the basis of yields, and that nominal yields derived from face values are of no use for this purpose.
(ii) A rise or fall in the price of a security will cause a change in the yield of that security.

(b) Timing of payments and yields

A yield is expressed as the interest accruing to capital in a year. Hence, if an investment is made of £1,000, and £100 in interest payments are made in each year, the yield is:

$$\frac{100}{1000} \times \frac{100}{1} = 10\%$$

But if the payment of interest is made in instalments then the yield will differ. For example, if the investor receives £50 after six months and a further £50 at the end of the year, then:

$$\frac{50}{1000} \times \frac{100}{1} = 5\% \text{ every six months}$$

The payment received after six months can be re-invested. Assuming it is re-invested in a similar investment, then further interest of 5% for the remaining half-year will be earned. The total interest payments at the end of the year are:

£50 (after six months) + (5% of £50) + £50 (end of year payment)

which equals £102.50.

And

$$\frac{102.5}{1000} \times 100 = 10.25\%$$

As the payment patterns change, such as quarterly in arrears or quarterly in advance, so will interest for the year change. This phenomenon is recognised in everyday life by the adoption of annual percentage rate (APR) figures which are quoted in respect of loan rates for borrowers or interest payments for credit card borrowers. The APR reflects the timing of the interest charged on the loans; this is rarely interest added solely at the end of the year. In the case of savings, an annual equivalent rate or AER will be quoted.

When reference is made to a yield, it is the yield as determined by the total annual interest expressed as a return on capital, ignoring the timing of the payments. Thus, in both of the foregoing examples, the notional or nominal yield is 10% but the true yields are 10% and 10.25%, respectively. Normally the simplistic approach is adopted by valuers and property investors, although the final yield chosen will reflect the timing of the payments. Thus the yield on a property let at £1,000 per quarter payable in advance, and offered for sale at £40,000, will be calculated by property valuers for valuation purposes as 10% [(4 × £1,000 = £4,000) and (£4,000/£40,000) × 100 = 10%]. The APR is in fact a little over 10.657%; valuers refer to this as a True Equivalent Yield (TEY) – see *Parry's*, pages 29–33, for yield conversion tables (see Chapters 9 and 12).

2. Principles governing yields from investments

The precise nature of "interest" and the relationship between, and the levels of, long-term, medium-term and short-term rates of interest are subjects for the economist and market analyst and are not dealt with in this book. The valuer is interested in why investors require investment A to yield 6%, investment B 3% and investment C 12%.

A reasonably simple explanation of the many complex matters that the investor must take into account in determining the yield required from an investment can be derived from the creation of an imaginary situation. For this purpose the following assumptions are made:

- the real value of money is being maintained over a reasonable period of years – that is, £1 will purchase the same quantity of goods in, say, 10 years time as it will now; and either
- there is no taxation, or the rates of tax are so moderate as not to influence the investor significantly, or
- the system of taxation is such that taxes bear as heavily on capital as on income.

In these circumstances the yield required by the investor would depend on:

- the security and regularity of the income;
- the security of the capital;
- the liquidity of the capital; and
- the costs of transfer, i.e. the costs of buying and selling.

The general principle is that the greater the security of capital and the greater the security and regularity of income, the greater the ease with which the investor can turn the investment into cash; and the lower the costs of transfer, the lower will be the required yield.

For example, if investment D offers a guaranteed income payable at regular intervals with no possibility of loss of capital, with capital repayable in cash immediately on request at no cost to the investor, then the investor will require the minimum yield necessary to induce him to make the investment. If investment E offers the same terms as D except that six months notice is needed in order to withdraw the capital, then a yield sufficiently higher than the minimum will be required to offset this difference. If investment F offers the same terms as E except that there are some transfer costs, the investor will require a still higher yield.

This imaginary situation must now be adjusted to reflect actual market conditions.

First, money does not retain its real value during periods of inflation. Thus, where an investor investing during a period of inflation is guaranteed a secure income of, say, £100 a year, yet finds the value of the pound is halved in 10 years, then this "secure" income will have a real value of only £50 a year. So, if "security" of capital means merely that if £1,000 is invested now and £1,000 can be withdrawn on demand, then the real value of the capital would be halved over a period of 10 years. If, in these circumstances, 10% for an income of £100 a year from an investment of £1,000 is a reasonable yield, the investor should be prepared to accept a lower initial yield from an investment which will protect the capital and income from the erosion of inflation. Thus, they might be prepared to accept a yield of 5% now on an investment of £1,000 if there is a reasonable chance that:

- the income of £50 a year now will have doubled to £100 a year in 10 years' time, thus maintaining its real value; and
- the capital of £1,000 will increase to £2,000 in 10 years' time, thus maintaining its real value.

An investment that offers the investor the opportunity of maintaining the real value of capital and income is described as a "hedge against inflation" and is said to be "inflation proof".

The second and third assumptions for the imaginary situation are related to the level and incidence of taxation. Over recent years taxable events and the rates of tax have changed frequently. Tax tends to divide between tax on capital and tax on income.

Until 1962 there was no tax on capital apart from estate duty. This led to surpluses arising from the sale of capital assets being free of tax. In 1962, a capital gains tax (CGT) was imposed on short-term gains, followed in 1965 by the establishment of CGT on all gains. This tax remains today, having been substantially revised in 2008. From 23 June 2010 the rates for individuals are 18% or 28%, for trustees and personal representatives 28% and for gains qualifying for Entrepreneurs' relief 10%; for companies, other provisions operate under corporation tax. In the past various other taxes have applied to gains from development land.

Estate duty was a tax imposed on the assets of a person on their death, but careful tax planning made it possible to keep this down to modest levels or even to avoid it altogether: it was regarded as an avoidable tax. Estate duty was replaced by capital transfer tax in 1975. This was a tax charged not only on assets at death but also on the value of gifts made during a person's lifetime. In 1986 capital transfer tax was replaced by inheritance tax which is more akin to estate duty. The rates of tax assessed on the value of the assets in the estate varied from 30% to 60%, but since 1988 a single rate of 40% has been imposed above an annually determined base on which no tax is levied.

As to tax on income, this has seen some significant changes in the rates of tax over the same period, but the taxes imposed (corporation tax on companies and income tax on individuals) have existed in some form for many years. Tax can be a significant factor in respect of investments and is subject to frequent change and fluctuation; in general the tax on income has tended to be more penal than the tax on capital, particularly on capital gains. For current rates readers should refer to www.direct.gov.uk.

The four principles listed above, adapted to meet conditions of inflation and levels of income and capital taxation, have governed yields in the investment market in the UK in recent years. This is apparent from examination of yields from different types of security during this period. British government securities, which in times of stable prices and moderate taxation have been described as "ideal securities", offer the minimum yield because they are practically riskless. In about 1955, the yield on government securities rose above that on ordinary shares in sound companies and for quality properties; this became known as the reverse yield gap; In the current economic climate, short-dated stock redemption yields have fallen to all time lows of less than 1% with undated stock at around 3.75%.

There are many reasons for the current financial position, which has seen the price of government stock rise and yields fall. Of particular importance is the security factor at a time when there is uncertainty over corporate profits, suggesting dividend cuts or in some cases short-term freezes on dividends. In addition, the historic expectation that share prices would keep pace with inflation appears to have taken, in some cases, a severe battering by the market. The market for stocks and shares is experiencing considerable readjustment to changing domestic and world economic conditions. Historic low prices for stocks and high yields reflecting the inflation-prone nature of stocks, compared to high prices and low yields from equities reflecting their inflation-proof characteristics, would appear to have been reversed. All-time low base rates and anticipated falling rates of inflation have caused investors to revise their investment expectations. The valuer has to observe and be familiar with every type of investment and with the macro and micro economic forces which influence buyers and sellers in the various markets. The general principles of risk referred to here are sound, but the effects of the 2007/08 credit crunch, following the impact on the financial markets of the sub-prime lending practices in the US and elsewhere, have led to significant repositioning of many investors across a range of markets. This has impacted on the UK property market. Under changing market conditions such as these, valuers must understand the underlying forces of market movement and must not only observe the affect on yields but also consider possible future movements in yields.

3. Yields from landed property investments

An investor in land and buildings (landed property) will be aware of the other forms of investment available and of the yields to be expected from them. Investors will, therefore, judge the yield they require from a landed property investment by comparison with the yields from other forms of investment such as insurance, building societies and stocks and shares. Although the principles governing yields discussed above are as applicable to landed property as to other forms of investments, certain additional features have to be considered.

First, there is the question of management. Investors have no direct management with, for example, government securities, as the dividend will be received by bank transfer or dividend cheque every six months. However, with most types of landed property, some direct management will be required. The actual cost of this management is allowed for in computing the net income, as discussed in Chapter 6. Apart from management, a landowner will incur costs in securing tenants, agreeing new rents, and may incur legal and surveyors' fees in disputes with tenants over various matters affecting the property. The yield must compensate for these costs and risks and where, in addition, management is troublesome, the investor will require the yield to be even higher.

The second special feature of landed property relates to liquidity of capital and costs of transfer. Shares can normally be bought or sold through a stock exchange very rapidly and at minimal cost. A transaction in landed property, however, is normally a fairly lengthy process and the costs of transfer, such as legal and agent's fees and stamp duty, are somewhat higher. The effect is reflected again in the need for the yield to be higher. The impact of such costs on the gross and net yields is referred to in Chapter 9.

The last of these special features is legislation. Legislation does, of course, impinge on many types of investment but its effects, direct or indirect, on landed property are frequently of major significance. The Rent Acts and various acts relating to planning provide excellent examples. The former limit or the limited amount of rent that can be charged for certain categories of dwelling-house, and the latter severely restrict the uses to which a property can be put. Existing legislation and the possibility of new legislation increase the risks of investing in landed property, and the yield required by investors will reflect all these additional risks.

There is more than one type of interest in landed property and different interests in the same property may have different yields. The yields considered below are from freehold interests in each type of property and merely indicate a typical range for the type of property at the time of writing. The general level of yields from all types of investment, or from landed property investments only, may change and there are often substantial variations between yields from landed property investments of the same type. The location, age and condition of the buildings and the status of the tenant are all factors that influence the yield. Other things being equal, the older the buildings and the poorer their condition, and the less substantial the tenant, the higher will be the yield. In an actual valuation the yield must be derived from market evidence.

(a) Shops

A normal range of yields is 4.75–15%, the lower rates being applicable to shops in first-class "High Street" positions, out of town centres and superstores where the occupiers are national retail companies, and the higher rates apply to shops in tertiary positions occupied by small traders. Position and type of tenant are vital factors in judging a shop as an investment and these matters are dealt with in Chapter 18. Retail trade fluctuates with the health of the economy and with consumer confidence. The availability of credit is a major factor affecting shopping behaviour. Retail yields move down when retailing is booming and rental growth is positive; they move up when retailers downgrade profit forecasts and rents are static or falling and voids have increased. The 2012 economic position is expected to continue for a number of years with a continuing impact on retail trading figures; recent increases in fuel costs together with social, political and environmental issues, including the effect of major earthquakes worldwide, will have an effect on consumer confidence and consumer choice and may affect rents and yields. The wide range of yields reflects the wide variation in the types of retail investment.

(b) Offices

A normal range of yields is between 4.25% and 12%. The lowest yield would be for a modern building in the West End of London let to a single tenant on FRI terms. The same building let in suites where the owner is responsible for the management, including the provision of such services as lifts and central heating, would yield a slightly higher rate even though the costs are covered by a service charge. The highest yield would be for older buildings, possibly let in suites and lacking modern amenities. In areas where the demand for offices was strong, rents increased steadily up until 2007, providing a secure investment in real terms with substantial capital appreciation. However, the recession after 1989 caused a reappraisal of office investments and a similar review occurred after 2009. The most important factors to be considered are: length of leases or unexpired term; and the quality of the tenant. If less than 10 years remain on the occupation lease the income is seen as less secure as the building may become, and may remain, vacant. The factors that affect office yields are location, quality of construction, supply of modern facilities (such as raised access flooring), layout of accommodation, economy of occupation, quality of tenant and length of letting. Energy Performance Certificates[2] for commercial space, coupled with a growing employer and employee wish to be associated with "green" buildings, may affect rents and increase the spread of yields between old and new office properties.

(c) Factories and warehouses

Until recently, factories and warehouses have not been a popular investment and the range of yields, 6.25–15%, to some extent reflects past unpopularity. However, significant changes have occurred recently that have blurred the distinction between factories and warehouses and other types of property. In the case of factories, a strong demand has arisen for space to be occupied by companies in the computer and electronic fields where the requirements are for standards closer to those of offices than the traditional, more basic, manufacturing space. These are commonly termed "high-tech" buildings and include data storage facilities. Similarly, alongside the traditional warehouse, there has grown up the development of retail warehouses, where the form of building is a single-storey building but with extensive car parking and with the occupier selling direct to the public. Thus, the building form is similar to a warehouse but the activity is

---

2 The Government proposal to ban the letting of all property with an EPC level of F or G from 2018 is a new market factor which needs to be reflected in the valuation of the less energy-efficient buildings. Currently it is only a possibility, but it may become a reality.

similar to that of a shop. Where the use is closer to office or retail, the yield will move to the yields appropriate to such uses. Business parks and distribution depots, including internet sales distribution centres, would fall into this category.

Yields will tend to be at the bottom of the range in the case of modern single-storey factories and warehouses in areas of good demand. In the case of factories and warehouses in areas of poor demand, particularly in areas suffering from general industrial decline accompanied by high unemployment levels, yields will be at the top of the range. For older buildings, which are frequently multi-storeyed and with low heights to eaves, the yield may be well in excess of 15%. Indeed, apart from buildings capable of conversion to small workshop units, they may even cease to be considered as investments and will change to other uses such as residential apartments or remain vacant pending redevelopment.

(d) Residential properties

This expression covers a very wide range of properties and some subclassification is required before even the broadest general statement on yields can be made. It should be noted that, whereas in the past residential properties formed a major part of the property investment market, this is no longer the case. The main impetus for this decline has been the growth in owner occupation, coupled with restrictive legislation imposed on the powers of landlords. Even where a property is let, it may be seen by a buyer as a speculation with the hope of a sale with vacant possession rather than as a long-term income-producing investment. Thus, the yields referred to in this section are intended to apply to those instances where the property is likely to remain an investment (see Chapter 16). The position has again changed with the removal of legislative restrictions on rents and most of the legislation providing tenants with security of tenure. A new "buy-to-let" market has emerged, with individuals and smaller property companies becoming significant owners. Some specialist players have been very active in developing units for the student and retirement markets.

With the tenement type of residential property the yield is high, from 12–18%. Properties of this type will normally be old, rent restricted, subject to a great deal of legislation imposing onerous obligations on owners relating to repair, cleanliness and other matters, and many tenants may be unreliable in payment of rent. Capital and income are not, therefore, secure; income may not be regular and the property may be difficult to sell. In recent years, such properties have tended to be acquired by public authorities or publicly financed organisations (such as housing associations) so that they are gradually ceasing to be investments found in the private sector. This sector of the residential market is in decline and rarely encountered.

Although the majority of housing is now either owned by a housing association, owner-occupied or owned by local authorities, a substantial number of housing units are held for investment purposes by individuals and companies.

Blocks of flats, if they were modern, were regarded as a good investment, yielding about 8–10%. Frequently they sold at much lower yields but these higher prices, known as "break-up" values, were based on the expectation that a substantial proportion of the flats would be sold to the sitting tenants, usually on long leases, at prices in excess of investment value, or would be sold with vacant possession at a later date. The valuation approach was often to calculate a percentage of vacant possession value rather than to apply a yield to the rent income. The percentage was generally in the range of 35–50%, but this increased substantially where the tenant was elderly and early vacant possession was expected. A few such investments are still to be found, but most have been bought and sold on long leases with the landlord receiving a low rent, retaining ownership of the freehold and managing the premises with a full service charge. A newer market has emerged for modern purpose-built student accommodation where yields of 7–8% might be encountered.

Let houses on terms other than assured shorthold tenancies yield in the range 5–15%. As with flats, only more so, such properties when bought outside the public sector are bought with a view to their ultimate sale so that the usual investment criteria are not applicable; very low yields apparently emerge, but these are not to be interpreted as simple investment yields.

In the case of new "buy-to-let" schemes, typically of one- or two-bedroom city apartment blocks, the unit price is the same for owner occupation as for purchase by potential landlords. These units are offered for let on assured shorthold tenancies where repossession or lease renewal is available at six month intervals. The investment yield to the landlord was often below 5% when there was evidence of strong capital growth. Market values in this sector of the residential market are as for similar units offered for sale with vacant possession. Fluctuations in tenant's demand and rents coupled with a fall in vacant possession values since 2008 has caused many buy-to-let units to have fallen in market value. This in turn attracts other investors who see an opportunity to buy for a better initial yield than previously.

Ownership of individual leasehold flats held as investments in the same way as individual houses have an additional risk factor due to the head lease requirement to meet the outgoings, including a service charge.

(e) Ground rents

A ground rent is a rent reserved under a building lease in respect of the bare land without buildings. In recent times, the practice has been to grant the building lease once the buildings have been built, with the developer holding an agreement to be granted the building lease during the period of development. Building leases are normally granted for the long term, 99 years at one time being fairly common, although modern leases are frequently for around 125 years or even 999 years, particularly in the case of commercial properties.

Secured ground rents are those where buildings have been constructed on the land. If the lease has many years to run the yield will be in the range of between 6% and 15%. Where the rent is a fixed amount throughout the term it is comparable in many ways to government securities and the yield will be similar to, but slightly higher than, that on 2.5% Consols and at the top of the range. The investor obtains better security in real terms where the lease provides for upward revision of the rent at reasonable intervals, and such an investment would provide a lower initial yield than where the rent is a fixed amount. In some cases, the ground rent is geared to capital value or market rent at review, which makes the investment more inflation proof, thus justifying an even lower yield.

The amount of the rent is also a significant factor. A single rent of a few pounds may be unsaleable unless the occupier is in the market, since the cost of collection will absorb most of the rent received. Usually these small ground rents will sell in "blocks".

Where leases are for less than 60 years the value is affected by the market's resistance to wasting assets, but there may be considerable marriage value (now called synergistic value by the RICS) available to the freeholder. This is the surplus value that sometimes emerges where two interests in a property are combined, especially the merger of a freehold and a leasehold interest to form a freehold in possession (see Chapter 9). In the case of houses with a rateable value of less than £1,000 in London and £500 elsewhere, the owner occupier lessee may force the sale of the freehold at a price that excludes marriage value (Leasehold Reform Act 1967), but in all other cases compulsory sale or lease extension (under the Leasehold Reform, Housing and Urban Development Act 1993) allows the freeholder not less than 50% of the marriage value.

Where the lease is a wasting asset the value of the freehold interest will reflect not only the reversion to vacant possession value but also "hope value" based on the assumption that sometime during its term the leaseholder will wish to acquire the freehold or an extended lease. The impact of leasehold reform is considered in Chapter 16.

4. Changes in interest rates

The landed property investment market normally responds to change less rapidly than other investment markets. The reason for this is probably the length of time taken to transfer ownership and the high costs of transfer. If, for example, it is felt desirable to hold a greater proportion of assets in cash than hitherto, stock exchange securities can be sold immediately even if some loss is incurred, but landed property cannot be sold with anything like the same speed. If, therefore, the change in preference is temporary, the landed property market may be less volatile than the other markets.

The landed property investment market tends to respond to longer-term changes. In recent years, the movements in interest rates have been more marked and more rapid than in earlier times. This reflects the closer integration of the property market into the general investment market brought about in the main by the growing involvement of pension funds and insurance companies ("the institutional investors").

Yields for property investments have been more stable and are historically seen as comparable to long-term interest rates available in the economy generally and not comparable to, or so immediately responsive to, short-term money market rates.

However, the 2007/2008 credit crunch reaction to the subprime mortgage problems in the US and UK had an impact first on the residential market but then on the commercial market. The impact on credit availability and the cost of credit filtered through to property yields faster than some valuers might have expected. In addition, there had been a rapid increase in retail investors, that is, in the quantity of individuals and volume of investment by individuals, not in retail units *per se* but in shares and units which provided individuals with an investment in property which previously could not have been afforded. This in itself would not have been a problem except that high numbers suddenly wished to realise their capital. So, if five thousand individuals had invested £10,000 each in a Real Estate Investment Trust (REIT) and decided they wanted to redeem their investment, then the fund had to sell £50,000,000 of property assets at short notice, or raise £50,000,000 in loans. This, replicated a dozen times across the market, together with other factors, created market uncertainty and yields rose, pushing values down and causing more retail investors to sell.

A downward spiral developed and an increase in volatility occurred in a market previously noted for its long-term benefits. The market is, therefore, now affected by the same short-term moves in interest rates as are found in other investment markets.

Understanding the details of market movements and interest movements has been noted elsewhere as being the role of the economist. The valuer's responsibility to clients is to be aware of market changes and to be responsive to such changes, making appropriate market-informed changes to both their assessment of rent and to yields in line with observable market movements.

Those looking for a more academic debate might be interested in Dunse, N., Jones, C., White, M. and Wang, L. (2007) Modelling Urban Commercial Property Yields: Exogenous and Endogenous Influences, *Journal of Property Research*, 24(4), 335–354.

Chapter 8

Investment mathematics as it applies to property valuation

I. Introduction

It is the valuer's duty to estimate the market value of a property. In making that estimate, the valuer must come to certain conclusions regarding the property, such as: its current net income and the likelihood of that income increasing or decreasing in the future; the possibility of future liabilities in connection with the property; and the yield prospective buyers in the market are likely to require as a return on their capital.

In undertaking a valuation using the income approach, the valuer makes use of the same financial principles that underpin bond and share pricing; financial mathematics provides the valuer with simple tools for converting estimates of future cash benefits and costs into estimates of present values (PVs) from which an estimate of market value can be deduced.

Although calculators and computers have substantially changed practice since *Parry's Tables* were first published in 1913, these tables are still extensively used. Whether a valuer uses printed tables, programmed calculators, industry standard software valuation packages or Excel, it remains essential to understand the principles of financial mathematics. This and Chapter 9 are fundamental to the understanding of the income approach or investment method.

References to valuation tables are to the 12th edition of *Parry's Valuation and Investment Tables* (2002, EG Books) by A.W. Davidson. These references are retained as the tables are still used by many practitioners and as a teaching tool, even though most investment valuations can now be completed with a calculator or by using investment valuation software.

This chapter looks at the various functions of £1, their mathematical construction, the formula on which they are based, what they represent, and when and how they are used in practice. All are based on the principle of compound interest and for this reason the first to be considered is the "Amount of £1" – which is compound interest. These functions are not currency sensitive and all references to £1 could be to $1, €1, yen, rupee or any other currency.

In this chapter, income is assumed to be payable annually in arrears and the effects of inflation are ignored, unless otherwise stated.

2. Amount of £1 (A£1)

(pp. 123–139 of *Parry's Valuation and Investment Tables*)

This is the amount to which £1 invested now will accumulate at different rates of interest over any given number of years on the basis that interest is payable annually in arrears and assuming that interest compounds. It can be calculated manually as shown in Example 8-1.

Example 8–1

To what amount will £1 invested at 5% accumulate to in three years?

Answer	Principal	Interest	Total
Amount at end of 1 year (£1 plus interest at 5% on £1)	£1.00	£0.05	£1.05
Amount at end of 2 years (£1.05 plus interest at 5% on £1.05)	£1.05	£0.0525	£1.1025
Amount at end of 3 years (£1.1025 plus interest at 5% on £1.1025)	£1.1025	£0.055125	£1.157625

This process is laborious and it is simpler to take the multiplier direct from the tables. This is given on page 130 as 1.1576, or calculated direct from the formula as:

$$(1+i)^n = (1+0.05)^3 \quad \text{i.e. } [(1.05) \times (1.05) \times (1.05)] = 1.15769$$

The amount of £1 is important because compound interest is the basis of all investment finance. It is sometimes used to calculate the loss of interest where capital sums are spent on a property which, for the time being, is producing no income.

Example 8–2

A site for residential development was purchased for £5,000,000 five years ago and £1,000,000 was spent at once on roads and other development costs. No further work has been possible and no return has been received for five years. What is the cost to the buyer at the end of the five years, assuming interest at 9%?

If the owner had not tied up £6,000,000 in the purchase and development of this land that sum could, for the purpose of this illustration, have been invested in some other investment producing interest at say 9%. The cost of the property to the investor is, therefore, the sum to which £6,000,000 might have accumulated in five years at 9%.

Capital sum invested	£6,000,000
A of £1 in 5 years at 9%	× 1.5386
Cost to buyer	£9,231,600

Another property example is where an owner borrows money to buy a development property on the basis that interest will be charged but will not be repaid until the development is completed (the interest, as shown here, of £3,231,600, is said to be "rolled up").

Example 8–3
A site was bought three years ago and a loan of £200,000 was obtained at a fixed annual interest rate of 8%, to be rolled up until the development is complete. The development will be completed in one year's time. Calculate the sum due for repayment at that time.

Capital sum borrowed	£200,000
A of £1 in 4 years at 8%	× 1.3605
Sum due for repayment (loan plus rolled-up interest)	£272,100

The formula for A£1 is derived as follows:
To find the amount to which £1 will accumulate at compound interest in a given time.
Let the interest per annum on £1 be i
Then the amount at the end of one year will be $(1 + i)$
The amount at the end of two years will be $(1 + i) + i(1 + i) = 1 + 2i + i^2 = (1 + i)^2$
The amount at the end of three years will be similarly $(1 + i)^3$
By similar reasoning the amount in n years will be $(1 + i)^n$
(The total interest paid on £1 in n years will be $(1 + i)^n - 1$

The amount of £1 formula is a general formula for compound interest. The rate of interest i must be the effective rate for the period n. For example, if a bank charges 1.5% per month interest on a credit card then the annual equivalent rate is $[(1 + 0.015)^{12} - 1] \times 100 = [1.1956 - 1] \times 100 = 19.56\%$. The reader is reminded that the expression $(1.015)^{12}$ means that the number inside the bracket is multiplied by itself 12 times, i.e. it is raised to the power of 12.

3. Present value of £1 (PV£1)

(pp. 89–109 of *Parry's Valuation and Investment Tables*)
   The PV£1 is the inverse of A£1; whereas the latter shows the amount to which £1 will accumulate at compound interest over any given number of years, the PV£1 shows the sum which invested now will amount to £1 in a given number of years with compound interest.
   The PV£1 is the reciprocal of the A£1 and so it is possible to obtain any required figure of PV£1 by dividing unity by the corresponding figure of A£1.

Example 8–4

What sum invested now will accumulate to £1 in six years' time, at 5% compound interest?

Answer

Let V equal the sum in question.

$V \times$ A£1 in 6 years at 5% = £1

$$\therefore V = \frac{1}{\text{Amount of £1 in 6 years at 5\%}}$$

$$\therefore V = \frac{1}{1.3400956}$$

so $V = 0.7462154$

It is quicker to take the figure from the PV£1 table or by calculating from the formula, thus:

$$PV = \frac{1}{(1+i)^n} = \frac{1}{(1+.05)^6} = \frac{1}{1.3400956} = 0.7462154$$

The PV of £1 is used to calculate the value at the present time of sums receivable in the future and for making allowances for future expenditure in connection with property.

The value today of the right to receive a capital sum in the future is governed by the fact that, whatever capital is invested in buying that right, it will be unproductive until the right matures. If £x is spent in buying the right to receive £100 in three years' time, it follows that for those three years the buyer's capital will be showing no return, whereas if the £x had been saved in some other security it might, during those three years, have been earning compound interest. The price which the buyer can fairly afford to pay is that sum which, with compound interest on it (calculated annually in arrears) at a given rate, will amount to £100 in three years' time.

Example 8–5

What is the PV of the right to receive £100 in three years' time on a 5% basis?

Answer

Let $V =$ sum which the buyer can afford to pay. The buyer will be losing compound interest on this sum during the three years and will have to wait before they receive the £100. V must be the sum which, together with compound interest at 5%, will equal £100 in three years' time.

$V \times$ A of £1 in 3 years at 5% = £100.

$$V = \frac{1}{\text{Amount of £1 in 3 years at 5\%}}$$

$$V = \frac{£100}{1.1576}$$

$$V = £86.38$$

The above method has been used to show the principles involved. The same result would be obtained direct from the PV of £1 as follows:

Sum receivable	£100
PV£1 in 3 years at 5%	× 0.8638376
PV	£86.38

The process of making allowance for the fact that a sum is not receivable or will not be spent until sometime in the future is known as "deferring" or "discounting" that sum. In the above example, £86.38 might be described as the "PV of £100 deferred for three years at 5%".

Since the valuer is concerned with the temporary loss of interest on capital invested in buying future sums of money, the rate per cent at which the value is deferred should generally correspond to that which an investor might expect from the particular type of security if currently owned. In other words, the rate should be a "remunerative" one.

Example 8–6

What is the market value of the reversion to a freehold property let for a term of five years at a peppercorn rent if the deferment or discount rate is 5% and comparable properties are selling with vacant possession (VP) at £700,000?

Since no rent is payable for the first five years (a "peppercorn rent" being a legal device for leases where no rent is to be paid, a peppercorn being the ingredient of a pepper mill), no value arises in respect of this period and therefore:

Value in five years time	
VP value	£ 700,000
PV£1 in 5 years at 5%	0.7835262
PV	£548,468

Where the sum is a future expense, or in the nature of a liability that cannot be avoided, the valuer is concerned with the interest rate at which a fund can be accumulated to meet the expense, which may have to accumulate at a net-of-tax rate. A future capital liability can be provided for, either by the setting aside of an annual amount in the form of a sinking fund or by the investment of a lump sum which will, with certainty, accumulate to the required amount in the future.

The first method will be discussed later in this chapter. The second method can be achieved through an insurance company by buying a single premium policy or by buying some other investment with a guaranteed rate of interest for the whole period.

Due to the need to guarantee the future payment there is an interest risk on the insurance company because of fluctuating market rates of interest. Therefore, the rate of compound interest on it allowed by insurance companies is low. It is therefore more sensible, as a rule, to defer sums to meet liabilities in the future at a low "accumulative" rate rather than at the "remunerative" or risk rate at which interest on capital is taken. The actual rate depends on the interest rates prevailing at the time and the financial market's view of future trends in interest rates. Historically, such rates have tended to be around 2.25–3.5%, net after tax, but might today be closer to 1% net of tax.

Example 8–7

Provision is to be made to replace a staircase in two years' time at a cost of £20,000; a chimney stack or flue in a further four years at a cost of £50,000; and the boundary fence two years later at a cost of £10,000. Calculate the sum that needs to be invested today in an assured savings account at 2.5% to provide for these future liabilities.

Calculation		
Staircase	£20,000	
PV£1 in 2 years at 2.5%	0.952	£19,040
Chimney stack or flue	£50,000	
PV£1 in 6 years at 2.5%	0.862	£43,100
Boundary fence	£10,000	
PV£1 in 8 years at 2.5%	0.821	£8,210
Say		£70,350

An exception to the low interest rule occurs where the future capital expense can be met out of income from the property itself, in which case there will be no need to provide for it by investment of a single premium at a low rate of interest and the sum can be deferred at the appropriate "remunerative" rate.

Where the future expenditure is of a kind that is optional – as distinct from a liability that cannot be avoided – it is probably better to allow for it at the remunerative rate, since it is assumed that an investor would not set aside a sum to accumulate at a low rate of interest, e.g. in a single premium policy.

In practice, cost inflation complicates these calculations as the sum required in the future is likely to be greater than today's estimated cost.

The formula for the PV£1 is as follows:

To find the PV of £1 receivable at the end of a given period of time.
Since £1 will accumulate to $(1 + i)^n$ in n years, the PV of £1 due in n years equals

$$\frac{1}{(1+i)^n}$$

If it is desired to take into account the payment of interest and its reinvestment at more frequent intervals, then i and n must be modified.
Let the number of payments in 1 year be m. Then the total number of payments is mn. As the annual rate of interest is i, the rate of interest for one period will be i/m and the amount of £1 in n years equals:

$$\left(1 + \frac{i}{m}\right)^{mn}$$

The PV of £1 will correspondingly be:

$$\frac{1}{\left(1 + \frac{i}{m}\right)^{mn}}$$

This assumes that the rate of interest is the nominal rate representing the total interest received over the year.

If it is assumed that interest paid quarterly is reinvested at the same annual rate, the effective annual yield will need to be determined. The effective annual yield represents the total interest that would have accumulated by the end of the year as a result of the re-investment at compound interest on a quarterly basis. The formula to determine the annual effective yield is:

$$\left[\left(1 + \frac{i}{m}\right)^{mn}\right]^{\frac{1}{n}} - 1$$

For example, if £1 is invested at 10% p.a. payable monthly for five years, the effective yield is:

$$\left[\left(1 + \frac{0.1}{12}\right)^{60}\right]^{\frac{1}{5}} - 1$$

which $= (1.6453)^{1/5} - 1 = 0.1047 \times 100$ (to convert decimal to percentage) $=$ 10.47% p.a.
(*Note*: not all calculators provide fractional power functions; 1/5 can be solved by using the decimal, which is 0.20.)

4. Amount of £1 per annum (A£1p.a.)

(pp. 141–156 of *Parry's Valuation and Investment Tables*)

The A£1p.a. is the amount to which a series of deposits of £1 at the end of each year will accumulate in a given period at a given rate of interest.

A calculation in this precise form rarely occurs in a valuer's work; but the A£1p.a. could be used to calculate the total expense involved over a period of years where annual outgoings are incurred in connection with a property that is not generating any income. Example 8-8 illustrates this use and distinguishes between A£1p.a. and A£1.

Example 8–8

A plantation of timber trees will reach maturity in 80 years' time. The original cost of planting was £2,000 per hectare. The annual expenses average £200 per hectare. What will be the total cost per hectare by the time the timber matures, ignoring any increase in the value of the land and assuming interest is required on outstanding capital at 5%?

The £2,000 per hectare spent on planting was a lump sum which will remain unproductive for 80 years. Its cost to the owner is the sum to which it might have accumulated if it had been invested during that period at 5%; this requires the use of the A£1.

The £200 per hectare on upkeep is an annual payment which will bring no return for 80 years. Its cost is the sum to which a series of such payments might have accumulated if placed in some form of investment earning 5%. Here it will be necessary to use the A£1p.a.

Original capital outlay per hectare	£2,000	
A of £1 in 80 years at 5%	× 49.56	£99,120
Annual expenditure		
A£1p.a. in 80 years at 5%	£200	
	× 971.229	£194,246
Total cost per hectare for period to maturity		£293,366

The derivation of the formula for the A£1p.a. is as follows:

The amount to which £1 p.a. invested at the end of each year will accumulate in a given period of time.

It is conventional to assume that interest is not paid until the end of each year, hence the first payment of £1 will accumulate for $n - 1$ years if the period of accumulation is n years.

The second payment will accumulate for $n - 2$ years, and so on year by year.

The amount to which the first payment accumulates will, applying A£1, be $(1 + i)^{n-1}$; the second $(1 + i)^{n-2}$; and so on.

Let the amount of £1 p.a. in n years be A.

Then $A = (1+i)^{n-1} + (1+i)^{n-2} + (1+i)^{n-3} \cdots (1+i)^2 + (1+i) + 1$; or, more conveniently:

$$A = 1 + (1+i) + (1+i)^2 \cdots (1+i)^{n-1}$$

These terms form a geometric progression for which the general expression is:

$$S = \frac{a(r^n - 1)}{r - 1}$$

Substituting therein, $S = A$; $a = 1$; $r = 1 + i$;

Hence $A£1$ p.a. $= \dfrac{(1+i)^n - 1}{i}$

5. Annual sinking fund (ASF)

(pp. 111–122 of *Parry's Valuation and Investment Tables*)

The ASF is the inverse of the A of £1 p.a.; it shows the sum that must be deposited annually at compound interest in order to produce £1 in so many years' time. The figures in the former are the reciprocals of those in the latter. Thus, any required figure of ASF can be found by dividing unity by the corresponding figure of A£1p.a.

The valuer uses the ASF to estimate the sum that needs to be set aside annually, out of income, in order to meet some future capital expense, such as a possible claim for dilapidations at the end of a lease, or a sum needed for the rebuilding or reconstruction of premises.

The provision of a sinking fund to meet future capital liabilities, although often neglected by owners, avoids the embarrassment of having to meet the whole of a considerable expense out of a single year's income and enables the owner to see precisely how much of the income from the property can be treated as spendable income. It occurs quite often in service charge agreements where a sinking fund provision is made to meet, for example, the costs of replacing lifts and plant when they reach the end of their useful lives.

Example 8–9
An investor recently bought a freehold property for £1,000,000 which, it is estimated, will yield a net income of £200,000 for the next 15 years. At the end of that time it will be necessary to rebuild at a cost of £1,500,000 in order to maintain the income. How should the owner provide for this, and what will be the effect on the yield of the investment?

Answer
The owner could provide for this by means of a sinking fund accumulating at, say, 3%.

Cost of rebuilding	£1,500,000	
ASF to produce £1 in 15 years at 3%	0.0538	
Sinking fund required	£80,700	p.a.

The owner's income for the next 15 years will therefore be £119,300 (£200,000 – £80,700), giving a return of 11.93% on the purchase price instead of the 20% which the investment appears to be yielding.

The provision can be made by means of a sinking fund policy taken out with an insurance company on terms and at rates of interest similar to those referred to in section 3 of this chapter for single premium policies, i.e. 2.25–3.5%.

If the sum to be set aside annually is considerable, it is possible that an owner may find opportunity for the accumulation of it in their own business, or in some other investment, at a higher rate of interest than that paid by an insurance company. It is probable that in many cases buyers will be inclined to take this fact into account in considering the investment value of property, when otherwise the cost of allowing for replacement capital on ordinary sinking fund terms becomes prohibitive. However, the insurance policy approach guarantees a certain sum that will be available at a future date, which is rare in any form of investment other than those for a short term of up to around five years. In any event the buyer will need to allow for any income tax or corporation tax that may be levied on the interest arising from other forms of investment. It is the interest *net of any tax payable* which will be needed for comparing alternative forms of investment with an insurance policy approach.

In the examples that follow, 3% net is used as a reasonable average figure where a sinking fund is needed.

The ASF can be used on a net-of-tax basis – that is, assuming that the accumulations of interest on the sinking fund are at net-of-tax rate. The allowance for income tax on interest on sinking fund accumulations, which is made by using a net rate of interest, should not be confused with the entirely separate adjustment for income tax on the sinking fund element of income, which is made when using dual rate Years' Purchase (YP), which is considered in Chapter 9.

The derivation of the formula for this table is as follows:

To find the sum which, if invested at the end of each year, will accumulate at compound interest to £1.

Let the ASF be S and the period n years; then the first instalment will accumulate to $S(1+i)^{n-1}$; the second to $S(1+i)^{n-2}$; and so on.

Hence
$$1 = S(1+i)^{n-1} + S(1+i)^{n-2} \cdots ; \text{ or, more conveniently, } S(1+i)^2 + S(1+i) + S;$$
$$1 = S + S(1+i) + S(1+i)^2 \cdots S(1+i)^{n-1}$$

The sum of the series will be:

$$1 = \frac{S[(1+i)^n - 1]}{i} \quad \text{and} \quad \therefore S = \frac{i}{(1+i)^n - 1}$$

As noted above, S (or ASF for Annual Sinking Fund) is the reciprocal of the amount of £1 p.a.

The sinking fund investment occurs at the end of each year. This is convenient when dealing with the income from real property, which is commonly assumed to be receivable yearly at the end of each year. For further discussion of this point see Chapters 9 and 12.

In comparing sinking fund figures with actual amounts payable under a sinking fund policy, it must be remembered that, under a policy, the premium is payable at the beginning of each year. Thus, by similar reasoning to the above, in this case:

$$S \text{ really} = \frac{i}{(1+i)^n - 1}$$

However, because generally rent is assumed to be payable yearly in arrears, the first formula is the one used by valuers.

6. PV of £1 per annum (PV£1p.a.) or years' purchase (YP)

(Single rate pp. 39–51, dual rate pp. 53–71, of *Parry's Valuation and Investment Tables*. Parry's also contains single and dual rate tables based on quarterly in advance assumptions.)

The PV£1p.a. shows, at varying rates of interest, the amount that might reasonably be paid for a series of sums of £1 receivable at the end of each of a given number of successive years. By applying the PV£1p.a. to the net income of a property, the valuer can estimate its present value. Where the income is perpetual, the PV£1p.a. or YP can be found by dividing 100 by the rate of interest or yield appropriate to the property, or by dividing unity by the interest on £1 in one year at the appropriate rate. Thus (as shown in Chapter 2), YP in perpetuity can be expressed as:

$$\frac{100}{\text{Rate per cent}} \quad \text{or} \quad \frac{1}{i}$$

But where the income is receivable for a limited term only the problem is more complex. For example, if a property producing a perpetual net income of £500 p.a. can be regarded as a 5% investment, a buyer can afford to pay £10,000 for it (£500 × 1/0.05 or × 20 YP). But if the income is receivable for six years only and 5% is still a reasonable return, then the buyer cannot pay £10,000 for it because, even though the income of £500 is still 5% on the purchase price for each of the six years, at the end of the term both capital and income cease.

There are two solutions to such a problem. The first of these is to consider the income flow for the next six years and to take each year's income in isolation, as below:

End of year 1, sum receivable	£500	
PV of £1 in 1 yr at 5%	× 0.9524	£476
End of year 2, sum receivable	£500	
PV of £1 in 2 yrs at 5%	× 0.9070	£454
End of year 3, sum receivable	£500	
PV of £1 in 3 yrs at 5%	× 0.8638	£432
End of year 4, sum receivable	£500	
PV of £1 in 4 yrs at 5%	× 0.8227	£411
End of year 5, sum receivable	£500	
PV of £1 in 5 yrs at 5%	× 0.7835	£392
End of year 6, sum receivable	£500	
PV of £1 in 6 yrs at 5%	× 0.7462	£373
		£2,538

The present value of the right to receive £500 for each of the next six years is £2,538. This is a laborious approach and would be cumbersome over a longer period. The second and tidier method would be to add the PV£1s together as follows:

Sum receivable each year		£500
PV of £1 in 1 year at 5%	0.9524	
PV of £1 in 2 years at 5%	+0.9070	
PV of £1 in 3 years at 5%	+0.8638	
PV of £1 in 4 years at 5%	+0.8227	
PV of £1 in 5 years at 5%	+0.7835	
PV of £1 in 6 years at 5%	+0.7462	5.0756
PV of £500 p.a. for 6 years		£2,538

Here the annual rent is multiplied by the sum of the PVs of £1 for each year. This can be simplified further as a formula can be derived to represent the sum of the PVs for any number of years. This is known by valuation surveyors in the UK as YP single rate (see pp. 39–52 of *Parry's Valuation and Investment Tables*) and internationally as PV£1p.a.

Hence, in the above example:

Rent receivable	£ 500
× PV£1p.a. for 6 years at 5%	5.0757
Capital value	£2,538

The first solution demonstrates that the sum of £2,538 is the accumulation of six individual PV calculations. In each year it can be shown that the investor earns interest at the rate of 5% a year on the sum used to acquire the future benefit, and that the future benefit represents the repayment of the original sum plus the accumulated interest. For example, the £373 paid for a future £500 in six years' time is equivalent to earning 5% a year on £373 (£373 × A£1 in 6 years at 5% = £373 × 1.3401 = £499.86, i.e. £500 if no rounding in PV and A£1). This rate of 5%

would be referred to as the Internal Rate of Return (IRR). The achievement of both a return of and a return on capital invested can also be shown in tabular form:

Capital paid or outstanding	Income received	Interest (i%) at 5%	Capital recovered
£2,538	£500	£126.90	£373.10
£2,164.90	£500	£108.25	£391.75
£1,773.15	£500	£88.66	£411.34
£1,361.81	£500	£68.09	£431.91
£929.90	£500	£46.50	£453.50
£476.40	£500	£23.83	£476.17*

*Accumulative error due to rounding first to four decimal places and then to two decimal places.

Note that in the table the capital recovered each year is increasing at 5% a year. In the last year the income of £500 is sufficient to meet the interest at 5% on the capital outstanding of £476.40, i.e. £23.83, leaving a balance of £476.17 (rounding error) to recoup the balance of the capital. This concept is at the heart of the investment decision-making process and loan repayment calculations including repayment mortgages.

An alternative approach to the problem of valuing a terminable income is to assume that the return remains constant throughout the time period and that the capital is recovered by allowing for an amount out of the terminable income to be set aside annually as a sinking fund sufficient to accumulate during the term to the capital originally invested. If this is done, the buyer having paid £x for the interest in the first instance, will receive the income during the term and set aside out of that income a sinking fund, so that at the end of the term the sinking fund will have accumulated to the original capital of £x. In this way, the terminable income has been perpetuated and is, therefore, directly comparable with the perpetual income of, say, a freehold. This concept differs from that of the IRR, where the capital cost is recovered year by year and the return is based on the amount outstanding from year to year.

The formula for finding the PV£1p.a. on this alternative basis is derived later in the chapter, but for the current purpose it can be noted as:

$$\frac{1}{i + \text{ASF}}$$

i being the interest on £1 in one year at the appropriate rate of interest and ASF being the sinking fund to replace £1 at the end of the term.

The next question is: at what rate of interest should the ASF be assumed to accumulate? Again, the proper function of the rate of interest is that of indicating the relative merits of different investments, for which the IRR is the most commonly used basis of measurement. However, as explained previously, sinking fund arrangements can be made by means of a sinking fund policy with an insurance company; here the interest is low because the investment is riskless and as near trouble-free as possible. If it is assumed that the sinking fund is arranged in this way rather than at the investment risk rate, then all of the risks attached to the actual investment in the property will be reflected, as they should be, in the rate of interest the buyer requires on the capital invested.

These two assumptions, (a) that a sinking fund is set aside and (b) that the rate of interest at which the sinking fund accumulates is the rate appropriate to a riskless and trouble-free investment, are in no way invalidated by the fact that many investors make sinking fund provisions in some way other than through a sinking fund policy or leasehold redemption policy, or that many investors make no sinking fund provision at all. Single rate PV£1p.a. and dual rate are two distinct ways of considering the capital replacement problem associated with a time-limited investment.

Example 8–10
What is the value of a leasehold interest in property producing a net income of £500 for the next six years, assuming that a buyer is seeking a return of 5% on his money and that provision is to be made for a sinking fund for redemption of capital at 3%?

Answer
Interest on £1 in 1 year at 5% = 0.05
  SF to produce £1 in 6 years at 3% = 0.1546
∴ YP dual rate for 6 years at 5% and 3% =

$$\frac{1}{i+s} = \frac{1}{0.05 + 0.1546} = \frac{1}{0.2046} = 4.888$$

Valuation

Net income	£500
PV£1p.a. or YP 6 years at 5% and 3%	× 4.888
Value	£2,444

Note
The YP dual rate at 5% and 3%, from *Parry's Valuation and Investment Tables*, is 4.8876.

Proof
The fact that the estimated purchase price does allow both for interest on capital and also for a sinking fund (return on capital and return of capital) may be shown as follows:

Interest on £2,444 at 5% (£2,444 × 0.05)   £122
Sinking fund to produce £2,444 in
6 years at 3% (£2,444 × 0.1546)   + £378

Income from property   £500

The YP used in the above example is called a dual rate YP because two different rates of interest are used. The first (5% in the example) is called the "remunerative" rate, and the second (3%) is called the "accumulative" rate.

As is shown below, the sum of the PVs of £1, the PV£1p.a. or YP, can be shown to have the same formula but the rate of interest for the sinking fund is the same as the remunerative rate. Hence there is a single rate adopted throughout and so the result is a YP single rate.

The terminable income of £500 p.a., when valued on a YP single rate, produced £2,538 whereas on the YP dual rate basis shown here the figure is £2,444. Here the two approaches to capital recovery have been shown; the choice is for the experienced valuer to make, based on the nature of the income and relevant factors, particularly the nature of the legal estate; this choice is considered further in later chapters, such as in Chapter 9.

The derivation of the formulae for YP dual rate and YP single rate is as follows:

(a) PV£1p.a. or YP single rate
   To find the value of £1 per annum receivable at the end of each year for a given time, allowing for compound interest.
   The PV of the first instalment of income is $\frac{1}{1+i}$, that of the second $\frac{1}{(1+i)^2}$, the third $\frac{1}{(1+i)^3}$, and so on.
   Let the PV be V and the term n years, then

$$V = \frac{1}{1+i} + \frac{1}{(1+i)^2} + \frac{1}{(1+i)^3} + \cdots \frac{1}{(1+i)^n}$$

   Hence, summing the series and changing the sign in the numerator and denominator:

$$V = \frac{\frac{1}{1-(1+i)^n}}{i} \quad \text{or} \quad = \frac{1}{i + \frac{i}{(1+i)^n - 1}}$$

(b) YP dual rate
   To find the PV£1p.a. dual rate or YP dual rate for a given number of years, allowing simple interest at i per annum on capital and the accumulation of an annual sinking fund at s per annum.

$$V = \frac{1}{i + \frac{s}{(1+s)^n - 1}}$$

This is the same as the second formula for YP single rate except for the fact that in the single rate formula the remunerative and accumulative rates are the same, as indicated by i, whereas in the dual rate the remunerative rate is shown by i and the accumulative rate by s.

In some circumstances, valuers may wish to adjust the dual rate sinking fund provision for the incidence of tax on that part of the investment income which is to be used for reinvestment in the sinking fund; if the sinking fund is to be adjusted for tax at $t\%$, the formula becomes:

$$V = \frac{1}{i + \left[\left(\frac{s}{(1+s)^n - 1}\right) \times \frac{100}{100 - t}\right]}$$

PV£1p.a. single rate is consistently used by valuers in the valuation of freehold interests in all types of investment property. YP dual rate is used by some valuers when valuing leasehold interests in property. The application of both to the valuation of investment property is considered in more detail in the next chapter.

Parry's Valuation and Investment Tables contain additional tables including Annuity £1 will purchase, Internal Rate of Return (IRR), tables for the valuation of life interests in property and various imperial/metric conversion tables.

Chapter 9

Income approach or investment method

This chapter reviews the traditional approaches to the income approach used by valuers to assess market value.

I. Valuation of freehold interests

Freehold owners may choose to let their property and accept rent rather than occupy it themselves. The letting may be periodic (e.g. from month to month or from year to year), or on a lease for a term of years. Normally the freeholder will seek to obtain the market rent, namely:

> The estimated amount for which a property would be leased on the *valuation date* between a willing lessor and a willing lessee on appropriate lease terms in an arm's-length transaction, after proper marketing where the parties had acted knowledgeably, prudently and without compulsion.
>
> (RICS VS 3.3)

Whenever market rent is provided the relevant lease terms must be stated. The terms will vary from a full repairing and insuring (FRI) lease to a fully inclusive lease as set out in Chapter 5.

Valuers use a variety of terms for market rent (MR) including market rental value (MRV), full rental value (FRV) and estimated rental value (ERV). MR is generally used in this edition.

At the start of the lease the tenant will generally be paying MR. This level of rent remains payable until the end of the lease, or until the tenancy is terminated, or until the rent is reviewed under its terms; this will be so regardless of any changes in the prevailing level of market rent. If, for example, a shop is let on a 15-year lease at an initial rent of £30,000 p.a., which is the MR, and if the lease provides for the rent to be reviewed after five and ten years, then, assuming that the MR of the shop increases by £3,000 each year, it follows that the rent and the MR coincide only in year 1, but differ for the following four years and coincide again in year 6 as the rent review operates and the pattern is then repeated for the next five years. Hence:

Year	1	2	3	4	5	6	7
MR	30,000	33,000	36,000	39,000	42,000	45,000	48,000
Rent payable	30,000	30,000	30,000	30,000	30,000	45,000	45,000

If valuers are required to value the freehold interest in year 1 they will be valuing a freehold interest let at market rent, whereas if valuation takes place in year 3 the rent will be £30,000 p.a. whilst the MR is £36,000 p.a. and so the valuers are valuing a freehold interest let at below MR. These two valuations are now considered.

(a) Freehold let at MR

The principle of the income approach is: Net Income × Years' Purchase (YP) = Capital Value or Net Income ÷ Yield expressed as a decimal = Capital Value.

Net income is the rent receivable less any outgoings borne by the landlord, other than income tax. Income tax is ignored as this is a generally applicable imposition and, in so far as it varies according to the personal status of the recipient, is a personal factor unrelated to the investment itself. The implications of valuing net of income tax are considered in Chapter 12.

Freehold interests are perpetual and, therefore, incomes from freehold property will also be perpetual. It is true that rent arising from buildings rather than land is unlikely to be perpetual since buildings eventually wear out, but the life of a building rarely has a predictable end and, unless the building's life is relatively short, say up to 25 years, then the likelihood of rent from the building ending at some distant date is insignificant to the valuation. For example, suppose that a building has a predictable life of 30 years and produces a rent of £10,000 p.a. whilst a similar building produces the same rent but will last for 100 years.

The 30-year building has a value of £10,000 p.a. × YP (PV£1p.a.) 30 years at say 7% = £10,000 × 12.41 = £124,100, and for the 100 year building the value is £10,000 p.a. at say 7% = 10,000 × YP (PV£1p.a.) 100 yrs at 7% = 10,000 × 14.29 = £142,900. The 30-years rent has a value equal to 86.84% of the 100-years' rent; in addition, the 30-year building will have a reversion to site value which will make up some of the difference. The YP for 30 years at 7% is 12.41, whereas YP in perpetuity at 7% is 14.29, and so in valuing any income to perpetuity at 7%, 86.84% of the total value will lie in the first 30 years' income.

The reason for this is that the PV of £1 for a short period is higher than for a long period, and the effect of compound interest is to magnify this effect. Hence the PV of £1 in one year at 10% is £0.909, two years is £0.826, three years is £0.751, and so on for each succeeding year, the value falling for each subsequent year at an accelerating rate, so that for 30 years it is £0.057 and for 100 years £0.00007.

The rent in the earlier years is therefore more valuable than the rent in the later years. An inspection of the PV£1p.a. or YP Single Rate Tables of *Parry's Valuation and Investment Tables* will indicate the effect of this mathematical consequence at different rates of interest, and is illustrated in the graph at the end of this chapter.

Quite apart from this mathematical issue, it should be noted that a building is normally demolished because it is at the end of its economic life, that is when the value of the site for a new development exceeds that for the standing building. For example, this 30-year building will probably be pulled down after 30 years because the site would have a value at that time greater than the MV of the existing land and building. Hence, in valuing the rent from a building, the income may be regarded as perpetual although it is recognised that the rent will not always arise from the existing building.

In some circumstances a building will have a predictably short life because, for example, of physical deficiencies or town planning restrictions. Here the problem is that the rent cannot be considered perpetual because the rent will fall. This situation is considered later.

The valuer is thus faced with a net income at MR which can be regarded as perpetual. At present, it is the MR that is capitalised, but the valuer, on the basis of recent market experience, might feel that the MR will change in the future. The change need not necessarily be upward even with the existence of inflation; obsolescence will also be a factor. With a new building the expectation would normally be that, with inflation, values will rise to reflect the change in the value of money, plus, possibly, the effect of growth in the economy. With older buildings the threat of obsolescence may override the expectation of any inflationary growth. There may also be other locational factors that will influence the valuer's expectation of future increases or decreases in value.

It follows that whilst the valuer might know that the MR will not remain at a fixed level, the valuer cannot know what changes will occur, no matter how clear their crystal ball. The best that can be done is to predict the general movements in rents and perhaps estimate average growth, above-average growth, or whatever. The predicted growth can be compared with that predicted for other investments. The comparison can then be translated into value effect by adapting the yield to be adopted, since the yield acts as a measure of comparison as well as a barometer of investment expectations. Thus, when valuing a building at a point in time the only known factors are the then current market rent, the physical factors of the building and its location; the future is unknown. Therefore, the only tool in the hands of the valuer is the yield, which is to be used to reflect the different potential of the subject building from the comparables.

For example, a prime investment where the property is well located, the tenant is sound and above-average growth is predicted, may attract a yield of 5%, i.e. investors buying such a property would expect a return of 5% on their capital, whereas a similar property but where only average growth is predicted may attract 7%. In effect, the investor would require a higher initial yield to compensate for the anticipated future shortfall. Thus, the choice of yield reflects the anticipated

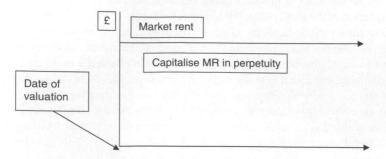

Figure 9.1 Freehold let at market rent.

future pattern of rent. The valuer can, therefore, take the net income at MR and regard it as perpetual and, to allow for the future changes in MR, adopt a yield appropriate to the predicted rental movement (see Figure 9.1).

Example 9–1

Value the freehold interest in shop premises in a prime London location. The premises were recently let to a multiple company at £100,000 p.a. exclusive on FRI terms for 15 years with five-yearly rent reviews. The rent receivable is MR.

Valuation

Rent reserved (and MR)	£100,000
Outgoings (NIL)	Nil
Net income	£100,000
PV£1p.a. in perp or YP in perp. at 5% (1/i:1/0.05)	20.00
	£2,000,000

Example 9–2

Value the freehold interest in shop premises in a secondary location. The premises comprise a shop on the ground floor with flat above, let recently to a single tenant at £30,000 p.a. exclusive on IRI terms for 15 years with five-yearly rent reviews. The rent is at MR.

Valuation

Rent reserved (and MR)		£30,000
less		
External repairs say	£3,000	
Management at 5%	1,500	4,500
Net income (and net MR)		£25,500
PV£1p.a. or YP in perp. at 10% (1/i:1/0.10)		10.00
		£255,000
Value, say		£255,000

Note that the less attractive investment features of Example 9-2, being (1) the inferior lease terms; (2) the shorter term of letting; (3) the lower quality of the location, and thus lower expectation of growth; and (4) the likely lower quality of the tenant; involve the use of a higher yield than in Example 9-1. If the new lease were for only a short period, say five years, then there is no certainty that the property would be re-let to the same covenant or one of a similar quality and there might be a void period between lettings. This could be reflected in the valuation by valuing in two parts:

Example 9–3

Value the same premises as in Example 9-2 but assuming only 5-year tenure.

Valuation

Rent reserved (MR) less		£30,000	p.a.
External repairs of	£3,000		
Management at 5%	1500	4,500	
Net income (and net MR)		£25,500	
PV£1p.a. or YP for 5 yrs at 10%		3.79	£96,645
Revert to net income		£25,500	
PV£1p.a. or YP in perp. at 10%	10.00		
PV£1 in 5.5 years at 10%	0.622	6.22	£158,610
			£255,255
Less costs of re-letting, say			£2,550
Value			£252,705

The extra half year in the deferment of the income to perpetuity reflects the potential void period. There are important matters to be considered in using the ARY approach. By definition, the choice of ARY by the valuer is a reflection of all the risks; these risks could be those to current income in terms of the tenant's covenant, liquidation, administration; risks to future income, falls in market rent, poor sustainable features in the property, non-renewal, voids and incentive payments for new tenants. If the ARY chosen captures all these risks, then there is no need to provide for specific risks such as voids. Some valuers will prefer to make certain risks such as voids and incentive payments explicit; if they do, then the ARY chosen should not be adjusted for these perceived risks for fear of allowing for the same risk twice. Those who favour an explicit approach have the problem that it is their view as to the length of a possible void and the cost of possible incentives in the future; but an argument in favour is that an adjustment to the ARY to reflect greater risks is an adjustment in perpetuity, that is for ever, whereas an explicit allowance for a void is an adjustment for a given period of time only. Those who favour DCF (Chapter 10) would argue that a DCF approach is explicit and can be based on the valuer's best interpretation of current property specific and market conditions.

If a valuer chooses to predict the future levels of rent at each review and thereafter and then includes them in the valuation, a higher yield would need to be adopted than that which was used here (which does reflect growth potential) as otherwise there would be double-counting. The problem then is the choice of a non-growth yield since yields excluding growth are not readily obtainable from comparables. The possibility of incorporating rental growth is discussed, together with comments on equated yields, in Chapter 12, where further aspects of this topic are considered.

It should be noted that although the valuations determine the sum which an investor can invest to achieve a desired return, they do not represent the whole cost of the investment as they exclude purchase costs such as professional fees and stamp duty. On all but small investments, where no stamp duty is payable, these on-costs average 5.80% for the buyer, made up of stamp duty land tax (SDLT) at 4%, legal, valuer's, surveyor's and other fees at 1.5% plus VAT at 20% on fees only, and 2.5% for the seller. The return on the actual sums spent on acquiring an investment will be slightly less than those adopted in the valuation, which represents the return on the net purchase price. This does not invalidate the valuation approach since the returns are based on the net purchase prices of comparable investments. As the on-costs tend to be a constant proportion there will be a fixed relationship between returns to net and gross sums invested.

For example, suppose that an investor pays £100,000 for an income of £10,000 p.a. The return to the net investment is:

$$\frac{10,000}{100,000} \times 100 = 10\%$$

If on-costs are 1.80% of £100,000 = £1,800*, the return to the gross investment is:

$$\frac{£10,000}{£101,800} \times 100 = 9.823\%$$

*The cost of acquiring this investment is lower as no stamp duty is payable below sales of £150,000.

Thus, investors seeking a 10% return recognise that they will achieve a 9.823% return on gross sums invested. To overcome this problem, the valuer will discount the valuation by the appropriate percentage of buyer's cost so as to return the true yield to the investor.

Example 9–4

Using the calculation in Example 9-1, the price to the investor wanting a 5% yield would be:

Purchase price	£x	
Costs of purchase at 5.80%	0.0580x	
Total cost	£1.0580x	
Rental income	£100,000	
Capitalised in perpetuity at 5% ($1/i = 1/0.05$)	× 20	£2,000,000

and £2,000,000 ÷ 1.0580x = £1,890,359

(b) Freehold let at below MR

A freeholder will be receiving a rent below MR if the rent was fixed some years ago and rental values have risen, or if the lessee paid a premium to the freeholder when they took the lease.

The approach "Net Income × YP = Capital Value" is inappropriate because, in addition to normal predictions as to underlying rental growth, there is certain knowledge that the rent will rise on a specific date when either the lease ends or a rent review operates. This change is different from normal rental value variations resulting from general market movements. It is impossible to draw from market transactions the appropriate yield to apply unless the valuer can find a similar situation where the rent bears the same proportion to MR, and the time for review or renewal is the same. In practice, the only situation that tends to follow these requirements is the valuation of freehold ground rents where the rent is nominal in relation to MR and the time for rent change is distant. The valuer must adopt a different general approach for this situation.

For example, suppose a valuation is required of a freehold interest where the rent receivable is £10,000 p.a. for the next three years and the MR is £40,000 p.a. For Years 1 to 3, the freeholder will receive £10,000 each year. At the end of Year 3 the rent will be revised to the then prevailing MR. It is not known what the MR will be at that time, but it is known that the current MR is £40,000 p.a. The valuation is made in two stages:

Stage 1. Present value of current rent

The period for which the rent is fixed is known as the term or "term to review" or "unexpired term". The freeholder will receive £10,000 each year. This may be payable quarterly, half yearly or annually, in advance or in arrears, but normal valuation practice is to assume annual in arrears. This is standard practice even though most leases require rents to be paid in advance and commonly quarterly or monthly in advance. This topic is further considered in Chapter 12.

Hence, following the normal convention, the valuer must value £10,000 receivable for each of the next three years. This can be done as follows:

Year 1		£10,000	
PV of £1 in 1 year	= say	£10,000	x

Plus			
Year 2		£10,000	
PV of £1 in 2 years	= say	£10,000	y
Plus			
Year 3		£10,000	
PV of £1 in 3 years	= say	£10,000	z
		£10,000	$(x+y+z)$

However, as was shown in Chapter 8, a simpler method is to use the PV£1p.a. or YP single rate.

The valuer needs to adopt the appropriate market yield to determine the multiplier, but the valuer will rarely have any direct comparables of similar properties let at one-quarter of MR for three years only. What the valuer does know is the market yield for valuing MRs from similar properties, from which an appropriate yield can be derived.

For example, suppose that if the property were let at its MR of £40,000 p.a. the yield would be 6%. What yield should be used in valuing the lower rent? One view is that the lower rent, being from the same property, has all the investment qualities of the MR, but in addition it is more certain of payment since the tenant is enjoying the accommodation at a cheap cost; thus a lower yield might be appropriate. On the other hand, if the tenant failed to pay the rent the freeholder could obtain possession and re-let at the higher rental, so that this added security might be said to be of no real importance. Further, if the tenant is of "a good covenant" the chances of default at any level are remote. But a low rent payable by a weak covenant is certainly more certain of being received than where the rent is at MR. A further issue is that if the low rent is fixed for a period beyond the normal review period then it has the disadvantage of providing a poorer hedge against inflation, a factor which will outweigh any additional certainty of payment; indeed, if the rent is fixed for several years, it takes on the characteristics of a medium fixed-interest investment where the yield may be more realistically derived from yields for such investments than from property.

From a practical point of view it will be found that, in cases where the term is for a short period, the value of the term is only a small part of the whole value. The choice of yield in such cases is of limited practical significance since the overall value is hardly affected by the yield chosen. Further, it is not possible to analyse comparable transactions to demonstrate the yield being used for valuing the term unless some assumption is made about the relationship between term and reversion.

The approach therefore is to weigh up the situation and make whatever adjustment appears to be appropriate – the valuer must make a subjective but rational judgment. It is suggested that as a general rule the same yield should be adopted as for the market rented property. However, if the reduced rent makes the investment a better quality investment or if the additional certainty of payment is especially attractive, for example where the tenants are financially strong, then a reduced yield is appropriate. On the other hand, if the rent is fixed for a period significantly longer than the normal review period, it should be increased. In every case, however, the yield is derived from the prevailing yield applicable to the valuation of the property if let at its MR.

In conclusion, the historic approach was to consider the lower rent as being, on balance, more secure and thus a lower yield than 6% would have been applied. The modern approach, except where the reversion is significantly distant, is to take the same yield as would apply had the property been let at its MR.

So, returning to the rent for the term, the valuation is:

Rent reserved	£10,000	p.a.
PV£1p.a. or YP 3 yrs at, say, 6%	2.673	
	£26,730	

Stage 2. Present value of rent for period after the term

The period following the term is called the "reversion". The valuer must first determine the rent that will arise at the end of the term. The valuer has two choices, to adopt a predicted rental value or to adopt the prevailing MR.

Adopting the former course, the valuer has two problems. First, there is no certainty as to what the rental value will actually be. Second, if a predicted future level is adopted, the general principles established in valuing a freehold interest let at MR are contradicted.

On the other hand, if the current MR is adopted there is a reasonable expectation that at least this rent is likely to be obtained. In addition, the yield adopted in valuing the MR reflects the likely trends in rental value. If the property were now let at a rent that was below the current MR, this would anticipate a growth in rent receivable at regular periods in the future which will occur in the property being valued.

Hence the valuer should adopt the current MR as the rent receivable on reversion, recognising that the yield applied reflects the possible changes in MR that may occur in the intervening period. The valuer can therefore project the calculations forward to the end of the term. At that time there will be a freehold interest let at MR, which can be valued as before:

MR	£40,000	p.a.
PV£1p.a. or YP in perp. at 6%	16.6667	
	£666,668	

But this is the value in three years' time. What is required is the value now. This can be found by applying the PV£1. The rate is that used to value the reversion since it is logical that an investor buying a future income showing 6% invests a sum today to grow at the same rate. Hence:

Value in 3 years' time	£666,668
× PV£1 in 3 yrs at 6%	0.8396
	£559,734

The valuation of the reversion is thus:

MR	£40,000
× YP or PV£1p.a. in perp. at 6%	16.6667
	£666,668
× P V £1 in 3 yrs at 6%	0.8396
	£559,734

Alternatively, as the PV£1p.a. or YP is the sum of the Pvs, so YP in perpetuity less YP for 3 years will equal YP perp deferred 3 years. Thus:

MR		£40,000	p.a.
PV£1p.a. or YP in perp. at 6%	16.6667		
less			
PV£1p.a. or YP 3 yrs at 6%	2.6730	13.9937	
		£559,748	

Note 1. The minor differences in value are due to rounding to four decimal places.

Note 2. The deduction of YPs is not possible on the dual rate principle, as will be shown later.

The valuer can short-circuit either of these approaches by using the YP of a reversion to a perpetuity from *Parry's Valuation and Investment Tables*. Hence:

MR	£40,000	p.a.
YP in perp. deferred 3 yrs at 6%	13.99367	
	£559,747	

Stage 3 Term and reversion (stage 1 plus stage 2)

The valuation of a freehold interest let at less than MR is the sum of the term and the reversion value (see Figure 9.2).

Term			
Rent reserved	£ 10,000	p.a.	
PV£1p.a. or YP 3 yrs at 6%	2.673		£26,730
Reversion to MR	£40,000	p.a.	
PV£1p.a. or YP in perp. def'd 3 yrs at 6%	13.994		£559,734
			£586,464

An alternative approach to the term and reversion is sometimes used where the initial rent is valued into perpetuity and the additional rent receivable at reversion is valued into perpetuity but deferred for the period of the term. Such an approach

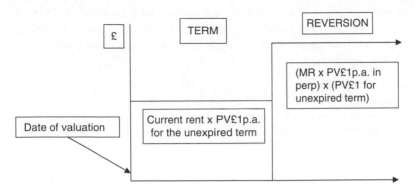

Figure 9.2 Freehold let at below market rent.

will give the same result as described above so long as a common yield is adopted throughout, but where the yield adopted in valuing the term should be different from the reversion, an artificiality is required for this layer method, which makes it less reliable. This layer or hardcore method is considered in Chapter 12.

The following examples illustrate the valuation of freehold interests let at less than MR.

Example 9–5

Value the freehold interest in shop premises in a secondary location. The premises are let at £25,000 p.a. net on lease with four years to run. The current MR is £60,000 p.a. net. The property is a 10% investment.

Term			
Rent reserved	£25,000	p.a.	
PV£1p.a. or YP 4 yrs at say 9% (rent less than MR and more secure)	3.240		£ 81,000
Reversion to MR	£60,000	p.a.	
PV£1p.a. or YP in perp. def'd 4 yrs at 10%	6.830		£409,800
			£490,800

However, as previously explained, the 1% adjustment to the term yield to reflect security is rarely seen in practice today. It was adopted because a rent payable below the MR gave the tenant a valuable lease which, if the tenant got into financial difficulty, could be sold at a premium, thus avoiding a void for the landlord. But the risk is that of an investment in property having regard to the quality of the tenant's covenant. That risk may not measurably change just because the rent is currently below MR. If this property is valued throughout at 10%, the term value falls to £79,248, giving a total value of £489,048. In both cases the valuer is likely to estimate MV at £490,000 and therefore the debate is academic, but it may not always be so.

Example 9–6

Value the freehold interest of a well-located factory let to a good covenant at £100,000 p.a. net on a lease with 12 years to run without review. The MR is £160,000 p.a. net and the property is an 8% investment (assuming a lease with five-yearly rent reviews).

Term			
Rent reserved	£100,000	p.a.	
PV£1p.a. or YP 12 yrs at 10% (rent less than MR but fixed for a long term)	6.814		£681,400
Reversion			
to MR	£160,000	p.a.	
YP in perp. def'd 12 yrs at 8%	4.964		£794,240
			£1,475,640

Note. The reversal of the yields reflects the inflation-prone nature of the term rent because of the absence of a review at the fifth and tenth years. Alternatively, an all-risk yield of 8.33% could have been used in both parts; this yield could only have been calculated either by trial and error or by using a computer program such as "goal seek" in Microsoft, but only if the value of £1,475,000 was known.

Example 9–7

Value the freehold interest in shop premises in a good location. The premises are let on a ground lease with 60 years to run at £1,000 p.a. without provision for review. The MR is £50,000 p.a. net. The property is a 7% investment (assuming a lease with five-yearly rent reviews).

Term			
Rent reserved	£1,000	p.a.	
PV£1p.a. or YP 60 yrs at 8.5% (the rent is very secure but is fixed for a very long term)	11.677		£11,677
Reversion			
to MR	£50,000	p.a.	
PV£1p.a. or YP perp. def'd 60 yrs at 7%	0.2465		£12,325
			£24,002

Note that the current rent provides a yield of 1,000/24,085 × 100 = 4.15% to capital invested. In practice it might be possible to show that similar ground rents sell on the basis of an initial yield of around 5%, or, put another way, they sell on a 20 YP basis. This is one case where a direct valuation of the interest may be derived from the present rent, i.e. capital value is £1,000 p.a. × 20 YP = £20,000. It is not as precise as the detailed approach but it may be considered sufficiently accurate in some circumstances. The use of this short-cut method can only be justified if there are comparables with similar facts because it is the reversion that dictates the true value.

An 8.5% yield has been used here to emphasise the need to reflect the inflation-prone nature of the term income. At the date of valuation it is the valuer's market knowledge which supports any adjustment to yields that may have to be made.

This valuation ignores any synergistic value. In finalising the opinion regard should be had to the potential to liberate this value, and some hope value may be added.

The general rules which can be taken from the above are:

1. That valuers should not try to guess the MR at a future date but should take the MR as at the date of the valuation.
2. The yield adopted to capitalise the MR must reflect the building's potential for growth in real and inflationary terms; the adaptability of the property; its potential to be relet at the end of any short unexpired term without much, if any delay; the quality of the tenant; and the length of the unexpired term.
3. The yield to be adopted for both the term and the reversion will usually be the same except where the investment becomes inflation prone due to a long term to reversion, in which event either different yields should be used or a single yield used which reflects the inflation proneness of the investment.
4. The end value should be reduced to reflect the buyer's costs as otherwise a lower yield will be achieved than would appear to be the case when analysed from comparables. This adjustment will, however, be used only where the investment is actually going to be sold and is not used when valuing the continuation of a discounted cash flow (DCF) calculation or for statutory valuations.
5. The potential to achieve synergistic or marriage value should be considered, which may justify a reduction in yield or an addition to reflect hope value.

(c) Freehold let at more than MR

A property may be let at a rent in excess of market rent because:

*   a competitive tenant has outbid the market to secure a property; typically an unrepresented retailer seeking a unit in a shopping centre;the market rent has fallen since the lease was negotiated; or the rent is indexed linked and the index has outpaced the market.

Over-rented property occurs most frequently when slump follows boom. The valuation problem is shown in Figure 9.3.

In valuing an over-rented property particular attention needs to be given to:

*   the tenant's covenant;the extent of the excess, e.g. 5% or 20%; the period of time during which the property may remain over-rented; the probable future movements in market rents.

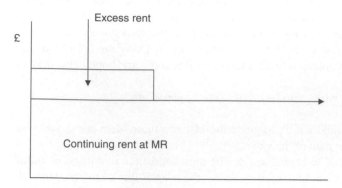

£

Excess rent

Continuing rent at MR

Figure 9.3 Over rented freehold.

The issue is whether there is any added risk attaching to the excess element of rent. If the tenant's covenant is exceptionally good then there may be no added risk. If the covenant is weak then there may be risk of default or an expectation that the tenant will request an agreement to pay only the market rent. If the excess is small then there will be little extra risk. If rents are rising then it may be only a matter of a few years before market rents are again equal to the rent reserved in the lease.

A standard valuation approach is to capitalise the market rent in perpetuity at the ARY which would be used if the property were let at its MR, then to capitalise the excess rent for the period up to the date when the rent could at law be renegotiated to the lower MR at a yield reflecting the higher risk. The two figures added together will be the market value. The choice of higher yield will depend on the valuer's assessment of those added risks. The period will depend on the unexpired length of the lease and on the nature of any rent reviews and break clauses.

The more correct approach is DCF because, when using an ARY, the valuer is reflecting future rental growth, if any. Hence part of the excess rent may be valued twice. Projecting changes in MR will identify when, in the future, MR may exceed rent reserved and thereby avoid double counting.

Example 9–8
Value the freehold interest in an office property let at £1,500,000 on FRI terms. The MR is £1,250,000. There is an upwardonly rent review in two years' time and the lease ends in seven years' time. The property is in a major city outside London and let to one of the top five firms of accountants, who are unlikely to default or to vacate at the end of the lease. Similar market-rented office premises are selling on a 7% ARY basis.

Valuation
Assuming there is little expectation of tenant default and, given current market conditions, little expectation of rental movement, then the valuation might appear as:

Market rent	£1,250,000	
PV£1p.a. or YP in perp. at 7%	14.29	£17,857,143
Excess rent	£250,000	
PV£1p.a. for 7 years at 10%	4.8684	£1,217,100
		£19,074,243
MV, say		£19,000,000

What the valuer is stating is that the freehold would normally have an MV of £17.8 million but, given the covenant strength, a buyer might be expected to pay about £1 million more for the fairly certain additional £250,000 for the next seven years. There is risk and uncertainty, issues which are considered in Chapter 12 and which will need to be made explicit in a valuation report.

2. Valuation of leasehold interests

Leasehold interests arise in two principal ways.

The first is where a freeholder of development land offers a ground lease under which the leaseholder carries out the development and enjoys the benefits of the development for the period of the lease in return for the payment of a ground rent. Such leases are usually for a long term of years, typically 99 or 125 years, to allow the lessees to recoup their capital investment in the buildings; the ground rent is low relative to the MR of the buildings.

The second is where either a freeholder or leaseholder offers an occupation lease in return for the payment of a rent which will normally be the MR of the premises. As previously mentioned, the rent may be less than MR because market rents have risen between the grant of the lease and the valuation date, or because an initial concessionary rent was agreed, possibly reflecting the payment of a premium.

The importance of the relationship between the rent payable and the MR is that a leasehold interest has no value unless the MR is greater than the rent payable. The reason for this is that as the MR represents the maximum rent at which the property can be let, a tenant will not be prepared to offer any additional sum to buy the lease. It must be remembered that when someone buys a lease, the buyer becomes responsible for the observance of all the covenants set out therein, including the duty to pay the rent; someone buying a lease for a capital sum and paying MR will be worse off than one paying MR for a new lease of comparable property.

A leasehold interest, therefore, has value only where the MR exceeds the rent payable. The difference between these two figures is termed a "profit rent".

For example, suppose that A holds a lease under which they pay £8000 p.a. when the MR is £15,000 p.a. The position is:

MR	£15,000	p.a.
Less		
Rent paid	£8,000	
Profit rent	£7,000	p.a.

Such a lease has a value because a buyer will be entitled to occupy the property under the lease terms and pay £8,000 p.a. for something which is worth £15,000 p.a., or alternatively it may be sublet at a rent of £15,000 p.a. and so, each year, £7,000 is retained as a "profit rent". This benefit will continue until the lease ends or the rent is adjusted, under a rent review clause, to the higher MR. The task of the valuer is to determine the market value of the profit rent, i.e. the value of the lease assuming there is no goodwill value in the premises being valued (see Figure 9.4).

Normally the valuation approach adopted is the investment method, notwithstanding that the buyer might intend to occupy. This is valid, particularly in the case of commercial properties, even though the buyer is not seeking an actual return on the capital invested as there will be an annual benefit which will boost the profit from trading in the property. This can be demonstrated from the above example, where the lessee pays £8,000 p.a. as rent, and comparing the position to an identical property which is available at the MR of £15,000 p.a. In the first property the tenant will pay £7,000 p.a. less than would be paid in the second property. Since all other conditions are identical, the profits from trading in the first property will be £7,000 p.a. greater than for the second property. When in the first property the lease ends or the rent is reviewed and the rent is increased from £8,000 p.a. to £15,000 p.a., there will be an immediate £7,000 reduction in the profits.

When valuing a leasehold interest, the basic principle of the investment method will still apply, namely:

Net income × YP = capital value

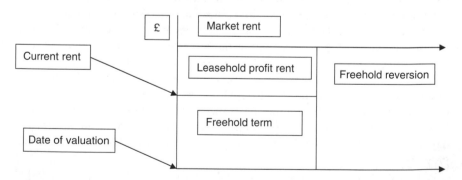

Figure 9.4 Leasehold Profit rent.

(a) Provision for loss of capital

Net income for this purpose will be the profit rent, as has been shown above. However, when an investor buys a leasehold interest there is one factor that will not be met in a freehold interest, namely that one day the buyer's interest must cease and the buyer will have no further interest in the property: if the buyer obtains a further lease in the property at market rent to follow the expired lease, that will be a new lease with no market value.

The investor knows that one day the interest will cease to exist, at which point there is no capital asset. The investment will therefore decline from whatever was paid to nothing. Such an investment is known as a wasting asset since the asset wastes away naturally and inevitably.

The investor faced with such a situation must therefore take steps to deal with the loss of capital. Various methods are available, but for many years the conventional approach has been to assume that part of the profit rent will be invested each year in an annual sinking fund. Since there is a need to guarantee the replacement of the capital, the investor must seek a medium which is as certain as possible to guarantee that the sums invested will reach a specific and certain figure at the end of the term. The only such medium is normally an insurance policy. Naturally the returns offered by insurance companies must be based on long-term rates of interest which ignore abnormal movements in interest rates and which give a reasonable guarantee that they can meet their obligations. Consequently, the returns offered will be low and an average yield of around 2.25–3% has been found to be realistic. For the purposes of our examples 3% is used. The appropriate rate will depend on market conditions at the time of a valuation but, in the case of *Nailrile Ltd* v *(1) Earl Cadogan* [LRA/114/2006] (a Leasehold Reform Act case), the Lands Tribunal determined in December 2008 that the appropriate sinking fund rate was 2.25%.

If the investor places some of the net income in such a sinking fund it can be shown that the original investment capital will be available at the end of the lease for further investment. In this way the investment can be continued indefinitely so that the return received after allowing for sinking fund costs will be perpetual. The effects of inflation on such an arrangement will be considered later.

An alternative approach is to write off the investment over its life. This is an accountancy convention often followed where companies such as industrialists employ plant and machinery which by their nature have a limited life. A typical approach would be to adopt straight line depreciation where the capital cost is divided by the predicted years of useful life and this figure is deducted as an annual cost of the item in the accounts. Such an approach does not lend itself readily to a valuation and fails to reflect the accelerating rate of depreciation suffered by a leasehold interest. Similarly, the other common approach of the reducing balance, whereby a stated percentage is written off each year, is also of no help since this approach writes off the value more heavily in the early years, which is the opposite of the depreciation pattern of a lease.

The use of the single rate approach is considered in Chapter 12. Dual rate is by no means uniformly accepted and there is a growing usage of single rate, where capital is deemed to be repaid on a diminishing balance basis from year to year in accord with present value principles. In some colleges and universities dual rate is almost off the curriculum; in some countries, such as Hong Kong, where all property is leasehold, it has never been used. The theory is set out below and the reader is left to consider whether dual rate is defunct after consulting the issues raised in Chapter 12 and elsewhere.

(b) YP dual rate

The investor in a leasehold investment, if providing for a sinking fund, can regard the net income after the sinking fund provision as perpetual but with no capital growth. The valuer may now value any leasehold profit rent with this in mind. Assume, for example, that a valuation is required of a leasehold interest where the lease has five years to run at a fixed rent of £8,000 p.a. The current rental value is £15,000 p.a. The valuation approach is as follows:

Stage 1. Determine profit rent

MR	£15,000	p.a.
less		
Rent paid	£8,000	
Profit rent (net income)	£7,000	

Stage 2. Assess annual sinking fund

As was shown in Chapter 8, the ASF for the redemption of £1 capital invested shows the annual payments needed to reach £1. In this case, the sum to be recouped will not be known until the valuation is completed, so assume the value of the interest is £V. Hence:

Capital sum required in 5 years	£V	
× ASF to replace £1 in 5 yrs at 3%	0.1884	
Annual sinking fund payment	£0.1884	V

Stage 3. Assess income net of sinking fund

Profit rent	£7,000	
less		
ASF to replace £V in 5 yrs	£0.1884	V
Net income	£7,000 − 0.1884	V

Stage 4. Value net income

Since the income net of the ASF is effectively perpetual, notwithstanding that the lease will end, the net income can be regarded as a perpetual net income as for a freehold in possession. However, as the income is produced by a leasehold interest rather than a freehold, the investor has to collect rent and pay rent on to the landlord; is burdened with the trouble and expense of replacing the investment at the end of the lease; and is subject to restrictions imposed by the covenants in the lease. Apart from these issues, the investment is the same as the freehold investment in an identical property (save for any capital growth) on which the yield will be known. Allowing for these adverse qualities, the appropriate yield for the leasehold investment should be higher than the freehold yield. The amount of additional return required depends on the facts of the case but, as a general rule, the addition will be around 0.5% on a freehold yield of 4%, rising to 2% on a freehold of 12%. There is, however, no fixed adjustment to be made: the adjustment is a matter of judgment coupled with knowledge of adjustments found by analysis of comparables.

If, in this case, the freehold yield is 7.5%, the leasehold yield would be, say, 8.5% and the valuation would be:

Net income	£7,000 –	0.1884V
× YP in perp. at 8.5%		11.7647
Capital value	£82,353 –	2.216V
BUT		
the capital value was taken to be		
£V, therefore	£82,353 –	2.216V = £V
	£82,353 =	V + 2.216V
	£82,353 =	3.216V
	£82,353/3.216 =	V
	V =	£25,607

This approach can be simplified by using the PV£1p.a. dual rate (YP dual rate) formula or using Parry's tables. These perform the function of stages 2 and 4 above and their use excludes the need for stage 3. The multipliers can be applied directly to the profit rent. Hence:

MR	£15,000	p.a.
Less		
Rent paid	£8,000	
Profit rent	£7,000	
× YP for 5 years allowing for a yield of 8.5% and		
annual sinking fund at 3%		
YP 5 yrs at 8.5% and 3% SF	3.6583	
	£25,608	

This shows how a conventional dual rate leasehold valuation would be set out, but this format cannot be used where there is a variable profit rent; this will be considered later in this chapter.

(c) Capital replacement and inflation

Mention was made earlier of the problem of inflation. The problem is that an investor who invests capital in a leasehold interest and who replaces the original capital at the end may find that, with the erosion of the value of money over the period, the replaced capital is worth less, and perhaps significantly less in real terms, than at the start. As a general rule, investors in freeholds will see their investments increase in value over time. The implications of this situation, which have led to some criticism of the approach described above, are considered in Chapter 12.

(d) The position of the owner/occupier

An occupier is not in the same position as an investor. When an occupier buys a lease of premises to trade from, the premium that could be paid should reflect the accumulated rent savings over the period to review/renewal less any allowance for advance payments.

Example 9–9

Value the lease of a shop to an owner/occupier, assuming an MR of £15,000 p.a. and a rent payable of £8,000 p.a. and an occupier's borrowing rate of 6%. As the rent saving is a constant £7,000 a year, the PV£1p.a. single rate can be used:

Year 1–5	Market rent	£15,000
	Rent payable	£8,000
	Rent saving	£7,000
	× PV£1p.a. for 5 years at a borrowing rate of 6%	4.2124
	Present value of saving	£29,487

The discount rate adopted needs to reflect the cost of finance to the occupier and not the yield required by an investor, which may be different. Note that the figure 4.2124 is the sum of the PVs for one, two, three, four and five years, and so each year's saving in the future has been discounted.

This is not a market valuation but an assessment of worth to an occupier buyer or to the current occupier. Further, it is not an assessment of value as required for accounting purposes.

(e) Annual sinking fund and tax

The valuation set out in (b) above can be analysed to show that it achieves its goals as follows:

Profit rent		£7,000
less		
ASF to replace £25,607 in 5 yrs at 3%		
(0.1884 × 25,607) =		£4,824
Remunerative income		£2,176
Net yield = (£2,176 ÷ £25,607) × 100 = 8.5%		
The investor achieves desired return.		
OR		
Profit rent		£7,000
Capital invested	£25,607	
Interest on capital at 8.5%	0.085	£2,176
Sum available for investment in ASF		£4,824
× Amt. of £1 p.a. for 5 yrs at 3%		5.30825
		£25,607

The investor's capital is replaced.

This analysis ignores income tax. The reasons for ignoring income tax in determining the return on an investment were set out at the beginning of this chapter. However, the ASF is a cost, and ignoring the effects of income tax in respect of this leads to a misleading value, as shown in the following analysis.

Assume that the investor pays income tax at the rate of 40p in the £. Hence:

Profit rent		£7,000
Capital invested	£25,607	
Interest at 8.5%	0.085	£2,176
Sum available for ASF		£4,824

Thus the profit rent is divided into £2,176 for return on capital and £4,824 for the annual sinking fund. Although income tax at 40% is payable on the £2,176 (as part of the £7,000 p.a. on all of which 40% tax is payable), this can be ignored for analysis purposes since the aim is to obtain 8.5% before tax. But the investor must pay 40% tax on the £4,824 as no tax relief is given on normal sinking fund investments. So the net of tax income which can be invested is:

ASF before tax	£4,824
less	
Income tax at 40%	£1,930
ASF after tax	£2,894

It is clear that 40% tax-paying investors will not be able to achieve their goal, since they will not be able to pay £4,824 each year into the ASF and meet their tax commitments. The true cost of the ASF is not what the investor must invest but the annual income which, after payment of tax, leaves £4,824 available for investment in the sinking fund.

This gross cost can be determined by multiplying the sum required by 100 ÷ (100 − 40), as is shown below.

Assume A earns £100 and pays tax of £40 (i.e. 40p in £). Then A's income after tax is $(100 - 40)$p. Thus for each £1 that A earns A will be left with 60p out of each 100p.

Thus A's net income (NI) will be $\frac{100-40}{100}$ of A's gross income (GI).
If

$$\text{Gl} \times \frac{100-40}{100} = \text{N1}$$

then

$$\text{Gl} = \text{Nl} \times \frac{100}{100-40}$$

The same will be true at any rate of tax. Thus, if tax is t, gross income can be reduced to net income by applying $\frac{100-t}{100}$.

Net income can be grossed up to gross income by applying $\frac{100}{100-t}$.

(f) Dual rate adjusted for income tax

Applying the tax factor to the leasehold income, where an annual sinking of £4,819 was seen to be necessary to replace the capital, the amount of profit rent needed to be earned each year to leave £4,824 after tax is:

£4,824 × $\frac{100}{100-40}$ =	£8,040	gross cost
less		
Tax at 40p in £1 = £8,040 × 0.4 =	£3,216	
Net income for ASF =	£4,824	

If the investment is now analysed to allow for the true cost of providing for the sinking fund, the result is:

Profit rent	£7,000
less	
Gross ASF to produce £4,894 p.a.	£8,040
Hence negative return on capital =	£1,040

Clearly, the investment has been over-valued if the price of £25,607 is to be invested by a taxpayer paying 40p in the £1 tax. The price of £25,607 will only leave £4,824 p.a. available for an annual sinking fund, after the return of 8.5% has been taken out, if the investor is a non taxpayer. Similarly, the cost of providing a sinking fund must vary with the individual taxpayer's marginal rate of income tax: the higher the rate of tax, the more gross income is needed before tax to produce the required net income for the sinking fund.

This presents a problem for the valuer called upon to prepare a "market valuation" when it is clear that the potential buyers in the market have differing rates of tax which have a direct effect on their sinking fund costs. The solution is for the valuer to adopt an "average rate" of tax, whilst recognising that investors whose personal rates differ sharply from this have an advantage or disadvantage which renders the valuation unreal for their own circumstances. If required, the valuer can easily produce a valuation for a particular investor's actual marginal tax rate.

Given that the valuer adopts an average rate of tax, what is an average rate? In any year, tax rates may vary from nil for tax-exempt persons to rates which historically have been as high as 98%, with many different rates between, and these rates tend to vary from year to year. Faced with such a variety of rates, the valuer needs to select a rate that is likely to be most commonly met for the remaining years of the lease. This rate is unknown, but a useful guide is the standard rate of income tax for individuals and the corporation tax rate for companies. The standard rate in recent years has come down steadily and in 2012 it is 20%, coupled with a single higher rate of 40% on higher incomes and a rate of 50% on individuals with an income of over £150,000 p.a. (50% rate will fall to 45% in 2013) small profits rate 10% change 25% in 2012 to 23% p.a. 2013 in same paragraph amend wording to read 'which for illustration purposes is taken at a rate of 25%. In similar fashion, corporation tax for small companies is 20% (2011), rising to 26% for profits above £1.5m (2011) but with a proposed reduction to 25% in 2012. On the basis that higher rate taxpayers will try to shelter their tax by using companies, bearing in mind that capital gains tax is currently 18%, it would seem to be appropriate to use the corporation tax rate which, for illustration, is taken at the proposed rate of 25%. However, if rates of tax showed a tendency to rise or fall, then the average rate should adjust to the trend.

Thus, the valuer adopts a predicted average level of tax for coming years, and a rate of 25% will be adopted for this purpose. In any event, a small variation in the rate will be significant only if the term of years is short. The valuer can now incorporate in the valuation approach a grossed-up cost of an ASF by applying $100/(100 - 25) = 1.334$ to the required annual sinking fund investment. This can readily be incorporated in the dual rate formula, given as:

$$\frac{1}{i + ASF}$$

where i is the gross of tax rate of return and needs no adjustment and ASF represents the annual sinking fund required to replace £1 of capital (after payment of tax).

It follows that $ASF \times 100/(100 - t)$ will represent ASF required before payment of tax, i.e. the gross ASF.

Hence, applying the facts of the example to the formula:

$$i = 8.5\% = 0.085$$

$$ASF = ASF \text{ to replace £1 in 5 yrs at } 3\% = 0.18835$$

ASF × 100/(100 − 25) = 0.18835 × 1.3334 = 0.251

YP 5 years at 8.5% and ASF at 3% (adjusted for income tax) = 1/(0.085 + 0.251) = 2.9749.

(Tables of YP dual rate adjusted for varying rates of income tax can be found in *Parry's Valuation and Investment Tables*).

Valuation

Profit rent	£7,000	p.a.
× YP 5 yrs at 8.5 and 3% adjusted		
for income tax at 25%	2.975	
Capital value	£20,825	

Analysis

Profit rent		£7,000
less		
Interest on capital at 8.5%	£20,825	
	0.085	£1,770
Gross ASF		£5,230
less		
Income tax at 25%		£1,307
Net ASF		£3,923
× Amt. of £1 p.a. for 5 yrs at 3%		5.309
Capital sum replaced		£20,827

The difference of £2 is accounted for by the rounding off of numbers.

This shows that the use of a YP dual rate adjusted for income tax allows investors to achieve their required return and also to replace their capital after meeting their tax liabilities (but only in respect of the sinking fund payments).

The need to use dual rate for leasehold valuations is strongly refuted by many valuers and the debate is considered in Chapter 12. The valuer and student must follow the debate and use the approach most frequently used in practice. However, it should be remembered that the maxim "as you devalue so shall you value" must always apply, and therefore, if the parameters for a valuation are taken from comparables, then different valuation methods may produce the same answers.

The object of a valuation is to arrive at the estimated price which would have been achieved in the open market on a given date. It may be that the market will comprise mainly owner/occupiers so that the single rate approach as set out in Example 9-8 should be used; alternatively, the entire market may comprise non-taxpayers such as charities. It should be evident from the discussion so far that a person or organisation not liable to pay tax would be in a very advantageous position in relation not only to the actual yield from any investment but, with lease-holds, to the rate of interest at which a sinking fund accumulates and to the amount which must be set aside; the gross ASF would equate to the net ASF. The shorter the lease, the greater will be the advantage. This advantage has been exploited by non-taxpaying investors, in some cases to the point where they comprise the entire market (see Chapter 12).

(g) Leasehold interest where income is less than MR

A leaseholder may be receiving less than MR, either because the property has been sublet in the past at the then prevailing MR and rental values have risen, or was sublet and a premium taken.

In these cases the valuer values the existing rent whilst it is receivable, i.e. during the "term", and then reverts to MR when this in turn is receivable, i.e. on "reversion". The same term and reversion approach is adopted as shown for freeholds but with a YP dual rate adjusted for tax being employed in place of YP single rate or in perpetuity, as the following example shows.

Example 9–10

A took a lease of shop premises 23 years ago for 42 years at a fixed rent of £10,000 p.a. 17 years ago A sublet the premises for 21 years at a fixed rent of £16,000 p.a. A valuation is now required of A's leasehold interest; the MR is £40,000 p.a. on shops let for 15 years with five-yearly rent reviews.

Term
A is receiving £16,000 p.a. from the under-lease. This will be payable for the next 4 years. Hence:

Rent received	£16,000	p.a.
less		
Rent paid by A	£10,000	
Profit rent	£6,000	p.a.

A will receive this profit rent for 4 years. If the freehold yield on similar shops is 5.5%, then the appropriate leasehold rate would be 6%.

Profit rent	£6,000	p.a.
× YP 4 yrs at 6% and		
3% adj. for tax at 25%	2.640	£15,840

Reversion
After four years, A will be able to re-let the shop at the prevailing MR. For the reasons set out above when considering the valuation of a freehold interest let at less than MR, the current MR is adopted as the rent receivable at reversion. Hence:

Reversion to MR	£40,000	p.a.
less		
Rent payable by A	£10,000	
Profit rent	£30,000	p.a.

After four years, A will still have 15 years to run and hence, in four years' time, the valuation will be:

Profit rent	£30,000	p.a.
× YP 15 yrs at 6% and		
3% adj. for tax at 25%	7.5937	£227,811

However, the value is required now, so that the value brought forward four years is:

Value in 4 years	£227,811	
× PV £1 in 4 yrs at 6%	0.792	£180,426

Note that, as before, the discount rate is the same as that adopted for the valuation of the reversion. Thus, bringing the valuation together, it becomes:

Term				
Rent received		£16,000	p.a.	
less				
Rent payable		£10,000		
Profit rent		£6,000	p.a.	
× YP 4 yrs at 6% and 3%				
adj. for tax at 25%		2.640		£15,840
Reversion to MR less		£40,000	p.a.	
Rent payable		£10,000		
Profit rent		£30,000	p.a.	
× YP 15 yrs at 6% and 3%				
adj. for tax at 25%	7.5937			
× PV £1 in 4 yrs at 6%	0.792	6.014		£180,426
				£196,266
Valuation of interest, say				£196,000

This is the only method of deferring the reversion. There are no tables in *Parry's Valuation and Investment Tables* which combine YP dual rate for a period of years × PV for the term (unlike YP reversion to perpetuity tables), and the YP for the period from the time of valuation minus YP for the term will give the wrong result (unlike YP in perp minus YP single rate for the term).

This valuation method has a built-in technical error because the first of the two sinking funds will continue to appreciate in value over the succeeding 15 years, thus producing a capital return in excess of the value estimate. This error can be eliminated by using an approach similar to that set out at the beginning of this section (known as the double sinking fund method) or by using Pannell's Method (see Chapter 12).

3. Synergistic[1] value, sometimes known as marriage value

The method of valuation of a freehold let at less than MR, and of a leasehold interest, have been considered. These interests will arise simultaneously in the one property since the lease held at less than MR creates one value and the freehold term and reversion creates a second value. If the owner of one of the interests buys the other interest, the interests can merge by operation of law so that the lease merges into the freehold to create a freehold in possession or, in valuation terms, a freehold at MR after the expiration of the sub-lease. It is interesting to consider the valuation implications of this situation.

Example 9–11

We use the same facts as in Example 9-9, where F is the freeholder and A is the leaseholder having an unexpired term of 19 years. The freehold yield when let at MR with five-yearly rent review is 5%.

Value of F's present interest term			
Rent received	£10,000	p.a.	
× YP 19 years at 7% (as the rent is fixed for a long term without review)	10.336		£103,356
Reversion to MR	£40,000	p.a.	
× YP in perp. def'd 19 yrs at 5%	7.915		£316,587
			£419,943
Value of leasehold interest			
As per Example 9-9			£196,266
Value of freehold plus value of leasehold			£616,209
Value of freehold after purchase of lease			
Rent received	£16,000	p.a.	
× YP 4 years at 5%	3.546		£56,735
Reversion to MR	£40,000		
YP in perp. def'd 4 yrs at 5%	£16.454		£658,162
			£714,897

Hence, if F buys the leasehold interest for £196,266, the existing interest valued at £419,943 will be lost but the freehold after merger will have a value of £714,897. Thus, there is additional value of £714,897 – £616,209 = £98,688, which represents extra value arising from the merger of the interests. This additional value is now called synergistic value by the RICS.

The reason for the additional value is that the profit rent element when owned by the lessee is valued at 6.0% dual rate whereas in the freeholder's hands it is valued

---

1 This is defined by the RICS as "An additional element of value created by the combination of two or more interests where the combined value is more than the sum of the separate parts (may also be known as marriage value)".

on a single rate basis at 5% in one case and 7% in another. This synergistic value phenomenon will arise in nearly all cases where interests are merged and can act as an incentive for owners of interests to sell to each other, or to join together in a sale to a third party, so as to exploit the synergies.

The question that arises is: who should take the benefit of the added value? Since the parties are interdependent they are of equal importance to the release of the added value, so as a general guide the sum is shared equally. Hence, in Example 9-10 above, F is able to offer the value of the lease (£196,266) plus half the added value (£49,344). In this way, both A and the leaseholder make a "profit" of £49,344.

This is the case where each party has equal bargaining strength. It does not apply where one party is forced to sell or sells by auction, although in the latter case some part of the merger value may be paid by a third party as "hope value", working on the assumption that a further sale may be made to the other party at an early date to achieve a profit. In this case, for the purpose of calculating synergistic value, the value of the sold interest will not be the sales price but its value excluding hope value, as otherwise there would be double counting.

During the negotiations between F and A the question may be raised about the use here of a dual rate adjusted for tax multiplier when valuing A's interest. The freeholder is not a buyer of a leasehold investment *per se*. After the merger the freeholder is in a position to sell the property at market value and thereby recoup any sums paid out on the buying back of the lease. There is no profit rent as such and no sinking fund, hypothetical or otherwise. Valuing dual rate unadjusted or indeed single rate reflects the greater value of this lease to the freeholder than to any other party. This fact is known to the valuers acting for both parties, who may agree a value of the lease on a single rate basis. What this does is to readjust the sums between lease value and synergistic value. However, F may again argue that this is double counting.

This type of value may also arise where owners of adjoining properties merge their interests to form a larger single site. This will be particularly so where the united interests create a development site but where the individual sites are incapable of, or are unsuitable for, development alone. In these circumstances a person buying the interests (known as "assembling the site") can pay over and above normal value by sharing the added value among the parties.

In some cases, site assembly will involve buying interests in adjoining sites and buying freehold and leasehold interests in particular sites, thus creating both forms of synergistic value.

4. Freehold interest where a building has a terminable life

In considering the question of MR in Chapter 5 it was stated that, as a general rule, the present MR of a property is assumed, for the purpose of valuation, to continue unchanged into the future. Reference was also made in section 1 of the present chapter to the fact that the majority of buildings are demolished when they reach the end of their economic life because at that time the site value for redevelopment exceeds the value of the existing buildings. However, there are circumstances where a building is approaching the end of its physical life (whether for structural or for limited planning permission reasons) and then it may be that some loss of rental value will be incurred.

Example 9–12

Freehold shop premises built about 60 years ago are let on lease for an unexpired term of 30 years to a good tenant at £28,000 p.a. following a recent rent review. The lease is subject to a schedule of condition, with no landlord's liability to repair.

It is considered that the building will have reached the end of its physical life at the same time as the current lease expires and that rebuilding will be required. The present cost of rebuilding is estimated to be £250,000 and the value of the site is also estimated at £250,000, equivalent to £15,000 p.a. based on a yield of 6% for a building lease with frequent rent reviews.

Analysis of sales of properties similar in nature in the same district, but more modern and whose structural life is estimated to be 75 years or more, indicates that they sell on an 8% basis.

What is the present market value of the property?

Valuation – Method 1

Applying the evidence of other sales, it would appear that the property in question could properly be valued at a higher rate of interest than 8% because the buildings have only a 30-year life.

Rent on lease	£28,000	p.a.	
YP in perp. at, say, 10%	10.0		£280,000

Method 2

Assuming a reversion to site value at the end of the lease:

That portion of present rent which will continue	£15,000	p.a.	
YP for 30 years at 6%	13.765		£206,472
Remainder of rent	£13,000	p.a.	
YP for 30 years at 8.5% and			
3% adj. for tax at 25%	8.847		£115,011
Reversion to site value	£15,000	p.a.	
YP in perp. at 5% def'd 30 years	4.627		£69,405
			£390,888

The yield reflects the expected growth in the site value over the next 30 years.

Method 3

Assuming continuance of rental value and making provision for rebuilding:

Rent on lease	£28,000	p.a.	
YP in perp. at 8%	12.5		£350,000
less			
cost of rebuilding	£250,000		
× PV £1 in 30 years at 8%	0.0994		£24,850
			£325,150

The yields and tax rates are being used for illustrative purposes only.

Method 1 has the merit of simplicity but the increase in the rate of interest is very much a matter of opinion. Methods 2 and 3 are open to objection on a number of grounds.

In Method 2 it may be argued that it is difficult to know what will be the value of the site upon reversion, after the useful life of the building, 30 years hence. The only evidence of the value of the site is present-day selling prices, which suggest the figure of £250,000 (MR £15,000 p.a.) used in the example, but this value may change in the future. A similar objection applies to the figure of £250,000 for the cost of rebuilding in Method 3, which, of necessity, has been based on present-day prices although these may be very different 30 years hence.

Another serious objection is the difficulty of predicting the life of a building with any degree of accuracy, particularly over a long period of years. Such an estimate of the life of the building as is made in the example must necessarily be in the nature of a guess unless considerable data is accumulated over a long period showing the useful life in the past of similar buildings. Even such evidence, over perhaps 40 or 50 years in the past, is not conclusive of what the useful life of a building will be in the future.

There is an inherent error in Method 2. The use of 8.5% to value the continuing rent for the next 30 years implies some growth in the rent. It follows that some part of the additional rent has been valued twice.

On balance it would appear that, in many cases, there is probably less inaccuracy involved in assuming the present income to be perpetual and using a higher rate to cover risks, than would be involved in the more elaborate method of allowing for possible variations in income.

On the other hand, the separation of land and building values and making provision for building replacement has the advantage of giving to an investor a more direct indication of the actual rate per cent he is likely to derive from an investment than does Method 1, provided it is used in conjunction with analysis of sales made on similar lines.

In cases where it can confidently be assumed that existing buildings will be worn out in a few years' time, Method 2 may be the soundest.

In the case of properties where the building clearly has a very limited life but the site is incapable of being redeveloped in isolation, whether for practical or for planning reasons, it may be impracticable to have regard to any future site value. In such a case, the existing income might be valued for the estimated building life as a terminable income with no reversion.

In practice it is very rare for a building to have a limited physical life. Parts of a building may wear out over its life but these can be replaced as and when necessary. Provided a building is kept in repair so as to prevent accelerated depreciation, a building's life is infinite but it will suffer from functional obsolescence over its lifetime so that its economic life is limited. However, there may be a limited life due to a planning restriction and a more temporary building may have been constructed to allow for that fact; this usually involves prefabricated building and thus Method 2 is the more logical valuation.

5. Premiums

A premium usually takes the form of a sum of money paid by a lessee to a lessor for the grant or renewal of a lease on favourable terms or for some other benefit. Alternatively, it is the price required for a lease that is being sold, e.g. "Lease of shop premises for sale. Premium £80,000". The following comments refer to a premium payable as a condition of a lease.

Where a premium is paid at the commencement of a lease, it is usually in consideration of the rent reserved being fixed at a figure less than the MR of the premises.

From the lessor's point of view, this arrangement has the advantage of giving additional security to the rent reserved under the lease, since the lessee, having paid a capital sum on entry, has a definite financial interest in the property, which ensures that he will do his utmost throughout the term to pay the rent reserved and otherwise observe the covenants of the lease. Premiums are only likely to be paid where there is a strong demand for premises on the part of prospective tenants or where their covenant is so weak that a landlord will not otherwise let to them. A rent deposit of say two quarters rent in advance is not the same as a premium, which is paid for a reduction in rent. On the other hand, in the case of leases of less than 50 years, the tax treatment of premiums is penal for lessors so that they may be discouraged from taking a premium (see Chapter 21).

The parties may agree first on the proposed reduction in rent and then fix an appropriate sum as premium, or they can decide the capital sum to be paid as premium and then agree on the allowable reduction in rent.

The calculation of the rent or premium can be made from both parties' points of view but, if so, different answers may emerge. The parties will then need to negotiate a compromise figure.

Where the premium is fixed first, in order to arrive at the reduction to be made in the rent it will be necessary to spread it over the term, either by multiplying the agreed figure by the annuity which £1 will purchase, or by dividing by the YP for the term of the lease.

Example 9–13

Estimate the premium that should be paid by a lessee who is to be granted a 30-year lease of shop premises at a rent of £20,000 p.a. with five-yearly reviews to 80% of MR (a "geared" rent review). The MR of the premises is £25,000 p.a.

Note

For illustrative purposes the calculations have been based on the dual rate table of YP at 8% and 3% adjusted for tax at 25% for the leasehold interest and 7% for the freehold interest.

Lessee's point of view:				
Present rental value			£25,000	p.a.
Rent to be paid under lease			£20,000	
Future profit rent			£5,000	p.a.
YP for 30 years at				
8 and 3% adj. for tax at 25%			9.257	
			£46,285	
Premium, say			£46,000	
Landlord's point of view: Present interest:				
MR	£25,000	p.a.		
YP in perp. at 7%	14.28		£357,143	
Proposed interest:				
Rent for term	£20,000	p.a.		
YP 30 yrs at 7%	12.41		£248,181	
Reversion to MR	£25,000	p.a.		
YP in perp. at 7% def'd 30 yrs	1.88		46,917	
			£295,098	
add				
Premium			*x*	
			£295,098 + *x*	

Present interest = Proposed interest
357,143 = 295,098 + *x*
x = £62,045
Premium, say = £62,000

Hence, the tenant's view suggests £46,000 and the landlord's view £62,000. A compromise would be needed to arrive at the premium. Since the premium would be a compromise it would be impossible to analyse it to show the assumptions made by the parties.

The need for a compromise has arisen due to the different treatment of the "rent loss" and the "profit rent gained" when they are in fact the same sum each year. Anything less than £62,000 requires the freeholder to accept a loss in total value, which is unlikely if the freeholder is well advised. Gearing the rent on review for so long a period is likely to be very expensive from the landlord's point of view if rents rise, because the actual value of the rent specified over such a term will be much greater than £62,000, but the landlord will have the use of the premium for further investment.

Consideration of mergers, premiums and surrender and renewal problems from both viewpoints has merits when preparing for negotiations, but freeholders should never accept less than the sum needed to fully compensate for the proposed deal unless they have an ulterior motive such as the need to raise cash from a non-banking source.

Example 9–14

A shop worth £50,000 p.a. is about to be let on FRI terms for 35 years. A premium of £100,000 is to be paid on the grant on the lease. What should the rent be throughout the term, assuming quinquennial geared rent reviews?

Lessee's point of view:		
Present MR	£50,000	p.a.
Deduct:		
Annual equivalent of proposed		
premium of £100,000		
£100,000 ÷ YP 35 yrs at 7% and 3% adj.		
for tax at 25% = £100,000 ÷ 10.8634 =	£9,205	p.a.
Rent to be reserved	£40,795	p.a.
Landlord's point of view:		
Present interest		
MR	£50,000	p.a.
YP in perp. at 6%	16.67	£833,333
Proposed interest		
Rent reserved	*x*	
YP 35 yrs at 6%	14.498	£14.498 *x*
Reversion to MR	£50,000	p.a.
YP in perp. def'd 35 yrs at 6%	2.168	£108,421
add		
Premium		£100,000
		£14.498x + £208,421

Present interest = Proposed interest
£833,333 = £14,225x + £208,421
14.498x = £833,333 − £208,421 = £624,912
Rent (*x*) = £43,103 p.a.

In this case the lessee's view suggests £40,795 p.a. and the landlord's view £43,103 p.a. A compromise rent around £42,000 p.a. might be agreed.

The reason for the difference is due to the freehold being valued single rate and the leasehold dual rate adjusted for tax. There is no purchase of a leasehold investment and therefore, arguably, no need for a sinking fund or for a tax adjustment. No actual profit rent is to be received, although notionally this will be reflected in higher business profits which will be taxed. In effect the premium is no more than the discounted cost of the rent foregone. Dual rate unadjusted for tax produces a revised figure of £41,346, which is closer to the freehold figure. In practice there is no value reason for a freeholder to accept less than the freehold figure because advisors for both parties would know that to do so would diminish the investment value to the freeholder.

In some cases, the parties may agree to a premium being paid in stages, perhaps a sum at the start of the lease and a further sum after a few years have passed.

The calculation of the premiums or rent follows the same approach as before, the future premium being discounted at the remunerative rate from the landlord's point of view (the landlord will receive it) and at the accumulative rate from the tenant's point of view (it is a future liability for the tenant).

6. Surrenders and renewals of leases

Where a lease is drawing where a lease is coming to an end tenants may approach their landlords with a view to an extension or renewal of the lease so as to be able to sell the goodwill of the businesses which they have built up,or to convert the lease into a more marketable form, or simply to secure their continued occupation.

The usual arrangement is for the tenant to surrender the balance of the present lease in exchange for the grant of a new lease for a similar term or for some other agreed period.

The Landlord and Tenant Act 1954, which gives tenants of business premises security of tenure, does not affect this practice. It has strengthened the hand of the tenant in negotiation, but tenants traditionally prefer to hold their premises under a definite lease rather than to rely merely on their rights under the Act, which in the end depend upon litigation, and an intending buyer of a business held on a short lease will be reluctant to rely merely on rights under this Act. As for landlords, a surrender and renewal may give them the opportunity to obtain terms in a new lease not obtainable on a renewal under the 1954 Act, and this could make such a proposal attractive. They may, therefore, be prepared to give way on rental arguments for the new lease in order to achieve better overall lease terms. The most important factor from a landlord's point of view is certainty of income for as long as possible; accepting the surrender of a lease and granting back a longer term will increase the capital value of the investment.

If the MR of the premises exceeds the rent reserved under the present lease, then the tenant will want to preserve their profit rent. It may be that a simple transaction will be agreed whereby the term is extended on the basis that the first rent review will be on the original term expiration date. Alternatively, the landlord may pay the tenant for taking a new lease at market rent, or there may be a new lease at a geared rent with gearing on review.

The valuer usually acts for one side or the other but may be called upon to mediate an appropriate figure of increased rent or premium.

The calculation is traditionally made from both the landlord's and the tenant's point of view with the tenant being credited with any value accruing to the current leasehold interest. For example, if the balance of the tenant's lease is seven years at a profit rent of £25,000 a year, then it is clear that the tenant has a valuable interest in the property which will be given up in exchange for the new lease. The tenant is therefore entitled to have the value of the surrendered portion of the term set off against any benefits which may be derived from the proposed extension.

From both parties' point of view, the principle involved is that of (i) estimating the value of each party's interest in the property, assuming that no alteration in the present term was made; and (ii) estimating the value of each party's interest, assuming that the proposed renewal or extension were granted. The difference between these two figures should indicate the extent to which the tenant will gain or the landlord lose by the proposed extension.

If a premium is to be paid, the above method will suggest the appropriate figure. If, instead of a premium, the parties agree on the payment of an increased rent, the required figure may be found by adding to the present rent the annual equivalent of the capital sum arrived at above, although the method used in the examples below is to be preferred.

The terms for the extension or renewal of a lease are seldom a matter of precise mathematical calculation. An estimate of rent or premium made from the freehold landlord's standpoint usually differs from one made from the tenant's, and the figure finally agreed is a matter for negotiation. A valuer acting for either party will consider all the circumstances of the particular case, not only as they affect their client but also as they affect the other party. Valuations will be undertaken from both points of view as a guide to the figure which the other side might be prepared to agree to in the course of bargaining. This practice is reflected in the following three examples.

The effect of income tax on the market value of the tenant's present and future interests is initially taken into account.

Example 9–15

A tenant holds a shop on an FRI lease for 40 years, of which six years are unexpired, and wishes to surrender the lease for a new 25-year lease at the same rent with geared five-yearly reviews. The rent reserved under the present lease is £10,000. The MR is £25,000 p.a. The rent at the reviews will be to 40% of MR.

What premium can reasonably be agreed between the parties?

Tenant's point of view:

Proposed interest:			
Profit rent	£15,000	p.a.	
YP 25 years at 8% and 3.0%			
adj. for tax at 25%	8.5785		£128,677
Present interest:			
Profit rent	£15,000	p.a.	
YP 6 years at 8% and 3%			
adj. for tax at 25%	3.4949		£52,423
On this basis, gain to lessee			£76,254

Freeholder's (landlord's) point of view:
Present interest:
Next 6 years

Rent reserved	£10,000	p.a.	
YP or PV£1p.a. 6 years at 7.5%	4.694		£46,940

Reversion

to MR	£25,000		
YP in perp., or PV£1p.a., def'd 6 yrs at 7.5%	8.639		£215,980 £262,920

Proposed interest:

Next 25 years proposed rent	£10,000		
YP, or PV£1p.a., for 25 years at 7.5%	11.15		£111,500

Reversion

to MR	£25,000		
YP, or PV£1p.a., in perp. def'd 25 years at 7.5%	2.19		£54,750 £166,250
On this basis, loss to freeholder			£96,670

Traditionalists will argue that depending on the negotiating strength of the parties, the premium should be between £96,670 and £76,254. But it should be noted that only at £96,670 will the freeholder be no worse off and no better off.

Example 9–16
Assume that in Example 9-15 it was agreed that the tenant should pay an increased rent throughout the new term in lieu of a premium; what should that rent be? It is assumed that the rent payable on review will bear the same proportion to MR as at the start of the lease (a "geared review").

Tenant's point of view:

Present interest as before	£52,423	
Proposed interest		
MR	£25,000	p.a.
less		
Rent reserved	x	
Profit rent	£25,000	$- x$
YP 25 yrs at 8% and		
3% adj. for tax at 25%	8.5785	
	£214,462 − 8.5785x	

Present interest = Proposed interest

£52,423	= £214,462 − 8.5785x
x	= £18,889

Freeholder's (landlord's) point of view:

Present interest			
as before		£262,920	
Proposed interest			
Rent under lease	x		
YP or PV£1p.a. for 25 yrs at 7.5%	11.15	11.15	x
Reversion to MR	£25,000	p.a.	
YP, or PV£1p.a., perp. def'd			

25 yrs at 7.5%   2.19   54,750

 11.15x + 54,750

Present interest = Proposed interest
£262,920   = 11.15x + £54,750
x   = £18,669

In this case the figures are very close and the rent will be fixed at about £18,700 p.a.

The freeholder's point of view rests upon the assumption that the freeholder will be satisfied with the arrangement so long as the freehold market value is not affected.

Where, as part of a bargain for a renewal or extension of a lease, the tenant is to make an expenditure upon the property which will benefit the value of the landlord's reversion, the tenant must be given credit for the value due to this expenditure and the sum must be taken into account when considering the cost of the new lease to the tenant.

Example 9–17

A warehouse in a city centre is held on a lease having five years unexpired at £32,000 a year. The MR is £40,000 p.a. The tenant is willing to spend £100,000 on internal improvements and alterations which will increase the MR by £12,000 p.a., provided the landlord will accept a surrender of the present lease and grant a new lease for a term of 30 years. The tenant is willing to pay a reasonable rent under the new lease, or a premium. The landlord is agreeable to the new lease provided the rent is fixed at £35,000 p.a., that a proper premium is paid, and that the new lease contains a covenant that the tenant will carry out the improvements. What premium, if any, would you advise the tenant to offer if the new rent is to be £35,000 p.a.?

Tenant's point of view:

Proposed interest MR	£40,000	p.a.	
Add rent due to outlay	£12,000		
	£52,000		
less			
Rent payable	£35,000		
Profit rent	£17,000	p.a.	
YP 30 years at 8% and 3%			
adj. tax at 25%	9.2571		£157,370
Deduct expenditure on improvement			£100,000
Value of proposed interest			£57,370
Present interest			
MR	£40,000		
Rent paid	£32,000		
	£ 8,000		
YP 5 years at 8% and 3%			

adj. tax at 25%	3.0199	£24,159
Gain to lessee		£33,211

Freeholder's (landlord's) point of view:
Present interest

First 5 years	£32,000	
YP, or PV£1p.a., for 5 years at 7%	4.10	£131,206

Reversion

Note. Since the estimated increase in rent due to improvements represents 12% on the sum expended, and since this is greater than the yield that might reasonably be expected from such a property as this when let at its MR, it would be worth the landlord's while, at the end of the present lease, to carry out the improvements at his own expense in order to obtain the increased income. The reversion can therefore be valued on the basis that the landlord would carry out the improvement work on the expiry of the lease, as follows:

b/fwd		£131,206
Reversion to MR after improvements	52,000 p.a.	
YP in perpetuity at 7%	14.286	
Deduct	£742,857	
cost of improvements	£100,000	
	£642,857	
PV£1 in 5 years at 7%	0.713	£458,357
Value of present interest		£589,554
Proposed interest		
Term first 30 years	£35,000	
YP, PV£1p.a., for 30 years at 7%	12.409	£434,315
Reversion to MR	£52,000	
YP or PV£1p.a. in perp. deferred 30 years at 7%	1.877	£97,604
		£531,919
Loss to freeholder		£56,038

Depending on the negotiating strength of the parties, the payment would be agreed at around £44,000 (i.e. near the average of lessee's gain of £33,211 and freeholder's loss of £56,038).

Suppose that it had been decided between the parties that no premium should be paid and that the rent under the new lease should be adjusted accordingly. What would be a reasonable rent in these circumstances?

Tenant's point of view:
Rental value after improvements
have been carried out   £52,000   p.a.

Deduct annual equivalent of cost of improvements	£100,000		
and present interest (as before)	£24,159		
	£124,159		
YP 30 years at 8% and 3.0% adj. tax at 25%	9.257	£13,412	
Reasonable rent from lessee's viewpoint		£38,588	p.a.
Freeholder's (landlord's) point of view: Value of present interest, as above			£589,554
Deduct value of proposed reversion to MR after improvements have been carried out	£52,000	p.a.	
YP or PV£1p.a. in perp. def'd 30 years at 7%	1.877		£97,604
Value of proposed term			£491,950
YP or PV£1p.a. for 30 years at 7%			12.409
Reasonable rent from freeholder's viewpoint			£39,644 p.a.

Again, a negotiated settlement might be reached at an initial rent of around £39,000 p.a.

7. Traditional valuation techniques and DCF

The valuation methods used in this chapter are generally known as the "traditional or conventional approaches" where an all-risks yield (ARY) is used to capitalise net income. The ARY approach reflects the valuer's assumed but unquantified perception of future rental change, and this is implicit in the yield. In the current market there is a greater lack of certainty and overlaying the traditional structures is the market concern of uncertainty which, with a lack of comparable data, has increased the challenge to the valuer in assessing MR and the ARY (see Chapter 12 Part 6, Valuing in periods of market instability). A number of authorities have suggested that a more explicit approach to future rental change should be adopted, but this requires assumptions as to the future which might be based on past performance, which is not a reliable guide to the future, or on econometric models which generally require to be outsourced by the valuer. The result might be an assumption as to growth or decline which could at best be guesswork. Explicit assumptions as to the future may be valid when analysing a market price to estimate a long-term return, or an internal rate of return (IRR), but it may not be easy to compare IRRs with returns on other forms of investment, such as shares, which are valued using a price earning (PE) ratio which is very similar in concept to a YP.

Dual rate and dual rate adjusted for tax have also been criticised by academics and practitioners; in some circumstances and in some markets there has been an increase in the use of single rate, but as with freeholds this remains an implicit approach as far as future rents are concerned and there are similar arguments for a more explicit approach to be used.

The explicit approach is based on the use of discounted cash flow (DCF). The RICS Valuation Standards 7th edition 2011 contains a guidance note, GN7 *Discounted cash flow for commercial property investments*. GN7 has been withdrawn from the Global and UK edition of 2012 but will be reprinted as a separate guidance note. DCF is a global tool for both investment analysis and investment valuation and international clients now look for a DCF approach in place of the ARY approach or in conjunction with it. DCF is considered in the next chapter, and other methodologies are considered in Chapter 12.

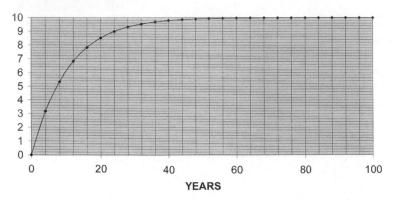

Graph showing years' purchase single rate at 10% for years 0–100.

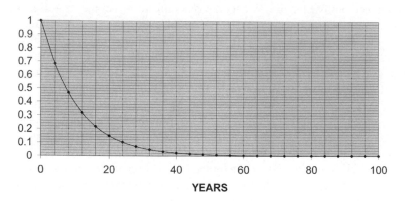

Graph showing present value of £1 at 10% from years 0–100.

Chapter 10

Discounted cash flow

1. Generally

The principle of the income approach or investment method of valuation is that future net cash flows from a property are discounted in order to determine its present (market) value. The value is the discounted cash flow. This principle is used in business to assess investment proposals generally and the technique is known internationally as discounted cash flow (DCF).

DCF techniques are used to assist in decisions relating to property investment and valuation. Their principal application has been in comparing or measuring investment opportunities to assist with the "buy" or "not to buy" question and occasionally with choosing between one or more property investments. The use of DCF by valuers was supported in the RICS Mallinson Report, published in March 1994 – *Report of the President's Working Party on Commercial Property Valuations*. A detailed consideration of the use of DCF was contained in the RICS Valuation Standards (2011) GN7 *Discounted cash flow for commercial property investments*; this guidance note has been withdrawn from the 2012 Red Book but is still available in its 2011 RICS guidance note format. It provides useful advice on using DCF for valuations and property investment advice. It states that:

> The explicit discounted cash flow (DCF) valuation method is of greatest application in the assessment of *investment value* to assist in buy/sell decisions or selection between alternative available investments. However, it can be used to estimate *Market Value* by adopting *assumptions* that are consistent with observed market prices, and then applying those *assumptions*, with appropriate adjustments, to the valuation of the subject property. Where there are no transactions, the explicit DCF model provides a rational framework for the estimation of *Market Value* not present in the ARY (capitalisation rate) approach, which relies on comparables for the identification of the ARY.
>
> DCF calculations involve the discounting of all future receipts and expenditures similar to direct capitalisation, but they can readily be used to allow for inflation, taxation and for changes in the amount of receipts and expenditure. In a DCF calculation explicit assumptions are made compared with the implicit

assumptions contained in the traditional all risks yield (ARY) capitalisation approach. DCF calculations involve the assessment of Net Present Value (NPV) and the Internal Rate of Return (IRR).

2. Net present value

Here the present value of the future receipts, from a proposed investment, is compared to the present value of the current and future expenditures. If the present value of receipts exceeds the present value of expenses the investment is worthwhile because the return must exceed the yield (the discount rate) adopted in the calculation. This excess is known as the net present value (NPV). If the same calculations are applied to different investment opportunities, the investment that produces the largest NPV is the most profitable. The NPV is useful in reaching a decision on the choices available, but it does not follow that the most profitable will be pursued. For example, it may be that the "best scheme" requires a large amount of capital to be invested, whilst other considerations make a scheme with a smaller capital outlay more attractive.

(a) The discount rate

If the future receipts and expenses are to be discounted, then the question is "at what rate?" When calculating the worth[1] of an investment to an individual investor, the discount rate will be the investor's target rate. When the method is used to determine a general market value, a discount rate must be adopted which is at or around the majority of individual target rates, i.e. a market-derived discount rate.

The discount rate is commonly arrived at by adopting a risk-free rate and making allowances for the risks associated with property investments in general and the subject property, i.e. adding a property risk premium (see RICS guidance note *supra*).

The general consensus is that a risk-free rate can be taken as the gross redemption yield on medium-term government gilts. This has ranged from 4% to above 10% over recent decades. To this must be added a yield to reflect the general risks of investing in property, such as illiquidity, rent risk, capital risk, depreciation, changes in law such as carbon footprint/green issues, and then a further addition is made to reflect the risks specific to the property in question such as strength of covenant, lease terms, condition, location, risks of voids. So the discount rate could be taken to be the risk-free yield, say, at the time of writing, 3.5%, adding for property market risks (say) 2.0% and for risks specific to the property in question (say) 1.5%, producing a discount rate of 7.0% (GN7 6.1–6.23). The examples

---

1 The RICS definition of investment value or worth is "The value of an asset to the owner or a prospective owner for individual investment or operational objectives. (May also be known as worth)" (RICS Red Book glossary).

shown here assume income is annually in arrears, but a DCF should always specify the cash flows correctly and the discount rate will need to be adjusted to the appropriate quarterly or monthly rate, as explained in Chapter 8.

(b) Rate of growth

The DCF calculation makes specific assumptions as to how the various elements of the calculation will change in the future. Given that inflation is the norm, the changes will be the rate at which these elements will grow in the future. Rates of growth may be different; for example, it might be assumed at one time that rents will grow faster than building costs, but under different economic conditions building costs may be increasing faster than rents.

There are methods of forecasting which are available to the Bank of England and the Government that assist in providing estimates of growth. These econometric models are not normally developed by valuers but can be obtained from appropriate sources. Alternatively, the valuer may rely on personal judgment as to growth rates. In the case of market rents, where the valuer knows the ARY appropriate to a property and also the discount rate, an implied rate of rental growth can be determined and used to estimate the predicted rent at each rent review.

The ARY equals the discount rate minus the ASF to recoup the growth in rent over the typical review period at the discount rate. From this, one can calculate the implied rate of annual rental growth. So, assume that the ARY is 6%, the discount rate is 10.0%, and that there are five-yearly rent reviews. Then:

6% = 10.0% – (ASF 5 yrs at 10% (0.1638) × growth over 5 yrs)
Let g = the growth over 5 years

Then:

$$0.06 = 0.10 - 0.1638g$$

$$0.1638g = 0.10 - 0.06$$

$$g = 0.04/0.1638$$

$$g = 0.2442$$

$$g = 0.2442 \times 100 = 24.42\%$$

Growth over 5 years = 24.42%

$$(1+i)^n = 1.2442$$

$$\therefore i = (^5\sqrt{1.2442}) - 1 = 1.2442^{0.20} - 1$$

$$\therefore i = 1.04467 - 1$$

$$\therefore i = 0.4467 \times 100$$

$$i = 4.467\%$$

Amount of £1 for five years at $4.467\% = (1.04467)^5 = (1.2442 - 1) \times 100 = 24.42\%$.

The argument here is that a market which is buying at 6% when a target rate of 10% is required, to meet the risks associated with a specific property investment compared to risk-free investments, must be buying for growth. Calculations such as those shown above indicate that, *ceteris paribus* (all other things remaining the same), the buyers are expecting a growth in rent at an average of, say, 4.5%; this rate being implied by the relationship between the yields.

(c) Cash flow period (see guidance note)

Income in the early years has a higher discounted value than that from later years. For example, the PV£1p.a. for 20 yrs at 6% is 68.82% of the PV£1p.a. in perpetuity at 6%; and the PV£1p.a. for 20 yrs at 10% is 85.14% of the PV£1p.a. in perpetuity at 10%; and so on. This demonstrates that a high proportion of value is likely to arise over the early years and so it is normal to prepare a cash flow for, say, 5, 10 or 15 years and then to assess the market value or exit value at the end of that period (GN,7 4.1). In estimating the exit value regard must be had to the ARY to be used to capitalise the income at that time; even if nothing else has changed, the property will be ten years older and that alone might require an adjustment on today's ARY of +0.25%.

Example 10–1
Value the freehold interest in shop premises which were recently let for 20 years at the MR of £20,000 p.a. with five-yearly reviews. ARY is 6%. Target rate is 10% and growth is 4.467%.

1.   *Conventional method*

Rent reserved and MR	£20,000	p.a.
YP in perp. at 6%	16.667	
Value	£333,334	

2.   *DCF*
Discount rate 10.0%. Growth rate 4.467% p.a.

Years	Cash flow (£)	YP at 10%	PV at 10%	Capital value (£)
1–5	20,000	3.7908		75,816
6–10	24,884	3.7908	0.6209	58,570
11–15	30,961	3.7908	0.3855	45,245
16–20	38,522	3.7908	0.2394	34,959
20–perp.	47,929	*16.6667	0.1486	118,704
NPV				£333,294**

*YP in perp. at 6% to establish exit value.
** Difference due to rounding of figures in calculations.

In the valuation of market-rented investments such as this, the two approaches give the same result. However, the DCF approach might be preferred if, at the end of the lease, it is assumed that the tenant will vacate, the premises will need to be refurbished, and this, coupled with a rent-free period on a new letting, will lead to a loss of one year's rent. These considerations are difficult to incorporate in the simple Income × YP model but can be accommodated in a DCF.

Example 10–2

Assume the same facts as in Example 10-1, but on expiry of the lease the freeholder will need to spend £80,000 on the premises at current prices. The work will take six months. It is expected to take three months to re-let and a three-month rent-free period is likely. Building costs are expected to rise at 3% p.a. The rental value will be unchanged by the work.

Years	Cash flow (£)	YP at 6%	PV at 10%	Capital value (£)
1–20	(as before)			214,590[1]
20–20.5	(146,640)[2]	–	0.1486	(21,790)
21–perp.	50,070[3]	16.667	0.1351	112,743
NPV				£305,543

1 This is the sum of the PVs from example 10–1 for the first 20 years.
2 This is today's £80,000 cost × A£1 for 20.5 years at 3%, which = £146,640.
3 This is the year 20 rent increased by 4.467% to allow for the works, re-letting and the rent-free period. It is capitalized at the ARY considered likely to be the long-term ARY at that time.

This example shows how expenditure can be allowed for. If there were several items it would be prudent to split the cash flow column into "cash in" and "cash out" columns for clarity, and similarly with the capital sums.

A further example where DCF can be used is in the valuation of a reversionary interest where explicit assumptions as to rental growth can be made.

Example 10–3

Same facts as in Example 10-1, save that the property was let 12 years ago, the current passing rent fixed two years ago is £17,000 p.a., with a rent review in three years, and the lease has eight years unexpired.

1.   *Conventional method*

Term			
Rent reserved	£17,000	p.a.	
YP or PV£1p.a. 3 yrs at 6%	2.6730		£45,441

Reversion to MR		£20,000 p.a.		
YP or PV£1p.a. in perp. def'd 3 yrs at 6%		13.9937	£279,874	
Value			£325,315	

2.   DCF

Years	Cash flow (£)	YP at 10%	PV at 10%	Capital value (£)
1–3	17,000	2.4869		42,277
4–8	22,802	3.7908	0.7513	64,941
8–perp.	28,369	16.6667	0.4665	220,569
NPV				£327,787

Note. The rent after four years is the MR of £20,000 p.a. increased at 4.467% p.a. and there is no void on expiry of the lease. The exit value is the year 8 rent capitalised at 6% in perpetuity.

As the values are close in this example, it may be said that there is no merit in changing from the conventional method. But as the DCF approach requires the valuer to be more explicit, it may encourage the valuer to consider the factors underlying value more carefully. In some cases where the conventional method is difficult to apply, such as where there are very few comparables or where the pattern of receipts and expenditures is irregular, the DCF approach can make the task of producing a value easier. The DCF approach is the recommended method to determine worth (see Chapter 12).

There are several DCF programs available which take away the labour of preparing the calculation and which allow for greater precision, such as allowing for rents payable quarterly in advance and for tax and other items not easily accommodated by the conventional method.

3. Internal rate of return (IRR)

As an alternative to the NPV method, an analysis can be carried out which discounts all future receipts and payments of a project at a discount rate whereby the discounted receipts equal the discounted payments. This discount rate will then show the actual rate of return on the capital invested in the scheme – the "internal rate of return" (IRR). At this point the NPV will of course be nil.

The approach to this method can be illustrated using the facts from Example 10-1 but, by trial and error, using different rates of interest to find the IRR.

A shop, as in Example 10-1, has been offered for sale at £300,000. Calculate the IRR.

Adopting the same assumptions regarding estimated growth and exit value, it is clear that the IRR must be more than 10% as, at 10%, the NPV is above £300,000. Try 12%.

Example 10–4

Years	Cash flow (£)	YP at 12%	PV at 12%	Capital value (£)
1–5	20,000	3.6048		72,096
6–10	24,884	3.6048	0.5674	50,897
11–15	30,961	3.6048	0.3220	35,937
16–20	38,522	3.6048	0.1827	25,370
21–perp.	47,929	16.6667*	0.1037	82,837
Total				£267,137

*This is the YP in perp. at the market yield of 6%.

At 10% the NPV is £33,294 above £300,000, and at 12% it is £32,863 below. By interpolation, try 11%.

Years	Cash flow (£)	YP at 11%	PV at 11%	Capital value (£)
1–5	20,000	3.6959		73,918
6–10	24,884	3.6959	0.5935	54,583
11–15	30,961	3.6959	0.3522	40,302
16–20	38,522	3.6959	0.2090	29,756
21–perp.	47,929	16.6667*	0.1240	99,053
Total				£297,612

*This is the YP in perp. at the market yield of 6%.

This is close to £300,000, so the IRR could be taken to be 11%. The IRR can be found by linear interpolation provided the two yields are no more than 1–2% apart.

At 10% NPV = £33,294 above: at 11% NPV = £2,388 below: range £35,682. Assuming a constant rate of change, then a closer figure can be calculated:

$$10\% + [(33,294/35,682) \text{ of } 1\%] = 10\% + 0.933\% = 10.933\%$$

This is not, of course, absolutely exact since the difference in discount rates is based on geometric rather than linear expansion. However, it is probably sufficiently accurate for most purposes.

This can be expressed as a formula that provides for linear interpolation:

$$R1 + \left[(R2 - R1) \times \frac{NPVR1}{NPVR1 + NPVR2} \right]$$

where R1 is the lower rate and R2 the higher rate.

Hence

$$10 + \left[1 \times \frac{33,294}{35,682} \right] = 10.933$$

An accurate approach is to use Microsoft Excel and the "Goal Seek" tool as set out below. As Excel is formula based, the figures it uses are not rounded. "Goal Seek" is an iterative process which rapidly recalculates the cash flow until it arrives at a rate of interest which precisely discounts the cash flow to the stated value.

Example 10–5
The cash flow is as set out in Example 10-4 with growth at 4.4670%, giving an increase over five years of 24.42155% (1.2442155):

> Using "Goal Seek" in Excel
> go to "tools"
> select "Goal Seek"
> place cursor in "set value" and click on answer, press "enter"
> place cursor in "to value" and type in required amount – in this case £300,000
> place cursor in "by changing cell" and click on the cell you wish to change – in

this case the rate of interest; "Goal Seek" will display the IRR in the cell allocated for this purpose; here it is shown to three decimal places as 10.929%.

Years	Years	Cash flow	YP at 10.929%	PV of £1 at 10.929%	Capital value
1–5	5	£20,000	3.7025299		£74,051
6–10	5	£24,884	3.7025299	0.5953594	£54,853
11–15	5	£30,961	3.7025299	0.3544528	£40,633
16–20	5	£38,523	3.7025299	0.2110268	£30,099
21–perp.	999	£47,931	16.666667	0.1256368	£100,364
					£300,000

This example shows two things. If the valuer is correct in the choice of the ARY at 6% and rental growth of 4.467% per annum, then the price is below its true value. A buyer would therefore realise a yield above the target yield. On the other hand, if the value is really £300,000 this would be because the ARY is above 6%, or the market's estimate of rental growth is below 4.467%, or a combination of the two.

The IRR is also known by valuers in the UK as the Equated Yield since it provides the yield that will be obtained at any chosen price after making explicit assumptions as to changes in rents and outgoings and other matters in the future. This yield contrasts with the yield or IRR determined in a similar manner, so that discounted income flow equals the NPV but without making any adjustment for changes in the future. This is known as the Equivalent Yield and is applied in the conventional investment methods.

4. Comparison of NPV with IRR approaches

The NPV compares the future receipts and expenditures on a discounted present capital value basis, whereas the IRR shows the return earned on an investment.

In the above examples, the application of each to the same investment shows that the investment is worthwhile on either basis. This will not always be so as sometimes it will be found that one method shows an investment proposal to be profitable whilst the other method suggests that it will show a loss, or that, in deciding between investments A and B, NPV method favours A whilst IRR favours B. The reasons for this will lie in the choice of discount rate or in the pattern of costs and revenue.

The resolution of this contradiction can be found in the standard texts on DCF, which provide various ways of overcoming the problem. As a general rule, however, the IRR approach will be favoured by a valuer who is using DCF for analysis since it will show the yield of the investment that enables comparisons to be made with yields on other known investment opportunities. On the other hand, given that the investment method of valuation and the NPV method are closely allied in their approach, the NPV method is preferred by valuers when comparing the value of an investment proposition with others.

5. Application of DCF and investment value (worth)

The principal use of DCF lies in relation to investment decisions whereby a choice between alternatives can be made. In the case of property development this may be a choice between development propositions on different sites, or between different development schemes on the same site. For example, a landowner may be able to consider developing a site with 100 houses over three years or 140 flats over four years or various other permutations. The analysis of each proposal by DCF will indicate which scheme is the more profitable.

Again, an investor may be offered the choice of different investments showing different rental patterns. Although each investment can be valued by the investment method, a further comparison between them can be made by using DCF analysis to show the IRR of each investment. In so doing, an allowance for projected rental increases can be incorporated. This approach has been developed by the determination of the "equated yield", which is considered further in Chapter 12.

More recently, an increasing number of investors require calculations of worth, which is now generally called *investment value*. This, as shown above, is defined as "the value of an asset to the owner or a prospective owner". These investment value calculations are an aid to the buy/sell decision. Typically, a cash flow is constructed over a five to 15-year period, with rental growth incorporated on the basis of the client's expectations or on the basis of agreed scenarios – "what if rents do not grow?", "what if there is no growth for five years and then there is a return to long-term growth of, say, 3%?". These scenarios can be played out on an Excel spreadsheet and will incorporate other factors such as rent review fees, lease renewal fees, capital expenditures, and an exit value at the end of the holding period. The cash flow will be discounted at the client's target rate or cost of capital rate. The *investment value* or *worth* is compared to the *market value*. If selling, then sell when MV exceeds worth; if buying, then buy when worth is greater than MV.

Example 10–6

An investor has been offered the freehold interest in an office building at a price of £14,285,000 to show an initial return on cost of £15,107,143 (£14,285,000 + 4% stamp duty land tax (SDLT) and 1.75% fees and VAT on fees) of 6.62%. The property is currently let on a full repairing and insuring (FRI) lease at a rent of £1,000,000. This is considered to be the market rental value. There is a rent review in three years' time. The investor requires an assessment of investment worth based on the following assumptions:

* No rental growth for three years followed by growth at 3% per year, i.e. no change in rent at the first rent review.
* Investor's target rate of 10%.
* Holding period of 10 years.
* Exit yield of 6.5%.
* Management cost, rent review fees and lease renewal fees to be ignored.

End year	MRV +3%	Rent	End year	Present value
1	£1,000,000	£1,000,000	0.90909	£909,091
2	£1,000,000	£1,000,000	0.82645	£826,446
3	£1,000,000	£1,000,000	0.75131	£751,315
4	£1,030,000	£1,000,000	0.68301	£683,013
5	£1,060,900	£1,000,000	0.62092	£620,921
6	£1,092,727	£1,000,000	0.56448	£564,474
7	£1,125,509	£1,000,000	0.51316	£513,158
8	£1,159,274	£1,000,000	0.46651	£466,507
9	£1,194,052	£1,159,274[1]	0.42410	£491,648
10	£1,229,874	£1,159,274	0.38554	£446,946
Exit value in year 10		£18,734,184[2]	0.38554	£7,222,777
Worth today				£13,496,296

1. The lease rent review on these growth assumptions for year 9 will be based on the market in year 8 as that is when the negotiations will normally be completed.
2. The exit value in year 10 is a simple Term and Reversion valuation: £1,159,274 × YP 3 years at 6.5% plus £1,229,874 × YP perp. def'd. three years at 6.5% produces the exit value which, discounted at 10% for 10 years, gives £7,222,777.
3. An alternative structure is to take time along the horizontal axis and information along the vertical axis.
4. Whilst this has been set out as an investment value or worth exercise, a similar format would be used for a DCF valuation. But all variables must be supported by market data, e.g. growth, which could be an indexed-linked annual increase, market discount rate, market-based ARYs for exit value calculations, etc.

This calculation has been presented as a table reconstructed from an Excel spreadsheet; the calculations have been rounded for illustration. Using an Excel spreadsheet allows the valuer to undertake a series of "what if" calculations to support the investor client's decision making. The conclusion here is that the market price is too high for this investor, given the investor's criteria. A worth calculation should always be accompanied by an estimate of MV.

DCF is a powerful tool in the hands of the experienced valuer. The less experienced can be confused by the use of both the ARY and the target or discount rate in assessing the implied rate of rental growth and in assessing the exit value before discounting at the discount rate. A frequent error, which must be avoided, is the misuse of the tool whereby growth is made explicit in the future rental figures, whilst at the same time the assessment of present value also allows for growth by using the ARY for discounting.

Currently, DCF is favoured in a number of countries for presenting an investment valuation; it is popular in the US and in a number of EU countries. This is due in part to lease practices which are on a gross basis but permit recovery of cost increases, such as utility cost increases, and in part due to the indexation of rents. Hence a cash flow will make the gross income, annual costs and annual net income explicit. It is a developing technique becoming more explicit with the level of market transparency. For example, in the US it is normal practice to incorporate lease renewal probabilities.

Further material on DCF can be found in other Elsevier (Taylor and Francis) books, including *The Income Approach to Property Valuation* by Andrew Baum, David Mackmin and Nick Nunnington (2011). But for members of the RICS the essential reading now is the RICS guidance note.

The use of DCF for development valuation work is considered in Chapter 11.

Residual approach or development method of valuation

1. Introduction

The residual approach is used in the valuation of development property that consists of bare land or land with existing buildings which are either to be altered and improved, an exercise commonly termed refurbishment, or to be redeveloped with new buildings.

The method works on the premise that the price a buyer can pay for such a property is the surplus after deducting the costs of construction, the costs of buying and selling, the cost of finance and the amount of profit required to carry out the project, from the estimated sale price or market value of the finished development. This can be expressed as:

> *Proceeds of sale less costs of development and profits = surplus for land in its existing state.*

These elements can be considered in turn in relation to an example.

Example 11–1

Value the freehold interest in a vacant house with permission to redevelop with a shop and offices. The shop will have a frontage of 6 m and an internal gross depth of 20 m. Two floors of offices will be built over the shop with separate access and will provide 90 m^2 NIA of offices per floor, 205 m^2 GIA. The offices are expected to let at £200 per m^2 and the shop at £60,000 p.a., and the completed property should sell on the basis of an 8.5% return.

2. Proceeds of sale

This is known as the Gross Development Value (GDV) and arises from the sale of the developed property. In the case of residential property this will be the price expected for each unit, based on comparison. In commercial developments the GDV will be the estimated market value or price that will be obtained on a sale, usually after the property has been let to create an investment, and will be estimated using the income approach.

The proceeds of sale are the whole of the anticipated moneys to be realised from the development. These will not be receivable until the work is completed, which may be some time in the future, but it would be wrong to discount the proceeds to their present value because the cost of holding the property is taken as a cost of the development, and therefore to discount the proceeds of sale would be a double adjustment.

What the valuer is seeking to establish is the size of the "development cake" which can then be cut up into the various slices needed for costs and profits and so establish how much of the cake remains as the land slice.

Where the buyer intends to occupy the completed development, or to let it on completion and then retain it as an investment, there will be no actual sale, but it is still necessary to determine the realisable value of the development in order to complete the residual valuation.

Applying this approach to Example 11-1; the proceeds of sale will be:

Market rent	
Shop	£60,000 p.a.
Offices 180 m^2 at £200 per m^2	£36,000
	£96,000 p.a.
YP or PV£1 p.a. in perp. at 8.5%	11.765
GDV	£1,129,411
GDV = say £1,130,000	

3. Costs of sale

The main costs incurred in a sale will be the agents' fees, including advertising costs, and legal fees in the conveyance; the general level of fees is in aggregate around 3% of the sale price.

In the case of investment properties it is usually best to let the property before selling; if so, the agents' and legal fees incurred in the letting should be deducted. Agents' fees will normally be around 10% of the agreed rents or around 15% if two or more agents are instructed. The aggregate fees will be around 20% of the rents obtained if promotion costs are included.

Where a buyer intends to retain the property, the costs of sale should still be deducted as the aim is to assess market value and not the value to an owner.

Sale costs	
Agents' and legal fees, say 3% of £1,130,000 =	£33,900
Letting costs	
Agents' and legal fees and promotion, say 20% of £96,000	£19,200
Total costs	£53,100

Where appropriate, VAT on these fees must also be allowed for (i.e. when the developer is unable to recover VAT – see Chapter 21).

4. Costs of development

The major items are the actual costs of building the property and funding costs. Certain other miscellaneous items may be met on occasion.

(a) Cost of building

In the preliminary stages, the costs of construction will need to be estimated. Estimates are normally based on the prevailing costs of building per m^2 of the gross internal floor area (GIA). As the details of the scheme become clearer, it may be appropriate to prepare a priced specification or a priced bill of quantities. In addition to the costs of building, the professional services of the design team are payable. The membership of the team depends on the nature and scope of the development but normally includes an architect, a quantity surveyor and one or more engineers; in the more complex schemes, structural, electrical, heating and ventilation engineers will be needed. The fees payable depend on the circumstances but usually vary between 8% and 14% of the building costs, with 12% as the average.

(b) Miscellaneous items

All sites are different and have unique features which may require various special costs. Typically, such items include costs of demolishing existing buildings, which will depend on the nature of construction and the salvage value; costs of obtaining possession, either by compensation to tenants on the site or even buying in minor interests; costs of agreeing compensation to neighbours, such as buying rights over the land (easements) or agreeing party wall rights and compensation; costs of providing above average quality boundary works; exceptional costs such as site clearance or filling of uneven land and site cleaning if contaminated, and diversion of services; and off-site costs such as highway improvements required as a condition for the grant of planning permission and landscaping. As can be seen, there are many possible problems that may need to be overcome. Where possible, an estimated cost for each item should be included. But as these items are sometimes difficult to predict and may not be known at the time of valuation, it is possible to allow some general sum by way of a contingency allowance. On the other hand, it could be said that these are part of the general risks of development, including the unknown, which are reflected in the allowance for profits.

(c) Costs of finance

Considerable sums of capital are needed for property development. Normally, this money is raised from banks or other lending institutions, or it may be loaned as part of an overall arrangement with an investing institution such as a pension fund, particularly in the case of medium to large scale commercial developments. The cost of borrowing the money, which will be repaid on the completion and sale of the development, is the interest charged at an agreed rate plus, in many cases, a commitment or finance arrangement fee of around 1% of the money to be provided. The rate of interest depends on the prevailing rates being charged and will also vary with the status of the borrower and the risks attached to the development scheme. The interest rates commonly range from 1% to 4% above the base rate of one of the clearing banks. These rates have fluctuated considerably in recent years within a range of around 6–20%. The valuer will use the rates current at the date of the valuation. Finance is extremely difficult to raise for speculative development at the time of this edition.

In some instances the developer may have raised money on a long-term basis at a favourable rate of interest which may be low compared with prevailing rates, or money may be provided from the developer's own resources. Nonetheless, the prevailing borrowing rate should be adopted in the valuation as this is the opportunity cost of the capital and it reflects the market for the site. The rate of interest chosen will vary according to the type of scheme and the size of the likely developer. A small scheme such as the one in Example 11-1 will attract small development companies who have, in general, a higher cost of interest than would be the case for a major developer with access to institutional funding. If the preferential rate is adopted, or none at all, this will produce a value to that person which is not necessarily the market value.

Once the rate of interest is known, the interest costs can be determined. The money to be borrowed relates to two items: building costs and land costs. As to land costs, these will be incurred at the start, so that the money is borrowed at the start and interest runs for the whole period of development. On the other hand, money required for building works will only be needed in stages and, as a rule of thumb, which experience shows is normally reasonably accurate, it may be assumed that the whole of the building money is borrowed for half the period of development. When the details of the development are known, it may be necessary to change this rule of thumb and adopt an S curve that reflects the more typical building cost drawdowns. In the case of housing development, the developer may obtain revenue from sales of houses as the development proceeds, with consequent savings in the money to be borrowed. The implications of this are considered in Chapter 15.

In Example 11-1 the costs of development are:

(a) *Building costs*
Shop, 120 m^2 (GIA)
at £500 per m^2 =   £60,000
Offices, 205 m^2 (GIA)
at £700 per m^2 =   £143,500
                                                       £203,500

add		
Professional fees, say 12%	£24,420	£227,920
(b) *Demolition costs*		
Estimated at		£7,000
(c) *Costs of finance*		
Assume development will take 1 year with interest at 8%		
Demolition costs £7,000 for 1 year at 8%	£560	
Building costs £227,920 for say		
6 months at 8%	£9,117	£9,677
Total costs of development		£244,597

As the value of the land is not yet determined, it is inappropriate to calculate interest on this element at this stage; this will be considered at the end of the calculation.

5. Development profits

As for any risky enterprise, a person undertaking a development will seek to make a profit on the operation. Target levels of profit will depend on the nature of the development and allied risks, the competition for development schemes in the market, the period of the development and the general optimism in relation to that form of development. Consequently, it is not possible to lay down firm limits of required profit levels. The profit is usually related to the costs involved, but sometimes to the development value. The profit is the gross profit to the developer before meeting the developer's general overheads and tax. Hence, developers may seek, say, 20% gross profit on the capital invested, namely building costs and land costs, or say 15% of the development value.

The profits are related to costs but, as the land costs are unknown, the profit on these cannot be calculated at this stage. The better approach may be to calculate profit as a percentage of the proceeds of sale and then to check this figure as a reference to the return on total costs.

Hence, in Example 11-1:

Developer's profit $= 15\% \times £1,130,000 = £169,500$

6. Surplus for land

At this stage the valuer has determined the net proceeds of sale, the total cost of development and the profits. The difference between these figures represents the sum available to spend on land costs, including interest and acquisition costs. In some cases building costs will exceed net proceeds of sale. If so, this shows that there is negative value for that development, but some other form of development might be profitable.

In Example 11-1, the surplus available for land costs is:

Proceeds of sale GDV		£1,130,000
less		
Costs of sale at say 3%		£33,900
Net proceeds of sale		£1,096,100
less Costs of development		
(i) Building costs	£227,920	
(ii) Demolition	£7,000	
	£234,920	
(iii) Interest on (i) & (ii)	£9,677	
(iv) Total letting costs	£19,200	
(v) Developer profit at 15% GDV	£169,500	£433,297
Surplus for land costs		£662,803

The land costs comprise three items. First, there is the price to be paid for the land, the very purpose of the valuation. Secondly, there are the professional fees and perhaps stamp duty land tax (SDLT) in relation to the purchase. These fees will generally be for an agent and for legal services in the conveyance. These, together with SDLT at 4%, are likely to be approximately 5.80% of the price paid unless the land value is below £150,000 in which case SDLT is zero, or if below £250,000 where SDLT is 1%, or between £250,000 and £500,000 where SDLT is 3%.[1] The third item is the interest on the money borrowed to buy the land to be repaid on sale, calculated for the period of development. All of these relate to the actual price of the land, which in turn depends on the three items. A simple way of apportioning the surplus between them is to express the land price as a symbol, say x, and then solve the subsequent equation; this in 11-1 becomes:

(a) Land price	$1.0000x$
(b) Fees on land purchase at 5.80% of price =	$0.0580x$
(c) Finance	
Interest at 8% for the period of development	
on (a) and (b) = $1.0580x \times 1$ yr at 8% =	$0.08464x$
Land surplus total	$1.14264x$
But the land surplus is	£662,803
$\therefore 1.14264x = £662,803$	
$\therefore x =$	£580,063
Hence, value of land is, say	£580,000

---

1 From 2012 on sales of residential property over £1m and below £2m SDLT is 5%, over £2m it is now 7% and over £2m purchased by certain bodies it is 15%.

An alternative way of calculating the net value of the land is as follows:

Surplus for land costs	£662,803
Less	
Allowance for interest × PV of £1 in 1 yr at 8%	0.92593
	£613,709
Less	
Acquisition costs at 5.80% (divide by)	1.0580
	£580,065
Hence, value of land is, say	£580,000

Thus the valuation is completed. The whole valuation is:

Proceeds of sale (GDV)			
MR			
Shop		£60,000 p.a.	
Offices, 180 m^2 at £100 per m^2		£36,000	
		£96,000	
YP or PV£1p.a. in perp. at 8.5%		11.765	
say			£1,130,000
less			
Costs of sale at 3%			£33,900
Net proceeds of sale			£1,096,100
less			
Costs of development			
(i) Building costs			
Shop, 120 m^2 at £500 per m^2		£60,000	
Offices, 205 m^2 at £700 per m^2		£143,500	
		£203,500	
(ii) Professional fees at 12%, say		£24,420	
		£227,920	
(iii) Demolition costs		£7,000	
		£234,920	
(iv) Cost of finance (interest)			
£7,000 for 1 yr at 8%		£560	
£227,920 for, say, 6 months at 8%		£9,117	
		£244,590	
(v) Agent's and legal fees on letting at 20% of £96,000		£19,200	
(vi) Developer's profit at, say, 15% GDV		£169,500	£433,297
Land surplus			£662,803
Less			
Allowance for interest, 1 yr at 8%			
× PV of £1 in 1 yr at 8%			× 0.92593
			£613,709
less Acquisition costs at 5.80% (divide by 1.0580)			1.0580
			£580,065
Hence, value of land is, say			£580,000

This example is comparatively simple and no allowance has been made for VAT, delays in disposals, phased developments or inflation. VAT would normally be recoverable if the developer has waived the exemption available to property and has registered the building for VAT purposes (see Chapter 21) or is zero rated (e.g. new residential developments). Where VAT is not recoverable then an appropriate allowance should be added to the development costs.

Where delays are expected in the disposal of the property, additional interest should be allowed for on both the development costs and the land. Thus, if it is assumed that the letting and sale of the property in the above example would take six months to complete, then the effect on the value would be as follows:

Net proceeds of sale		£1,096,100
less		
Development costs as above		
(including interest)	£244,590	
Interest for 6 months at 8%	£9,784	
Agents' fees on letting	£19,200	
Developer profit at 15% GDV	£169,500	£443,074
Land surplus		£653,020
less		
Allowance for interest, 18 months		
At 8% = × PV of £1 in 18 months at 8%		× 0.8903
		£581,389
less		
Acquisition costs at 5.80%		1.0580
		£548,571
Hence, value of land is, say		£549,000

Where there is a phased development, then each phase can be valued separately, but a further allowance must be made for interest on the land value for the second and subsequent phases to reflect the additional holding cost of the land. If it is assumed that three shops with offices over are to be developed but with work starting on the second shop six months after the start of the first, and starting on the third shop six months later still, the calculations will be as follows:

Site value for shop 1, say		£580,000
Site value for shop 2	£580,000	
× PV of £1 in 6 months at 8%	× 0.962	
	£557,960	
Site value for shop 3	£580,000	
× PV of £1 in 1 yr at 8%	× 0.926	£537,000
		£1,674,960
Hence, value of land is, say		£1,675,000

No allowance has been made for inflation because of the normal valuation assumption that all values and costs are taken at present values. In times of significant inflation this approach may not be correct, and this will be further considered in Chapter 15.

In the calculation above of the land value of £580,000, the profit was taken as £169,500, being 15% of the GDV. The return on total costs equates to:

$$169,500/[(443,074 - 169,500) + 653,020]$$

$$= 169,500/926,594$$

$$= 18.29\%$$

If the valuer wishes to use return on cost rather than return on GDV to arrive at the profit allowance, then this can be done by substituting the relevant figure for the £169,500 used in the example [i.e. say 18.29% × (244,590 + 9,784 + 19,200) = 18.29% × £273,574 = £50,036] and the resultant land surplus will include the next tranche of developer profit [i.e. £1,096,100 − (£273,574 + £50,036) = £1,096,100 − £323,610 = £772,400 land surplus including developer profit]. This figure can then be reduced by dividing by 1.1829 to give the land surplus excluding developer profit, i.e. £653,047 [the difference from the figure in the example, which is £653,020, is due to rounding of figures in the calculations.]

This valuation contains many figures, all of which are based on estimates of cost or value or derived from such estimates. As with any estimate, one person's opinion may differ from another so in that sense they are all variable. Given a calculation based on a large number of variables, the actual range of answers that can be produced is wide. This variability is the method's weakness, but it is one which is acceptable so long as the estimates are prepared by experts with specialist costing skills in order to reduce any possible errors.

The residual method is disliked by courts and tribunals in compensation cases (see Chapter 26) because it is not tested by "haggling in the market". However, in the market, the residual method will continue to be the main cornerstone of many opinions of value, particularly those involving land for development or redevelopment. Nevertheless, the method is weak because of the speculative nature of the main items, a small variation in any one of which can make a large difference to the answer. Care must therefore be taken to see that the items in the valuation are as correct as possible and as verifiable as possible.

Because of the mathematical nature of the calculations, this method of valuation is ideally suited to solution using Excel or specialist software. Many property development programs have been produced which not only produce a value for the land but also produce cash flow charts and schedules, as well as feasibility studies and sensitivity analyses which allow for rapid recalculations and objective testing. This approach will be considered in Chapter 15. The RICS has published a Valuation Information Paper, VIP 12, on the *Valuation of Development Land* (RICS, 2008); this is currently under review.

Chapter 12

Developments in valuation approaches

1. Introduction

In recent years critical attention has been given to valuation methods both from within the valuation profession and from other quarters, particularly the financial sector. This has occurred for various reasons. There have been sharp variations in property values which have led to property transactions attracting considerable attention. There has been a growing involvement by institutional and overseas investors who have invested significant funds in property. Public interest in and awareness of property has been fostered by press commentaries, with most of the national newspapers carrying regular property articles. The importance of property to the economy is now clearly recognised: a fact that was highlighted by the collapse of the property market after 1973, 1990 and 2007 when many companies were at risk, even though they were not in business as property companies. Indeed, many companies have substantial property interests as a natural consequence of their activities, and some companies have a property portfolio rivalling that of the major property companies and institutions in value and quality.

The importance of property to the economy has again been highlighted by the "credit crunch" which began in 2007 and developed into recession in 2008/2009. The cause may in part have been a function of banks' competitiveness in raising their accepted residential loan to value ratios, thus raising, in the residential mortgage market, their income multipliers and increasing the number of higher risk loans – subprime lending activity. The impact of subprime lending and the reduction of bank-to-bank lending in the light of the subprime position of some banks has combined to have a negative effect on the residential and commercial property markets. This in turn has had a multiplier effect on the economy in terms of a whole variety of goods and services which flow from an active property market, such as legal work, brokerage work, surveyors' work, removals, DIY, construction and mineral extraction connected to construction. A slowdown in sales activity affects market prices in both the residential and commercial property markets and raises questions, in the minds of some, about the process of estimating market values during periods of limited market activity.

It is natural for the work of valuers to be closely scrutinised, given the importance of property to the economy. A valuation may be a vital factor in deciding whether shares or bonds will be purchased, companies remain solvent or expensive projects are carried out. Most importantly, a valuation may affect members of the public at large who could suffer seriously if the valuation is misleading. Consequently, the methods by which valuers arrive at a value have, increasingly, been examined, especially the investment method of valuation. This is evidenced by criticism of methods by academics and practitioners in books, journals and other publications.

This body of critique has led to many suggestions as to alternative approaches to certain valuations, or considerable adaptation of the general principles. However, until these methods receive wide acceptance, such as to oust the methods set out in previous chapters, they cannot be set down as the appropriate method to adopt. Indeed, given the variety of views expressed, it would be difficult to select any one against the others. This chapter will be limited to a consideration of those aspects which give rise to criticism of investment valuations, and some brief comments on the solutions are offered so that readers can judge their validity for themselves.

2. Relationship of rent in arrears and in advance

Valuation methods were developed on the assumption that rents were paid and received annually in arrear. This premise has persisted even though rent has nearly always been payable on a quarterly or monthly in advance basis.

Where rent is payable either annually in arrear or annually in advance, the difference between Years' Purchase (YP) (rent in arrear) and YP (rent in advance) is obviously $1 - $ (Present Value (PV) for nth year). The YP (rent in advance) can be derived from $1 + $ YP (rent in arrear) for $(n - 1)$ yrs, e.g. YP 3 yrs at 10% (rent in advance) = YP 2 yrs at 10% + 1 = 1.7355 + 1 = 2.7355. This compares with YP 3 yrs at 10% (rent in arrear) of 2.4869, a difference of 0.2486 or 10%.

Over time, the difference between the YPs grows as the PV falls. For example, YP 50 yrs at 10% (rent in arrear) is 9.9148 whereas YP 50 yrs at 10% (rent in advance) is $1 + $ YP 49 yrs at 10% (rent in arrear) = 1 + 9.9063 = 10.9063, an increase of 0.9915 on 9.9148, again a difference of 10%. In fact, the difference will always be exactly 10% at a yield of 10% and it can be shown that the percentage increase at any rate will always be the rate adopted. Hence, for example, YP 30 yrs at 6% (rent in advance) will be 6% greater than YP 30 yrs (rent in arrear), YP 40 years at 7% (rent in advance) will be 7% greater than YP 40 years at 7% (rent in arrear), and so on.

However, rent paid annually in advance is probably as untypical as rent paid annually in arrears. Where the rent is paid quarterly in advance the method is as follows.

Example 12–1
Value a rent of £1,000 p.a. receivable for 10 years and payable quarterly in advance. The rent is regarded as arising from a 12% investment.

(a) Conventional approach (YP SR, rent annually in arrears)

Rent	£1,000 p.a.
YP or PV£1 p.a. for 10 yrs at 12%	5.65
Capital value	£5,650

(b) Valuation approach reflecting rent in advance
One method is to treat the income as being four payments a year with an effective rate of one-quarter of the annual rate, subject to adjustment as described above of $(1 - PV)$ or adopting $1 + YP (n - 1 \text{ year})$. Hence:

Rent per quarter		£250
YP [(4 × 10) − 1] quarters at 3% per		
quarter (= YP 39 periods at 3%)	22.808	
Add 1 =	1.000	23.808
Capital value		£5,952

But this alternative approach is not a true comparison because a yield of 12% means that in one year the investor earns 12% on their capital. If the interest is added at periods of less than a year it will normally be on a compound basis. Thus, where the annual yield is 12%, and interest is added quarterly and is compounded, the quarterly rate is not 3% as this would produce an annual yield of 12.55%.

Amt. of £1 for 4 periods at 3% =	1.1255
less	
£1 invested	1.0000
Interest after 1 year =	0.1255

Converted to a percentage, this is 12.55% (0.1255×100).

The problem derives from what is meant by the yield (see Chapter 7). If a yield of 12% is meant to represent not only the nominal yield but also the effective yield (APR or AER), then the quarterly equivalent is not 3% but 2.874%. If this quarterly effective rate were adopted, then there would be no difference between interest being paid quarterly or annually in arrears. This fact is understood by the market and all valuations are carried out on the same conventional basis assuming annual in arrears rent. This means that a property investment will provide a better rate of return than is implied by the capitalisation rate used in the valuation; this is understood to be the case by valuers and investors. When one property investment is compared with another, the same "error" occurs in both and therefore it is effectively self-cancelling, and since this yield is calculated on the same "error" basis the result is in any event correct.

The Investment Property Forum suggested that all valuations where rent is payable quarterly in advance should be presented on a quarterly in advance basis. They acknowledged that this change does not in itself change market values; it merely presents the valuation differently. If valuers wish to present their calculations on a quarterly in advance basis, then they need to analyse all market data on the same basis. What they must not do is simply adopt the quarterly in advance figures presented in *Parry's Valuation and Investment Tables*. For example, the YP annually in arrears at 10% for 10 years is 6.1446, the YP for 10 years quarterly in advance at 10% is given as 6.5240; simply changing assumptions without consideration of the nominal and effective yields will suggest an increase in value. Switching from annual to quarterly on some software packages could lead to a similar error. Valuing on a quarterly basis requires the valuer to establish first the correct quarterly in advance yield to provide the same value as the previous annual in arrears value. Possibly for this reason, the market has continued to value on an annual in arrears basis, thus maintaining continuity in comparability of the market yields used for valuation with other investments where an annual rate is quoted. For comparison, the True Equivalent Yield needs to be known and is generally requested by investors.

3. Taxation of incomes

In the investment method of valuation income is capitalised on a before-tax basis. The main reason for this is that tax payable depends on the status of the recipient of the income. Hence, the net income reflects this status, as indeed will the net yield to any investor. If net yields are adopted as a measure of comparison they will no longer act as a means of comparison between respective investments since they will reflect, in addition to the qualities of the investment, a unique feature of an individual taxpayer or investor. (But note that in the dual rate application to leaseholds, this reasoning is sometimes refuted.)

However, valuations may be made on a net of tax basis and these will be of use in determining the net worth of an investment to an individual. Such approaches are referred to as "true net" approaches. Consider first a freehold interest let at its market rent.

Example 12–2
Value the freehold interest in a property let at the market rent of £10,000 p.a. The yield is 10%.

Market rent (MR) (and rent received)	£10,000	p.a.
YP or PV£1p.a. in perp. at 10%	10.0	
Capital value	£100,000	

But allowing for tax at the investor's tax rate of 40%, the valuation is:

Income net of tax	£6,000
YP or PV£1p.a. in perp. at 6%	16.667
Capital value, say	£100,000

A gross income of £10,000 becomes a net income of £6,000 with tax at 40% and a gross yield of 10% with tax at 40% becomes a net yield of 6%. Each approach produces the same value. This will be true of any valuation of a freehold let at market rent and at any rate of tax. The same will be true of a leasehold interest sub-let at MR and valued on a dual rate basis. as shown in Example 12-3.

Example 12–3

Value the leasehold interest in commercial premises sub-let at an MR of £12,000 p.a. The leasehold interest has 10 years to run at £2,000 p.a. This is a 10% investment. The investor pays tax at 40% on rents.

MR (and rent receivable)	£12,000	p.a.
less		
Rent payable	£2,000	
Profit rent	£10,000	p.a.
YP 10 yrs at 10% and 3%		
(Tax at 40%)	4.075	
Capital value	£40,750	

Now value the investment on a net of tax basis.

MR (and rent receivable)	£12,000	p.a.
less		
Rent payable	£2,000	
Profit rent	£10,000	p.a.
less		
Income tax at 40% = 0.4 × £10,000 =	£4,000	
Net profit rent	£6,000	p.a.

As before, the net yield becomes 6%; the net income £6,000 and there is no need to gross up the ASF as the profit rent is now net of tax. Hence:

Net profit rent	£6,000	p.a.
YP 10 yrs at 6% and 3%	6.792	
	£40,750	

Dual rate adjusted for tax is a hybrid as it uses a gross remunerative rate for comparison and a net accumulative rate, adjusted for tax, on that part of the gross profit rent used for the SF.

So, valuations of a freehold and a leasehold interest let at MR give the same result whether valued gross or net of tax. From this it may be concluded that such investments present no advantage to one taxpayer as against another. But this should not be confused with the advantage to a non-taxpayer, as against a taxpayer, which a short leasehold interest offers for quite different reasons, as explained below.

The valuation of a freehold or leasehold interest where the present rent receivable is not at market rent (MR) – a reversionary investment – will, however, produce different results as between gross and net approaches.

Example 12–4

Value on a 10% basis the freehold interest in commercial premises with an MR of £10,000 p.a. which are currently let for the next five years at £4,000 p.a.

Term			
Rent receivable		£4,000	
YP 5 yrs at 10%		3.791	£15,164
Reversion			
MR		£10,000	
YP or PV£1p.a. in perp. at 10%	10.0		
× PV of £1 in 5 yrs at 10%	0.6209	6.209	£62,090
			£77,254

On a net of tax basis, assuming tax at 40%, the valuation becomes:

Term			
Rent receivable		£4,000.00	
less			
Income tax at 40% of £4,000		£1,600	
Net rent receivable		£2,400	
YP or PV£1p.a. for 5 yrs at 10% gross, 6% net		4.212	£10,109
Reversion			
MRV		£10,000	
less			
Income tax at 40% of £10,000		£4,000	
Net MR		£6,000	
YP or PV£1p.a. in perp. at 6%	16.667		
× PV £1 in 5 yrs at 6%	0.74726	12.45	74,700
			£84,809

The net of tax valuation gives a higher value. Indeed, as the rate of tax is increased it will be found that the value rises on a net of tax basis. In five years' time, when let at MR, the property will be valued at £10,000 p.a. × 10 YP or £6,000 p.a. × 16.667 YP = £100,000. Clearly the difference between gross and net of tax valuations arises because there is a term at a lower rent and because the deferment of the reversion is at a net of tax PV, even though there is no income to be taxed during this period other than the term income.

It can be seen that, in valuing the term, although the rent is reduced by 40%, the YP is not increased correspondingly. In determining the present value of the reversion, the value at reversion is the same but the present values differ because of the different rates. If 10% were to be used for the deferment in the net valuation, because there is no income to be taxed, this element would be the same for gross and net. A net of tax valuation of a term and reversion is more complex than a valuation in perpetuity.

This net of tax valuation is incomplete as no allowance has been made for capital gains tax (CGT); allowing for CGT at 18%, the valuation becomes:

Example 12–5

Term (as before)		£10,109
Reversion to net MRV £6,000		
YP in perp. at 6%	16.667	
	£100,000	
less		
CGT		
Net value of interest $= x$		
$\therefore$ Gain $= (100,000 - x)$		
$\therefore$ Tax $= 0.18\,(100,000 - x) =$	£18,000 − 0.18x	
$\therefore$ Reversion net of CGT	£82,000 + 0.18x	
$\times$ PV £1 in 5 yrs at 6%	0.747	£61,254 + 0.1345x
Capital value		£71,363 + 0.1345x

But capital value $= x$

$\therefore x = £71,363 + 0.1345x$ and, as $x - 0.1345x = £71,363$, so $0.8655x = £71,363$
and $\therefore x = £82,453$

Note: The CGT element has ignored any annual exemptions from CGT that might apply.

This calculation suggests that there will be a different worth for investors with differing rates of tax. If a full allowance is to be adopted to analyse an investment to an individual, it is probably more appropriate to use a DCF approach as discussed in Chapter 10. An allowance for tax in determining market value is inappropriate since the value would then depend on the choice of tax rate. Of course, if it can be shown that a group at a certain rate of tax is dominating the market for a specific type of investment, then it may become appropriate to adopt a net of tax approach that applies that group's tax rate. Such a situation has arisen in the case of short leasehold investments, as described below.

It has been stated frequently that rates of tax vary. At one extreme an individual or company may pay tax at a high rate whilst, at the other extreme, charities, and in particular pension funds, pay no income tax whatsoever – hence their description as "gross funds".

In Chapter 9, the valuation of leasehold interests was described, including the need to allow for the gross cost of an annual sinking fund (ASF). Such a cost represents a high proportion of profit rent when the term is short. However, if a gross fund invests in such interests, it is not faced with a tax liability, so that the ASF costs do not need to be grossed up to allow for tax. The effect of this is that it can afford to pay significantly more for a short leasehold interest (say up to 15 years) than a taxpayer or, alternatively, if it can buy such an interest at the net of tax value then it will obtain a higher than anticipated yield since the grossed up element remains as a return to capital.

This has led to gross funds tending to be a major force in this part of the investment market. However, it cannot be said that all short leasehold investments should be valued ignoring the tax element; rather that it may be appropriate to value grossing up the cost for tax and also ignoring this factor, with the price payable lying in between the two resultant figures as the seller and gross fund buyers share the benefits of the latter's tax-free status on the value. In practice, many gross funds adopt a single rate approach or a DCF approach to the valuation of short leasehold investments, and it would be appropriate to consider these different approaches when valuing investments likely to appeal to gross funds.

The RICS, in its brief guide for users of valuations,[1] notes:

> Strictly speaking, the relative tax impact should be reflected in valuations but the convention is that it is not, certainly not in an all-risks yield valuation. The tax impact is just another of the many factors wrapped up into the all-risks yield. It may be reflected in a DCF market valuation; and arguably DCF assessments of worth should reflect the specific investor's tax position.

In the case of short leaseholds the RICS does recognise that the market tends to reflect tax and its suggested approach is net of tax profit rent capitalised single rate at a net of tax rate. No reference is made here to dual rate.

4. Valuation of freehold interests

Two aspects of the valuation of freehold interests cause the greatest controversy; these aspects are the yield to be adopted and the effects of inflation.

(a) Nature and function of yields

The choice of, and the function of, the yield appropriate to a particular investment are both areas for comment. The fundamental issue turns on the function of the yield itself.

As a matter of basic principle, the conventional approach treats the yield as being a measure of comparison between various investments available, and the yield chosen reflects all the different qualities between the investments in question and others that are available. Thus, it reflects the potential for future growth, the

---

1 Property Investment in the UK, RICS, 2009.

strength of covenant, the likely performance in an inflationary or recessionary economy and any other factors that are thought to be relevant. It is referred to as an "all risks" yield.

Another school of thought argues that the yield itself should be analysed to reflect these different and distinct qualities. That view has been expressed in various articles, books and theses where a "real value approach" is advocated.

It is not the role of this book to attempt to evaluate the different methods advocated, but it is important for the student in particular to be aware of them.

(b) Equated yields

The valuation of a freehold interest calls for the determination of the appropriate yield for the class of investment under consideration. Thus, it may be that the yield on first-class shops is 5%, on modern factories 8%, and so on. What is meant by this is that where such a property is let at its MR on a normal market basis, then that is the yield to be used in the valuation.

This is sufficient for valuation purposes, but the converse of the method may be inadequate if an analysis is to be made of investments allowing for differing rent review patterns or projections of inflation in rents.

Indeed, there may be difficulties in valuing some reversionary properties. For example, if by analysis of similar properties let on 21-year leases with seven-yearly reviews the yield is known to be 5.33%, what is the appropriate yield to adopt if one further similar property is found to have been let on a 21-year lease with three-yearly reviews? In the absence of any comparables on this basis, valuers must use their best judgment. The yield is probably lower, but by how much? Should it be 0.25%, 0.5%, 1.0% or more? It is fair to say that in many cases the valuer's judgment, in conjunction with the investor's judgment, will determine the price offered in the market. But in some instances, and particularly where the investor is a fund or an institution, a more reasoned approach would probably be required.

A solution to the valuation problem, and also to that of how to allow for projections of inflation, can be found in a technique that has been developed and is referred to as the equated yield approach.

The expression "equated yield" has been given different definitions, so it should be made clear that in this instance the definition is taken to be the discount rate which needs to be applied to the projected income (allowing for growth), so that the summation of all the incomes discounted at this equated yield rate equates with the capital outlay.

This definition follows the internal rate of return (IRR) concept of DCF as set out in Chapter 10. Thus, an investment can readily be analysed allowing for actual and projected rent at whatever the review dates may be, against the market price, to show the IRR or equated yield. Alternatively, given the actual and projected rents and review dates, and given the equated yield appropriate to the investment, the price that should be paid to maintain these factors can be determined.

Hence, in the valuation problem posed above, if the analysis of the investments where the rent reviews were seven-yearly is established, the appropriate price/value for three-yearly reviews can be determined and thus the initial yield for valuation purposes.

Example 12–6
Similar properties have been letting for £10,000 p.a. on leases for 21 years with seven-yearly reviews. Rental growth is anticipated at 12% p.a. The properties have been selling at £187,500 (= 5.33% return). Value a similar property recently let at £10,000 p.a. for 21 years with three-yearly reviews. (12% annual rental growth is unrealistic but is used here to provide a more substantive illustration than would be the case with a more conservative 3%).

Step 1: Find equated yield of properties sold at £187,500

Try 16%

Years	A £1 at 12%	Expected rent (£10,000 × A £1)	PV £1 at 16%	PV × YP 7 yrs at 16%	PV of rents
0–7	–	10,000	1.0	4.0386	40,386
8–14	2.2107	22,107	0.3538	1.4289	31,589
15–21	4.8871	48,871	0.1252	0.5056	24,710
22–perp.	10.8038	108,038	0.0443	0.8306[1]	89,736
					186,421
Less price aid					187,500
NPV					−1,079

Note 1: 0.8306 = YP in perp. at 5.33% (18.75) × PV 21 yrs at 16% (0.0443).

Try 15%

Years	A £1 at 12%	Expected rent	PV £1 at 15%	PV × YP 7 yrs at 15%	PV of rents
0–7	–	10,000	1.000	4.1604	41,604
8–14	2.2107	22,107	0.3579	1.5639	34,573
15–21	4.8871	48,871	0.1413	0.5879	28,731
22–perp.	10.8038	108,038	0.0531	0.9956[1]	107,563
Less price paid					212,470
					187,500
NPV					+24,970

Note (1): 0.9956 = YP in perp. at 5.33% (18.75) × PV 21 yrs at 15% (0.0531)

$$IRR = 15\% + \frac{24,970}{24,970 + 1,079} = 15.95\%$$

Step 2: Apply IRR to the property to be valued

Years	Expected rent ($£10,000 \times A £1$) at 12%	PV £1 at 15.95%	PV × YP 3 yrs at 15.95% (2.2477)	PV of rents
0–3	10,000	1.0	2.2477	22,477
4–6	14,049	0.6415	1.4419	20,257
7–9	19,737	0.4115	0.9250	18,257
10–12	27,729	0.2640	0.5934	16,454
13–15	38,956	0.1693	0.3805	14,823
16–18	54,730	0.1086	0.2441	13,359
19–21	76,890	0.0697	0.1567	12,048
22–perp.	108,023	0.0447	0.0447x	4,828.628x

117,675 + 4,828.628x

Less price paid   10,000x

NPV   0

$$10,000x = 4,828.628x + 117,675$$
$$10,000x - 4,828.628x = 117,675$$
$$\therefore 5,171.372x = 117,675$$
And x = 22.755 YP (= YP in perp. at 4.394%)

Note: x = YP in perp. at the initial yield.

Hence initial yield on property to be valued is 4.394%.
The price that should be paid is £227,583.

It can be seen that the valuer making a subjective judgment as to the appropriate reduction to the 5.33% yield would need to have reduced the yield by close to 1%, but this would be a matter of chance. On the other hand, the calculations have produced a mathematical answer which is not necessarily acceptable as a true valuation. This is because one would need to assume that rents are unaffected by the rent review pattern. In practice, if identical properties are offered on leases with either three-year or seven-year review patterns, it is most unlikely that the initial rent would be £10,000 p.a. for both properties. The more likely situation is that, if properties let on 21-year leases with seven-year reviews command an initial rent of £10,000 p.a. and sell for £187,500, the valuer would be required to determine the approximate initial rent for a lease with three-yearly reviews so as to maintain the value at £187,500.

The calculation is the same as for Step 2 above, save that the initial rent is x, the value after 21 years is £187,500 × Amount of £1 for 21 years at 12% (10.804) × PV £1 in 21 yrs at 15.95% (0.0447) = £90,551.03, and the price paid is £187,500. This produces a result of:

$$11.769x + £90,551.03 = £187,500$$
$$x = \text{say, } £8,237.66$$

Thus, the initial rent should be, say, £8,240 p.a.

There are several tables available that provide a simple means of adjusting rents to match changes in rent review patterns. However, these are rarely applied in practice. The most common practice is to apply a rule of thumb, which is usually to add 1% to the rent for every year that the review pattern exceeds the normal.

The relationship between the equated yield, the all risks yield and the growth rate, given a specific rent review frequency, was considered in Chapter 10.

(c) Effects of inflation

Until recently, the impact of inflation has been experienced widely. It has led to widespread increases in the prices of many goods, including property. The term inflation is often used imprecisely, so that any increase in price is treated as an effect of inflation. Thus, in property, as rents increase they are said to rise with inflation. If inflation is measured as the increase in the retail price index, then the growth in rents of many types of property has sometimes been in excess of inflation and sometimes behind it. Whatever may be the cause of changes in rents, such changes have led people to examine the investment method of valuation and to question whether adequate or even any allowance is made in the valuation.

Taking the simplest example of a property recently let at its MR, the valuation method is to multiply the rent passing by a YP in perpetuity. As explained in Chapter 9, the yield from which the YP is derived reflects the likely future increases in rent, be they real or nominal. In this way, the valuation can be said to reflect fully the likely effects of inflation, used in its widest sense. Nonetheless, it has been argued that a more explicit allowance should be made for growth. No generally accepted approach for making such an allowance has emerged, although the DCF method or methods derived from DCF techniques enable explicit allowance to be made quite readily. As the recent past has demonstrated, there is no guarantee that rents will rise in the future and care must be taken not to confuse short-term changes with long-term trends. In so far as commercial property is concerned, technical changes make property prone to obsolescence, which will have an adverse effect on future rents and values.

(d) Valuation of varying incomes

It was shown in Chapter 8 that the valuer may be required to value a freehold interest where the rent receivable is currently less than the MR. The method adopted was to capitalise the present rent until the MR could be obtained, the "Term", and then to capitalise the rental value receivable thereafter, the "Reversion". An alternative approach is sometimes adopted whereby the rent currently receivable is capitalised in perpetuity, and the incremental rental receivable at the expiry of the lease is capitalised separately. The present rent represents a hard core of rent which will continue into perpetuity whilst future rents are incremental layers or top slices. Hence, it is commonly referred to as the hardcore or layer method.

The valuation approach by the hardcore or layer method involves applying a YP in perpetuity to both the continuing rent and the incremental rent, the capital value of the incremental slice then being deferred by applying the appropriate present value. The problem that arises, and the major criticism of the method, is the choice of the appropriate yield for each slice. Where the term and reversion would be valued at the same yield, then that same yield can be applied throughout the hardcore method, and an identical result will emerge since each pound of future rent is being discounted at the same rate.

Example 12–7

Value the freehold interest in commercial premises let at £50,000 p.a. for the next three years. The MR is £80,000 p.a. The yield at MR is 8%.

Conventional approach

Term			
Rent received	£50,000 p.a.		
YP 3 yrs at 8%	2.577		£128,850
Reversion to MR	£80,000 p.a.		
YP or PV£1p.a. in perp. at 8%	12.5		
× PV £1 in 3 yrs at 8%	0.7938	9.923	£793,840
Capital value			£922,690

Layer method

Continuing rent	£50,000 p.a.		
YP or PV£1p.a. in perp. at 8%	12.5		£625,000
Incremental rent (top slice)	£30,000 p.a.		
YP or PV£1p.a. in perp. at 8%	12.5		
× PV £1 in 3 yrs at 8%	0.7938	9.923	£297,690
Capital value			£922,690

However, where it is felt that the rate for the term should be different from the reversion, perhaps lower because the covenant of the tenant makes the rent receivable more certain of receipt, or higher because the rent receivable is fixed for an extended number of years, then the layer method is likely to produce a different result. For example, where the term rent is valued at a lower rate, then the results will only coincide if the incremental rent is valued at an appropriate marginal rate.

Take the rents in Example 12-7, but assume that the term rent is valued at 6%.

Conventional approach

Term			
Rent received	£50,000 p.a.		
YP 3 yrs at 6%	2.673	£133,650	

Reversion to MR		£80,000 p.a.	
YP or PV£1 p.a. in perp. at 8%	12.5		
× PV £1 in 3 yrs at 8%	0.7938	9.92	£793,840
			£927,490
Layer method		£50,000 p.a.	
Continuing rent			
YP in perp. at 6%		16.667	£833,350
Incremental rent		£30,000	
YP or PV£1 p.a. in perp. at 18.9%	5.29		
× PV of £1 in 3 yrs			
at 18.9%	0.593	3.137	£94,110
			£927,460

If the incremental rent is valued at approximately 18.9% and the continuing rent at 6%, the same result is obtained. In practice, the likelihood is that the valuer, whilst raising the yield in valuing the increment to reflect that it is the top slice, is unlikely to adopt what appears to be a very high rate. This leads to the criticism that, in the hardcore method, variable yields, if adopted, cannot be checked by direct comparison with the yields of similar investments unless identical investments can be found.

Similarly, where the term rent would be valued at a higher yield, the layer method tends to be difficult to apply. Assume again the facts in Example 12-7, but assume that the rent of £50,000 p.a. is fixed for 15 years, following which the property can be let at £80,000 subject to regular reviews. Clearly, the appropriate yield to apply to the £50,000 p.a. could be significantly higher than 8%, based on medium-term fixed interest investments plus a property risk premium. On the other hand, it would be unrealistic to value £50,000 into perpetuity at such a high rate.

Another problem with the hardcore or layer method is encountered when the current lease is on internal repairing terms and a deduction is needed for external repairs. The assumption is that this will be deducted in perpetuity, whereas in the term and reversion method the assumption is that the reversion will be to an effective net rent.

The main argument in favour of the hardcore method is that it is simpler to apply, particularly where the rent varies at renewal stages. A further argument is advanced that the yield for the term and reversion should be the same in all circumstances since this is the yield for the investment as a whole. If this is accepted, then the case for the hardcore method becomes one of convenience only. This latter view is now the more generally accepted one in the market. The choice between the methods must be made by the individual valuer, who must be able to justify the approach and the yields chosen.

5. Leasehold interests

Considerable criticism has been expressed about the dual rate method of valuing leasehold interests. The criticism concerns two areas, one being the nature of capital recoupment and the other the inherent errors to be found in the valuation of varying profit rents.

The valuation of a leasehold interest recognises that a lease is a wasting asset. Customarily, valuers have provided for capital recoupment through a low safe sinking fund, the dual rate approach. The low safe sinking fund rate has been shown in this edition, for illustration purposes, at various rates and can be assumed today to be between 2.25% and 3.5%, although the range given in *Parry's Valuation and Investment Tables* does go as high as 4%. The lowest figure currently used in *Parry's* is 2.5%. The current prevailing Bank of England Base Rate of 0.5% would suggest that accumulations net of tax might now be at or below 1%. Current tax rates would also suggest a need to reconsider the tax adjustment rate; for illustration purposes some dual rate adjusted for tax examples are shown in this edition at 35%, and other examples at 25%.

The questions that have been raised about dual rate include the following. Why replace capital in a low safe rate sinking fund when it can be demonstrated that few investors actually take out such policies? Why infer that this is what buyers do, or should do, when it is doubtful whether a leasehold redemption policy per se can be obtained from any insurer? Why assume investors are willing to buy a more risky type of investment but are so risk averse that they must recoup capital in the safest SF available in the market? Why is it only the historic cost that is recouped, rather than a replacement of cost plus an allowance for inflationary growth? Why is it an average tax rate that is used rather than a rate typical of the investor class most likely to be buying such interests? As inferred above, it is the valuer who selects what he or she considers to be an appropriate sinking fund rate and average tax rate and, by so doing, the valuer assesses the investment worth of the leasehold at a given tax rate rather than a market value based on analysis of market sales. Investors take investment decisions on the basis of NPV calculations at their target rate, or on the IRR – buyers are not interested in sinking funds but are interested in the yield expressed as the IRR. If these criticisms are valid, then considerable doubt is thrown upon the method of valuation described in Chapter 9. Before considering the overall effects of the criticisms, it is useful to consider the criticisms in turn.

(a) Replacement of capital

The purpose within the valuation method of replacing the capital that must ulti- mately disappear in the case of a leasehold investment is to isolate this factor, which is unique to leasehold interests as against freehold interests, and so to allow a ready comparison between freehold and leasehold interests. The alternative is to reflect this unique feature in the yield. This could lead to adjustments in the yield which would be partly subjective in the absence of direct comparables. The adjustment in the case of short leaseholds would need to be considerable if it is accepted that investors who buy at prices allowing for a sinking fund accept such prices as being reasonable. Market sentiment suggests that the single rate yield is likely to be between 2% and 7% above the freehold yield in the same property, but this will all depend on the perceived risks.

Example 12–8

Value the leasehold interest in commercial premises held for five years at £1,000 p.a. They are underlet for the term remaining at £11,000 p.a. The appropriate yield is 7%.

Conventional approach	
MR (and rent receivable)	£11,000 p.a.
less rent payable	£1,000
Profit rent	£10,000 p.a.
YP 5 yrs at 7% and 3% (Tax at 35%)	2.78
	£27,800
Value without ASF at 3%	
Profit rent (as before)	£10,000 p.a.
YP or PV£1p.a. for 5 yrs at 23% single rate	2.80
	£28,000

If it is felt that 23% is unrealistic and a lower single rate is adopted, the value will rise, which might suggest that short leasehold interests are being undervalued by the conventional approach. Such a view is supported in practice where gross funds are in the market, but mainly because of the taxation effect on both the income and the sinking fund yield. Also, in short-term leases the inflation effect is not so marked and these investments can be directly compared with dated gilt-edged securities. Here the valuation has been undertaken at a yield based on the equivalent yield in a freehold in the same property, which might be a reasonable assumption if, apart from the wasting nature of the leasehold, the lease is directly comparable. In most cases it is not directly comparable, as explained later.

If a longer term lease is considered, then the increase in yield is lessened. For example, if the term were 50 years at 7% and 3% adj. tax at 35%, the dual rate YP would be 11.956, which is approximately equal to YP (PV£1p.a.) 50 years at 8.2% single rate.

A further justification for incorporating a sinking fund is that it follows the accounting convention of writing off wasting assets. Indeed, it would be possible to write off the value of the lease by one of the other depreciation methods such as straight line instead of sinking fund.

The valuation would then become:

MRV	£11,000 p.a.
Less rent payable	£1,000
Profit rent	£10,000 p.a.
Less depreciation over term of lease (5 years) =	$0.2x$
	$£10,000 - 0.2x$
YP or PV£1p.a. 5 years at 7%	4.10
	$x = £41,000 - 0.82x$
	$x = £22,527$

A further criticsm is that dual rate fails to provide an investor with a true measure of the investment yield. The dual rate argument is very specific in describing the remunerative rate as the rate of return after allowing for the recoupment of capital in a sinking fund. This is not the same type of yield as the equivalent (no growth) or equated (with growth) yields used by investors for the purpose of comparing alternative investments; these yields represent the IRR, which is the only objective investment measure.

Thus, whilst some valuers continue to value using dual rate, investors analyse sale prices or asking prices single rate in order to assess by comparison whether the additional yield is sufficient to compensate for the additional risks associated with leasehold property investments. Single rate analysis adheres fully to the concept of the time value of money and the recovery of capital outstanding on a year-by-year basis.

Example 12–8A

Analyse the sale price of £28,000 in Example 12-8 to show that a buyer would be able to achieve a return on his/her money of 23% and obtain all the capital back by the end of the lease.

Year	Capital owed	Profit rent	23% return on capital	Capital recouped in each year
1	£28,000	£10,000	£6,440	£3,560
2	£24,440	£10,000	£5,621.2	£4,378.80
3	£20,061.2	£10,000	£4,614.08	£5,385.92
4	£14,675.28	£10,000	£3,375.31	£6,624.69
5	£8,050.59	£10,000	£1,851.64	£8,148.36*

*The discrepancy between £8,148.36 and £8,050.59 arises from the use of 23%, which is fractionally incorrect; the actual IRR is 23.0588%.

Here the investor receives £10,000 and this provides a return of 23% on the capital still not recovered from the investment; by the final year the investment has repaid all capital invested. Single rate here is illustrating the normal concept of present value; each year's profit rent has been discounted at the yield rate to its present value. What the investor does with any of the money is entirely a personal choice, but as most investors are not individuals but pension funds and charities the total net proceeds, after meeting the fund's liabilities, will be reinvested strategically each year.

The other point to note about the single rate approach to capital recovery is that this is the stock market's approach to analysing and pricing time-limited stocks and bonds. Single rate is also the basis for calculating all loan repayments of capital and interest including mortgages, which, from the bank or building society's point of view, are time-constrained investments in borrowers – a loan of £28,000 for five years at 23% would require annual repayments of £10,000 a year.

The tax issue remains, but as analysing or valuing at any specified rate of tax has to be, by definition, a calculation based on an individual's tax position, or like individuals' or corporate tax position, it cannot be an objective analysis but an analysis on a specified tax rate. Valuing net of tax has already been questioned in section 3 above as inappropriate since tax rates vary with the investor's tax status. Obviously, investors who are taxpayers will assess all investment opportunities on a gross and net of tax basis in order to be satisfied that a purchase at, say, £28,000 will yield an acceptable net of tax return. Here, a 40% tax rate would reduce the investment income to £6,000 and the IRR net of tax to 2.345%, which probably explains why leasehold investments are not considered attractive to taxpayers. Further references are given at the end of this chapter.

(b) Replacement of capital in real terms

It is true that the conventional dual rate method allows for an ASF to replace only the original capital invested. In times of inflation there will be an erosion of capital value in real terms. Clearly, if £10,000 had been invested in a lease 20 years ago which now expires, then the £10,000 now available under the sinking fund is not putting the investor in the same position as would have been the case if the sinking fund had matched inflation.

The YP dual rate can be adjusted quite simply if it is decided to replace the capital in real terms. It would be necessary initially to determine either the predicted rate of inflation or the predicted rate of growth in an equivalent freehold interest. The latter approach is more logical since the aim is to put freehold and leasehold interests on an equal footing. Once the rate of inflation or growth has been determined, then the ASF should be calculated to replace not £1 invested but the amount of £1 at the chosen rate over the period of the lease.

If this is done, the leasehold interest is more secure than it would otherwise have been, so that the remunerative rate should be reduced from what it would otherwise have been, possibly to the level of the yield on an equivalent freehold interest. The YP difference is shown below.

Example 12–9

Assume a lease with 40 years to run and a freehold yield of 9% and leasehold yield of 10%.

Conventional YP
YP 40 yrs at 10% and 3% (tax at 35%) = 8.305

Inflation-proof YP
Assume inflation at 7% p.a., leasehold rate now 9%

$$\frac{1}{i + (\text{ASF to replace A of £1 for 40 years at 7\%}) \times \left(\frac{100}{100-35}\right)}$$

$$\text{Formula} = \frac{1}{0.09 + (0.0122^1 \times 14.96742^2 \times 1.538^3)}$$

$$= \frac{1}{0.09 + 0.3054}$$

$$= 2.529\,\text{YP}$$

[1] ASF for 40 years at 3%
[2] Value of £1 in 40 years at 7%
[3] Tax adjustment at 35

This difference in YP suggests that the conventional method greatly over-values leasehold interests. There is, however, a fallacy in the argument that the ASF under-provides without an allowance for inflation.

The fact is that a leasehold interest is a wasting asset, by which is meant that it ultimately wastes away to nil. However, in the period between the start and finish of a lease the value may rise considerably before falling away. The reason for this is that, if rental values increase faster than the YP for succeeding years falls, then the capital value of the leasehold interest will rise in nominal terms. The effect of inflation is therefore reduced. If investors sell their interest before the capital value starts to fall, they will obtain more than they paid at the start.

Indeed, an inspection of the YP dual rate tables will show that it is only in the last few years that there is a significant falling away of the value of the YP figures such as to outweigh the inflation on rents, provided these are reflected in the profit rent. Thus, it is only in the case of short leases that ignoring the effects of inflation in the sinking fund can be criticised, but the impact of any allowance is far less over such short periods.

Example 12–10
As for Example 12-9, but the lease has four years to run.

Conventional YP
YP 4 yrs at 10% and 3% (tax at 35%) = 2.138

Inflation-proof YP

$$\frac{1}{0.09 + (0.239 \times 1.311 \times 1{,}5388)} = 1.749$$

In practice, neither on a dual rate basis nor on a single rate basis do valuers make any specific adjustment to reflect the impact of inflation, relying, where available, on the market evidence of all risk yields.

(c) ASF at low rates of return

Very few investors actually take out leasehold redemption policies. Instead, it is said that they either invest their redemption funds in alternative investments showing a higher return than the low rates of around 2–3% obtainable, or make no such provision at all. This may perhaps be so, but such criticism misses the essential point, promoted by dual rate proponents, of the ASF allowance; that it is a notional allowance to allow comparison between freehold and leasehold investments, as explained before, namely to retain a close relationship of 1–2% in the yields. Given that the notional allowance makes such a comparison possible, it remains that only by means of insurance policies can investors, if they so choose, obtain a guarantee of a future specific capital payment, and the rates offered under such policies are naturally low. It is true that, in the case of short terms, there are investments that guarantee more than 2–3%, and it is in these cases where the dual rate YP is less likely to be adopted.

The issue of recoupment by reinvestment does not arise with single rate in that the investor is assumed to be accepting a return on capital outstanding from year to year, which is notionally at the same rate as the remunerative rate. The risk with any leasehold lies in the continuation of the investment income. Once that is acquired, investors either will or will not recoup their capital; the issue is how risky that particular leasehold is in terms of possible non-payment of rent. A sinking fund per se will only deliver if there are sufficient funds from the profit rent to meet the annual premiums. Inability to meet the annual premium due to non-payment of the rent by the occupying tenants could nullify the ASF or result in an actuarially reduced payment. The risk is the same whichever method of valuation is used.

The use of single rate for analysis is considered to be a better reflection of the additional risks seen by the market in leaseholds, as it is not superficially linked by the valuer to an equivalent freehold rate.

(d) Why use dual rate YP?

The answer to this question would appear to be because, by convention, valuers have done so since the beginning of the 20th century for the reasons explained previously. Nonetheless, as has been indicated above, there is a body of opinion that argues for the abandonment of the dual rate approach and its replacement by a single rate YP (PV£1p.a.) or by DCF.

Some of the reasons for doing so have already been noted, namely:

- Single rate provides for replacement of capital on a year-by-year basis, in keeping with the concepts of the IRR as embedded in present values.
- IRR is the only objective tool of investment analysis of leaseholds; dual rate analysis requires an initial assumption of the ASF rate and, where required, of a tax rate.

- Single rate requires a valuer to assess risk in order to determine an appropriate yield or capitalisation rate; dual rate assumes all leaseholds are 1–2% above the equivalent freehold rate.
- Dual rate net of tax for leaseholds makes the comparison of leasehold remunerative yields with freehold all risk yields difficult as each is based on different investment concepts.
- Investors compare investment risks and make investment decisions based on IRR. Valuation is supposed to mirror the market, which would require valuation of leaseholds to be on a single rate basis.

Fundamentally, the valuation of any leasehold investment is complicated by the unique characteristics of each lease. The following examples illustrate just a few of the structures that a valuer might encounter:

- A simple investment where the head lease and the sub-lease(s) are co-terminating and the profit rent is fixed for the whole of the unexpired term. This is seen by many to be similar to a stock investment and might be valued single rate with a property premium on top of the redemption yield of government stock.
- A situation where a leaseholder is in possession and where the profit rent is changing on an annual basis as MR changes and/or there are changes in the head rent due to rent reviews. This might be very similar to a freehold and could be valued at a rate comparable to the freehold rate.
- A lease where the sub-lease rents have been agreed to change to specific rental sums in the future but the head rent reviews are to MR for the land only.
- A long, old lease where the head rent is subject to rent reviews, say, every 25 years, but the occupation leases are subject to rent reviews every five years, creating a stepped pattern of profit rents.
- A relatively new building lease where the head rent is geared, say, at 5%, to the occupation or sub-leases.

Only by a careful examination of the leases can a valuer derive a visual picture of how the profit rent will change in the future. Adhering to the simple idea that profit rent × YP dual rate = capital value, or that profit rent × YP single rate = capital value, can obscure the real nature of the investment. DCF is favoured by some as it allows the valuer using implied rates of rental growth to set out the cash inflows and outflows over the whole of the remaining lease term. DCF is undertaken using the single rate approach embedded in the present value concept. This can be illustrated by considering two identical profit rents with unexpired leases of 40 years and with no head rent reviews.

Here, a simple capitalisation ignores the fact that the profit rent in B grows faster than in A. These appear to be two similar profit rents, which should have similar values, but only by using DCF will their true nature be identified. DCF is the investment market's key tool for analysis of all investments and should therefore be used as a check when valuing leasehold investments.

	A		B
Market rent	£30,000	Market rent	£100,0000
Rent payable	£20,000	Rent payable	£90,000
Profit rent	£10,000	Profit rent	£10,000
YP for 40 years at 10% and 3% adj.tax at 35%	8.6066		8.6066
Value	£86,066		£86,066
Position in 10 years with 30 years to run and with long-term average rental growth of 3%	£40,317		£134,392
Rent payable	£20,000	Rent payable	£90,000
Profit rent	£20,317	Profit rent	£44,392
YP 30 years at 10% and 3% Adj. tax at 35%	7.8474		7.8474
Value	£159,436		£348,362

In these examples it can be seen that with growth in market rents there has been no loss in value, even though these are wasting assets. Both could be sold after 10 years to achieve a capital gain. The critical point is that the gain for each is very different, due to the lease structures. The gain in case A is £73,370 and in B £262,296 – respective gains of 85% and 305%. The valuer who relies on the tradition of adding 1–2% to the freehold rate, plus a sinking fund adjusted for tax, may be failing to reflect fully on the specific issues associated with a given leasehold valuation. This illustrates the need for extreme caution when valuing leasehold interests in property.

A further problem that has emerged with the use of dual rate is that associated with variable profit rents. The following section illustrates how this problem can be corrected. Single rate approaches do not run into the same problem.

(e) Valuation of variable incomes

The dual rate method, when used to value leaseholds with variable profit rents, has an inherent error. The method envisages more than one sinking fund being taken out to provide for the redemption of capital, whereas in practice a buyer would take out a single policy to cover the whole term. The concept of a variable sinking fund is not in itself incorrect, and a valuation could be made on this basis if a method were employed which would enable the varying instalments to compound to the required amount over the full unexpired term.

The normal dual rate approach used in the valuation of varying incomes for a limited term does not allow for the correct compounding of the sinking fund accumulations. This type of valuation is made in two or more stages, so that a separate sinking fund instalment will be based on the number of years in the appropriate stage of the valuation and will not be based on the full length of the lease. Consequently, the accumulated sums provided by these sinking funds over the full length of the lease will not match the capital sum required.

The following examples illustrate the error that occurs:

Example 12–11

A ten-year profit rent of £1,000 is to be valued on an 8% and 2.5% basis.

Profit rent	£1,000 p.a.	
YP 10 years at 8% and 2.5%	5.908	
Capital value	£5,908	

Example 12–12

If the term of 10 years in Example 12-11 is considered in two stages of, say, four years and six years, the valuation will take the following form:

Profit rent		£1,000 p.a.	
YP 4 years at 8% and 2.5%		3.117	£3,117
Profit rent		£1,000 p.a.	
YP 6 years at 8% and 2.5%	4.227		
PV in £1 in 4 years at 8%	0.735	3.107	£3,107
Capital value			£6,224

In these two examples, the profit rent of £1,000 p.a. has been capitalised in each case for a total period of 10 years at the same remunerative rate of 8%; each example should therefore produce the same result. The difference of £316 is caused by the error that arises in Example 12-12. This dual rate valuation is made in two stages and an incorrect sinking fund instalment is implicit in the valuation, creating a mathematical error. The degree of error here is approximately +5.25%. If the 10-year term is divided into three stages, the error increases to over 8%.

(f) The effect of tax

Repeating the calculations on a net of tax basis produces a similar error but of a different magnitude.

Example 12–11(t)

Profit rent	£1,000 p.a.	
YP 10 years at 8% and 2.5% (tax at 50%)	3.87	
Capital value	£3,870	

Example 12–12(t)

In this example, the 10-year term is again considered in two stages:

Profit rent		£1,000 p.a.	
YP 4 years at 8% and 2.5%			
(Tax at 50%)		1.78	£1,780
Profit rent		£1,000 p.a.	
YP 6 years at 8% and 2.5%			
(tax at 50%)	2.54		
PV £1 in 4 years at 8%	0.735	1.87	£1,870
Capital value			£3,650

These examples show that, in making the allowance for income tax, a compensatory "pull" on the degree of the error occurs which reduces the variation from +5.25% to −5.5%.

In many instances, the effect of the tax adjustment will be more marked and will usually over-compensate for the "normal" error to such an extent that a much greater inaccuracy will be introduced. In some cases, an error that is well in excess of 10% may be involved.

(g) The double sinking fund method

The following method, developed by A.W. Davidson, one time Head of Valuations at the University of Reading, enables this form of valuation to be made without involving any significant error.

Example 12–13

The same leasehold is again considered in two stages, as in Example 12-12(t); the dual rate YP is not used.

Let the capital value = P			
Profit rent		£1,000	p.a.
less ASF to replace P in			
10 years at 2.5%	0.0893P		
adj. for tax at 50%	× 2.0	0.1786P	
Spendable income		£1,000	− 0.1786P
YP 4 years at 8%		3.312	
		£3,312	− 0.5915P
Spendable income		£1,000	− 0.1786P
YP 6 years at 8% × PV £1 in 4 yrs at 8%	4.623		
	0.735	3.398	
		£3,398	− 0.6068P
		£6,710	− 1.1983P
*Plus repayment of the capital replaced by			
the single rate SF			
P × PV of £1 in 10 years at 8%			0.4632P
		£6,710	− 0.7351P

$\therefore P = 6,710 - 0.7351P$

$1.7351P = £6,710$

$P = £3,867$

*As the valuation has been reduced by allowing for the capital to be replaced twice, a figure equal to the deferred capital value has been added back to the valuation.

The spendable income has been valued on the single rate basis. This means that the "capital" has been replaced by allowing for a sinking fund at 8% in the capitalisation. Thus, two "capital recoupments" have taken place, one at 2.5% and one at 8%.

The capital value of £3,867 is higher than that produced in Example 12-12(t) of £3,650; removing the inherent error has increased the value by 5.9% to produce the same value as in Example 12-11(t), which is correct.

A simplified approach has been produced by P.W. Parnell which is probably sufficiently accurate for most cases. This again applies a single rate YP, the result being adjusted by applying the ratio of the YP dual rate to the YP single rate.

Example 12–14

The same facts are adopted as in Example 12-13, using the Parnell approach.

1st 4 years			
Profit rent		£1,000 p.a.	
YP 4 yrs at 8%		3.312	£3,312
2nd 6 years			
Profit rent		£1,000 p.a.	
YP 6 yrs at 8%	4.623		
PV £1 in 4 yrs at 8%	0.735	3.398	£3,398
			£6,710
			0.576
$\times \dfrac{\text{YP 10 years at 8\% and 2.5\% (tax 50\%)}}{\text{YP 10 years at 8\%}} = \dfrac{3.868}{6.710} =$			£3,865

The answer is similar to that produced by the double sinking fund method, and is much simpler.

(h) Summary

So where does this leave the valuation student? It is fairly simple and definitive to use dual rate as a means of distinguishing a leasehold from a freehold valuation. It is standard practice to raise the remunerative rate by 1–2% above the equivalent freehold rate for that type of property in that location. So one might see a freehold shop valued at 6%, and the leasehold in the same shop at 7% and 3%.

But it is important to be aware of the lease structure, and the valuer must be able to set out the rationale for the yield used. In most cases a valuer will be unable to use directly comparable leasehold transactions as evidence for either dual rate or single rate, due to the limited availability of leasehold investment sales.

When dealing with premiums, mergers and surrender and renewals, it appears that using single rate for freeholds and dual rate for leaseholds produces solutions that can be reconciled. But there is a problem in that these negotiations, whilst based approximately on market value equivalences, are potentially assessments of worth. In a merger, neither party would need to set up a sinking fund because the two interests are merged and the freehold can be sold to recover market value. In a surrender and renewal, neither party is actually buying a leasehold investment so again there is no need for a sinking fund. In neither situation is there any need for any tax adjustment because there is no capital acquisition of a lease per se, nor is there a need for a sinking fund because there has been no capital outlay. Any reconciliation made here by taking an average is unlikely to meet with the freehold valuer's approval because to do so would lower the value of the freeholder's investment. Here it would seem that the freehold valuation is likely to prevail. In the case of a merger, adopting single rate or dual rate unadjusted for tax for the leasehold is likely to increase, for this purpose, the value of the lease and reduce the element of merger value.

In later chapters, dual rate and dual rate adjusted for tax will be used for lease-holds. This does not imply that the authors believe it is the right and only method, but simply that it remains a convention in parts of the UK market, particularly in some statutory situations based on case law precedent – a convention, incidentally, which is unlikely to be encountered in any other property market in the world. But note that a number of RICS publications make no reference to dual rate.

6. Valuing in periods of market instability – valuation certainty

As indicated at the beginning of this chapter, questions on valuation methods tend to be raised by those outside the valuation profession during periods of market instability. The profession is also challenged at such times by the fall in sales activity and the resulting reduction in comparable evidence. Property investment, being dependent on the availability of credit, means that in periods where there is little or no debt capital available, such as those starting in 2007, not only are sales volumes affected but other market issues emerge concerning: "who might be active buyers", particularly for the higher multimillion pound property investments; and "what level of rent" might be regarded as market rent during an economic down-turn. None of these issues affect valuation methods as such, but they do impact on practice and processes.

Valuation is based on comparison with known market sale prices; lower activity reduces the quantity of comparable evidence. These periods of reduced activity, as witnessed in the 1930s, in the early 1970s and 1990s and currently, are generally associated with economic downturns which impact on market rents and market prices for all types of property.

A concern over the security of the rent payable and the uncertainty about future reversionary rent was one argument used to support a lower capitalisation rate for the term than the market rent yield used for the reversion from the 1930s to the 1980s. In these conditions, tenant covenant grows in importance. There is an enhanced need for the valuer to assess the potential for voids and to allow for appropriate void allowances and for lease incentives.

In the leasehold market, arguments in support of dual rate may re-emerge as low interest rates are used to stimulate investment and the only "safe" places for money appear to be government-backed banks.

Every issue requires careful consideration, but the critical issue is the reduced evidence arising from comparable sales. The RICS *Valuation Standards* in GN 1 *Valuation certainty* notes five matters that may affect valuation certainty, namely:

- status of the valuer;
- inherent uncertainty;
- restrictions on enquiries or information provided;
- liquidity and market activity; and
- market instability.

The problem for the valuer is that in periods of greater uncertainty clients still require a single figure of value to be provided. Valuations are still required for sale and purchase, for lending purposes and for accounting and other purposes. In the case of market instability the RICS suggests that "Although valuers should remain able to make a judgement, it is important that the context of that judgement is clearly explained." The use of value ranges is not recommended, nor is the use of terms such as "in the region of". Elsewhere in GN1, emphasis is placed on being explicit and transparent.

Bywater[2] notes that investors and lenders normally require information on the following matters in a valuation report:

1. Property information
2. Tenant information
3. Planning information
4. Market information
5. Market data
6. Valuation methodology
7. Valuation factors

It is here and in the other areas of the report specified in VS 6 that the valuer provides information to the client on the uncertainties that may exist and may be implicit in the ARY. In this guidance note Bywater also discusses the main areas of investment uncertainty, namely:

1. *Economic, financial and political uncertainty*; e.g. impact of job losses in the banking sector in some locations, the effect of issues in Europe and their impact on the euro and £GBP.

---

2 Bywater, N (2011) *Reflecting uncertainty in valuations for investment purposes. A brief guide for users of valuations*. RICS.

2. *Legal and regulatory uncertainty*; e.g. enforcement of downward rent reviews in periods of recession
3. *Physical uncertainty*; e.g. accelerated obsolescence
4. *Occupational uncertainty*; e.g. rationalisation of multi-branch banks
5. *Leasing uncertainty*; e.g. voids at ends of leases
6. *Market uncertainty*; e.g. impact of rising food prices on retail rent recovery
7. *Valuation uncertainty*; e.g. lack of recent comparable sales evidence

(Words in *italics* from Bywater, examples from authors)

Again, the emphasis is on being transparent and, by reporting to an investor or lender in detail under these headings, the valuer provides them with the clarity and transparency referred to above. Bywater's paper also emphasises the usefulness of cash flows to make explicit what is implicit. Developments in the use of the DCF approach in the United States, as part of the income approach, show how probability factors can be incorporated into the cash flow. For lenders, Bywater recommends the use of Swot analysis to alert the lender to the Strengths, Weaknesses, Opportunities and Threats relating to the property being used as security for a loan.

The International Accounting Standards Board in their 2008 paper, *Measuring and disclosing the fair value of financial instruments in markets that are no longer active,* picks up this theme in the broader context of financial instruments. Two interesting points are picked out in their summary:

8. A thorough understanding of the instrument being valued allows an entity to identify and evaluate the relevant market information available about identical or similar instruments. Such information to be considered includes, for example prices from recent transactions in the same or a similar instrument, quotes from brokers and/or pricing services, indices and other inputs to model-based valuation techniques. An entity uses such information to measure the fair value of its financial instruments by assessing all available information and applying it as appropriate.
9. When the market for a financial instrument is no longer active, an entity measures fair value using a valuation technique (commonly referred to as 'mark-to-model'). The selected valuation technique maximises the use of observable inputs and minimises the use of unobservable inputs in order to estimate the price at which an orderly transaction would take place between market participants on the measurement date. Regardless of the valuation technique used, an entity takes into account current market conditions and includes appropriate risk adjustments that market participants would make, such as for credit and liquidity.

This text is about real estate valuation methods. However, it could be that in some areas of the property market where comparables are few, if any, a valuation approach, but not a direct market method, might be to consider the advantages and disadvantages of property in general, or of a specific property, to invest in, and the price of government stock. In some conditions property, although more risky, takes on similar characteristics to government stock. A freehold with a secure tenant occupying on a lease with at least 10–15 years unexpired with regular upward-only rent reviews might be seen, with no immediate signs of market rental growth, to be comparable to fixed income government stock with 10–15 years to redemption. A market value based on limited market evidence could possibly be compared to a valuation based on a risk-adjusted discount rate derived from the sale of government stock.

Valuation methods do not have to change when valuing in periods of inactivity or instability, but application is more difficult and disclosure to the client as to the robustness of the opinion is essential. As the RICS suggests – "an expression of less confidence by a valuer is not an admission of weakness".

7. Valuing sustainability

Reflecting sustainability in property valuations is not, strictly speaking, a development in valuation methods but a development in the application of methods. The RICS has published Valuation Information Paper 13[3] on sustainability and commercial property valuation. This contains, in section 5, details on "Assessing a building's sustainability characteristics" and, in section 6, "Reflecting sustainability characteristics in the valuation". Here it is noted that:

> If at the date of the valuation, the market does not differentiate, in terms of either occupier or investor demand, between a building that displays strong sustainability credentials and one that does not, there will be no impact on value.

However, given that sustainability is high on the agenda of all political parties, there is a professional expectation that the valuer will be sufficiently aware of these "green" matters to be able to record and make a judgment upon "sustainability" as part of his or her inspection and reporting. Also, the valuer will need to assess any impact that sustainability is having on rents and yields. Sustainability matters do not affect the method of valuation used by a valuer, but will impact on due diligence and may affect the valuer's assessment of the inputs to a given valuation. In VIP 13 there is a list (6.4) of a number of questions that "may be relevant in considering the impact of various sustainability issues"; a note of a few indicates the increased care a valuer needs to exercise in considering the impact of sustainability:

- What impact will the building's sustainability criteria have upon rental growth and rent obtainable at current levels of value?

3 Valuation Information Paper No. 13, *Sustainability and commercial property valuation*, RICS, 2009.

- Are the building's sustainability characteristics such that it is likely to suffer more or less voids and delays on lettings than comparables?
- Will the building be economic to run in terms of outgoings both from an occupier perspective and, in the case of multi-let buildings, in terms of service charges?
- Does the building represent increased or decreased overall risk to the investor …? If so, should the risk premium in the discount rate be adjusted?

The inferences of sustainability are evident. Sustainability is becoming an increasingly important factor in a valuer's deliberations over assessment of market rent and market value. The valuer must maintain a high level of appreciation of the criteria, legislative changes and owner, occupier and investor expectations.

Valuers will need to increase their knowledge and understanding of the Building Research Establishment Environmental Assessment Method (BREEAM), which is a popular UK method for assessing environmental aspects of buildings. Theory would argue that in time the market may begin to distinguish in rent and capital terms between a "pass" building and an "excellent" or "outstanding" building. The impact of the Carbon Reduction Commitment Energy Efficiency Scheme will need to be monitored. It is a question of supply and demand; currently the pressure from government is on suppliers of space to provide more sustainable, and more energy efficient, buildings. Greener buildings and green leases are part of the market and, as the numbers increase, the less green properties, in theory, will become harder to let and may depreciate faster in terms of rent and capital values.

The RICS has also issued guidance for those involved in residential valuations[4]. This is important reading for residential valuers and agents and especially for those engaged in valuations for mortgage and loan security. Here again one can note that "The role of the valuer is to assess Market Value in the light of evidence normally obtained through analyisis of comparable transactions", and "therefore as part of establishing Market Value and Market Rent, residential valuers should seek to keep abreast of trends and ensure that they collect appropriate and sufficient sustainability data when inspecting property, as this will enable them to analyse and apply them to any property valuation."

This book is about valuation but, as indicated earlier, there is a need for valuers to have a sound knowledge of buildings, their construction, their operational costs and now their sustainability, as these factors can have a profound impact on value.

8. Conclusion

The purpose of this chapter has been to raise some of the issues relating to valuation practice. The comments here do not seek to accept or reject the views being

---

4 RICS Information paper *Sustainability and residential property valuation* (2011).

expressed, nor is it suggested that all points of debate have been covered. Hopefully, the reader has realised that methods of valuation are under scrutiny from outside the profession and under review from within the profession, and that this may result in methodological changes. However, until such changes are accepted in the market, the authors believe that the current practice of valuation set out in the preceding chapters will continue, albeit with greater emphasis on current rents paid and the initial yield rather than reversionary expectations of uplift. But they expect an increase in instructions requiring cash flows to be used both in valuation and investment advice; and, whilst all markets remain affected by concerns of economic recession, fears of tenant failure, administration or liquidation, valuers will need to provide fully transparent reports. Hopefully, the observations in this chapter will encourage the reader to question convention.

9. Further reading

Research Report on Valuation Methods by Andrew Trott (RICS/Polytechnic of the South Bank, 1980).

Valuation of Commercial Investment Property: Valuation Methods (RICS Information Paper, 1997). *Calculation of Worth* (RICS Information Paper, 1997).

Property Investment Appraisal by Andrew Baum and Neil Crosby (Blackwell, 3rd edn, 2008).

Dual rate is defunct? A review of dual rate valuation, its history and its irrelevance to today's UK leasehold market by David Mackmin, *Journal of Property Investment and Finance*, 2008: 26(1), 80–95.

RICS (2010) *Is sustainability reflected in commercial property prices: an analysis of the evidence base*. RICS Research.

RICS (2010/11) *Cash flow forecasting*, RICS Guidance Note.

RICS (2010/11) *Valuation certainty user guide*, RICS User Guide.

RICS (2011) *Sustainability and the valuation of commercial property* (Australia).

RICS, *Sustainability and residential property valuation*, RICS IP 22/2011.

RICS, *Sustainability and commercial property valuation*, RICS IP 13/2009.

Chapter 13

Some practical points

1. The valuation process

The primary purpose of valuation is to assess market value (MV). The RICS has formally adopted the International Valuation Standards (IVS) definition of MV in the 2012 Red Book (*RICS Valuation – Professional Standards*) (see Chapter 1) and the full definition must always be referred to when undertaking a market valuation. There is, however, nothing complicated about this definition as it conforms to the commonly held understanding of MV as being the price at which an interest in a property would have sold on a given date as between a willing seller and a willing buyer, both being fully informed regarding the interest in question.

The valuation process is captured in the Valuation Standards which, together with guidance notes, information papers and practice standards, are essential reading for members of the RICS and IRRV and firms regulated by the RICS, who must comply with them when undertaking valuations that require a written valuation. There are exceptions, noted in VS 1.1 as listed below, but the principles should still be followed.

The exceptions to the red book requirements are listed in VS 1.1 and should be read with their full commentary. The exceptions apply when:

- the advice is expressly in preparation for, or during the course of, negotiations or possible litigation; the valuer is performing a statutory function or has to comply with prescribed statutory or legal procedures;the valuation is provided solely for internal purposes;the valuation is provided in connection with certain agency or brokerage work;a replacement cost figure is provided for insurance purposes, whether separately or within a valuation report.

The standards largely describe the process, namely:

1. Ensure compliance (VS1) requirements are met, including RICS valuer Registration Scheme requirements.
2. Agree the terms of engagement (VS 1.2 and VS 2); the latter contains a list of twenty terms that must be agreed including the property, the date, the purpose,

the basis and the fee. The basis of value will be one or more of market value (MV), market rent (MR), worth and investment value, and fair value, which is mainly for accounting purposes as defined fully in VS 3.

3. Be sure that the valuer or firm is appropriately qualified (VS 1.5), has the knowledge and skills (VS 1.6), confirms there are no conflicts of interest and meets all the other requirements of VS 1.

4. Meet the application requirements under VS4, e.g. for loan security purposes.

5. Complete all appropriate inspections and investigations and verify all information as per VS 4, which contains important check lists, but also follow other information papers and practice standards such as the RICS *Contamination, the environment and sustainability* GN 13/2010, which also contains helpful checklists for property inspections.

6. Prepare a valuation report that meets the minimum content of VS 6.

In brief, the valuer needs to be sure that: he or she is qualified and competent to complete the valuation; the appropriate professional indemnity insurance cover is in place; a full contract and terms have been agreed – to do what, when, where, how, by when and for how much?; any special assumptions[1] have been noted; all application requirements are met; all data needed for the valuation has been collected from inspection and investigatory work, and has been verified; and that a written report can and will be produced by a given date. In this process the valuer will typically select one of the methods of valuation described in this book to produce a final figure of value. Some of the other UK-specific Valuation Standards are considered in Chapter 19.

Although many of the valuer's problems involve consideration of MV, there are often other problems which require a valuer's attention.

It is often necessary to consider future trends of value; to consider whether prices at which properties have been sold are reasonable or likely to be maintained; or to examine the possibilities of changes in market rents. For example, the value of a short lease will change over a comparatively short time because of the wasting nature of the asset. Valuers may need to report, with supportive data, their fears that the market is unreasonably high due to a shortage of supply which might be amended in the short term, or unreasonably low due to the absence of buyers caused by a particular event, for example the effect of the credit crunch in 2008.

When valuations are made for certain purposes, e.g. for taxation or in connection with compulsory purchase , the valuation, although based on MV, may be regulated by statutory provisions as to the date at which the valuation is to be assumed to be made and as to the factors which may or may not be taken into account in making it. For these purposes, it is normally permissible to use postdated evidence

---

1 A special assumption has been redefined by the RICS as "An assumption that either assumes facts that differ from the actual facts existing at the valuation date, or that would not be made by a typical market participant in a transaction on the valutation date" (RICS Red Book glossary).

because the object of the exercise is to establish the correct value of the property as at the relevant date and not what price a valuer might have placed upon the property in advance of that date. This concept of acceptable post-dated evidence has been accepted by the courts and also in the Privy Council case of *Melwood Units Property Ltd* v *Commissioner of Main Roads*.[2] The same principle of utilising postdated evidence has also been accepted in connection with rent reviews, and for the same reason. However, the further that the valuer moves away from the valuation date the more likely it is that the postdated evidence will not be relevant because of changes in the market between the relevant dates.

Where properties are purchased for investment, the valuer may be asked to advise on policy, to suggest what reserves should be created for future repairs, or in respect of leaseholds what redemption policy, if any, should be arranged; and to advise generally on the many problems involved in good estate management.

The knowledge required to deal with all these matters is best gained by experience, but the valuer will find it both necessary and of great advantage to keep abreast of current affairs (international, national and local) and of yields in other investment markets. The valuer then values in the light of this information and after making a careful study of sale prices, yields and rents; in this respect it is useful if the valuer can have sight of those properties that are the subject of recorded transactions.

It is usually possible to obtain evidence of yields for different kinds of properties from the property market itself, but this may not always be the case. Sometimes the valuer has to look at similar investments which might be quoted on the Stock Exchange. A good example of this relates to the valuation of ground rents. These are often compared with undated gilts issued by the British Government which, at the time of writing, has an average yield of 2%+ for undated stock and 1.0% for index-linked. It is then usual to add to this yield to reflect the relatively high cost of collecting relatively small ground rents, the costs of selling and the illiquid market (the risk factor in a ground rent is virtually non-existent because of the security offered by the overriding property), to arrive at the ground rent yield.

The valuer needs to obtain as much information and evidence as possible so as to render the valuation as correct as possible. Even so, it should be emphasised that a valuation is only an expression of one valuer's opinion and it is not surprising if other valuers hold different views. For this reason a valuation is a best estimate of the price that will be achieved, but the best estimate must be based on the best available evidence.

Some interesting observations on valuation in general, including limits of accuracy, were made in *Singer & Friedlander* v *John D Wood & Co*[3] and these have led to a general proposition that a valuation may be accurate if it falls within a reasonable range of the true value. This acceptable bracket or margin of error has been crucial in determining negligence by valuers in a stream of cases. The

2 [1979] 1 All ER 161.
3 [1977] 2 EGLR 84.

bracket is often taken to be 10% of the true value (above or below), but the Court may regard this as too small or too large, having regard to the circumstances of the case.[4]

2. The analysis of rents and prices

In estimating the MV of an interest in property, the valuer must consider all the evidence available. The property may be let at a rent, in which event its reasonableness or otherwise must be considered in relation to comparable market rents. The capitalisation rate will be determined by reference to the yields achieved on the sale of other similar properties and investments. Accurate records of rents, sale prices and their analyses are essential; analysis of current sales to determine market yields is essential if an opinion of value is to be supported.

An example is given in Chapter 5 of an analysis of rents made to determine the rental value of offices. Similar methods, usually related to the NIA or GIA of premises, can be used for the making and keeping of records in connection with a variety of properties (see Chapter 17 for analysis of lease incentives).

Although in many offices manual records are still maintained, computers have added considerably both to the ease of maintaining records and to the ways in which information can be analysed, manipulated and retrieved. National databases have been organised by the Estates Gazette (EGi) and other internet service agencies.

In analysing sale prices, the usual steps in making a valuation are reversed. The purchase price is divided by the estimated net income to find the figure of Years' Purchase (YP), from which can be determined the yield that the purchase price represents, or more directly by expressing the income as a return on purchase price. In the case of perpetual incomes, the "yield" can be found by dividing 100 by the figure of YP. In the case of incomes receivable for a limited term, the single rate and dual rate yields have to be found by inspection of the valuation tables or use of computer programs.

Example 13–1
Freehold property producing a net income of £20,000 p.a. has recently been sold for £200,000. Analyse the result of this sale for future reference.

Analysis:
Purchase price £200,000 represents 10 YP
Net income £20,000
Rate at which purchaser will receive interest on money

$$\frac{100}{10YP} = 10\%$$

---

4 See, for example, *Arab Bank plc* v *John D Wood Commercial Ltd* [1998] EGCS 34 and also the article 'Close Brackets' in the *Estates Gazette* of 4 April 1998, p. 133, which considers the result of research at Reading University into margin of error cases where the bracket varies widely.

Alternatively this could be found here, as the income is assumed to be perpetual, in the following way:

$$\frac{£20,000}{£200,000} \times 100 = 10\%$$

The 10% found here is the all-risks yield or capitalisatiion rate. This is not the only analysis; if the £20,000 is not the MR then the analysis will involve a term and reversion calculation.

Example 13–2
Shop premises held on lease for an unexpired term of 52 years at a rent of £15,000 p.a. have been sold for £130,000. The market rent is estimated to be £27,500 p.a.

Analysis:
The profit rent is £12,500 p.a. The purchase price is £130,000 which, divided by £12,500, represents 10.4 YP. Reference to the dual rate tables at 3% and tax at 25% for a term of 52 years indicates a remunerative rate of 8.5%. But on a dual rate basis a range of remunerative rates will be found, dependent on the SF rate and tax rate chosen. On a long lease the variations are small but on a short lease the variations will be more pronounced; hence the plea to use the IRR or single rate for analysis of all property investment sales.

A useful tool for analysis is the Microsoft Excel "Goal Seek" program. It can be found under "tools" and it saves making calculations by interpolation or trial and error.

Valuers short of rental evidence but with evidence of sale prices will sometimes seek to analyse the sale price to find the rent by assuming a typical market yield. This is not considered to be an acceptable means of securing evidence of market rent. The object of all sales analysis is to assess yield, not rent.

Comparisons of market rents are generally restricted to a particular locality; this is obviously so in the case of shops, where wide variations in rent can occur within a distance of a few hundred yards. On the other hand, with experience, the valuer may be able to identify a level of values that will be similar in locations of similar qualities within a region or even nationally. For example, in towns close to a common factor such as a motorway, the levels of office rent will tend to display regional rather than narrowly local characteristics. Similarly, shops in major shopping locations will tend to have a comparable Zone A rental value.

In the case of industrial or warehouse properties the area for comparison purposes may be similarly extended, but with care, as evidence of value of a particular type of property in one town may have limited significance in relation to another town say 30 miles away. Comparisons derived from nearby property will always be more significant.

In assessing the yield for a particular type of investment property, evidence over a much larger area can be used. Indeed, there are two broadly defined markets found in practice, a national and a local market. Prime properties, such as first-class shops or offices, or well-designed modern warehouses or factories, tend to produce a similar yield nationally. On the other hand, secondary properties are more prone to local factors, and yields need to come from local comparables.

It is in the analysis of past sales and the application of the evidence to a particular case that the skill and experience of the valuer are called into play. The valuer must consider carefully the extent to which the property to be valued is similar to those that have been sold. The correctness or otherwise of the rent must be determined, and trends of value since the evidence was accumulated must be considered.

In some cases the prices at which properties have changed hands may be above or below MV. The seller may have been anxious to dispose of a property quickly and accepted a lower price than might have been expected, or a high price may have been paid by a buyer in urgent need of a certain type of accommodation.

When analysing sales for the purposes of Leasehold Reform valuations, care must be exercised in excluding any part of the value attributable to the right to acquire the freehold or an extended lease. Indeed, it may sometimes prove impossible to obtain "untainted evidence" from the market because of the existence of these rights. In some cases valuers should use their own knowledge and experience to distinguish between that part of the price which represents the true value of the leasehold interest concerned and that which effectively represents the lessee's share of the marriage value which is being paid to the outgoing lessee. This is a special situation where open market evidence is not to be used for a statutory valuation without adjustment.

It follows that, whilst the analysis of transactions may show a level of value (or rents) for a particular type of property in a particular area, this level may not accord with the valuer's own expectations based on his or her knowledge and experience. In such a case, the valuer should advise the client that the level of value may not be sustained if there are in existence some facts or factors that may not have been taken into account. An example is where a property has been subject to a sale and leaseback, the agreed rent may have been above MV, and therefore the analysis of the transaction could lead the unwary valuer to over-value a similar property.

The duty of the valuer is to question evidence before applying it, and not to slavishly follow other people's assessments without understanding the basis of their assessment.

3. Special purchasers or buyers

A valuation is prepared adopting certain assumptions as to the market conditions. In some instances the valuer will be aware that some people are in the market for whom the property in question will be of special interest. A special purchaser is defined by the RICS as "A particular buyer for whom a particular asset has special value because of advantages arising from its ownership that would not be available to general purchasers in the market". Typical examples are where synergistic value exists (see Chapter 9).

The valuer must always consider whether there are special purchasers since they are likely to pay the highest sum and one which is in excess of the value to the general market ("an overbid"). Even when the existence of the special purchaser is known, it is often difficult to determine the price to which such a purchaser will go.

In Chapter 9 the price that might be agreed in marriage value situations was discussed. But in other cases the only way of discovering the overbid is by marketing the property in a manner designed to draw out the final bid. The valuer must therefore be able to comment on the appropriate method of offering the property to the market. For example, it may be appropriate to sell at public auction, or by public tender, or by a closed tender (when the property is offered only to a selected group) or, of course, by a general offer for sale by private treaty.

The case of special purchasers creating valuation difficulties is commonly found where leasehold interests are offered for sale in respect of prime shops. In general, the major retailers will be established in these prime locations and they will be determined to remain while the centre continues to thrive. The chance of a unit becoming vacant is unlikely. So when a trader decides to move out and offers the balance of their leasehold interest, they may find that there is considerable competition to buy the lease. Valuers may find that their estimate of market value by traditional investment methods is far less than the price obtained. The most appropriate method might be a profits approach, perhaps valuing the profits for several years ahead, beyond the term of the lease offered, since traders may be prepared to buy the right to be able to earn such profits. Whatever method is adopted, it will often be found that the price paid defies analysis on any normal valuation basis since the price paid is the cost the retailer is prepared to pay to get into the centre. These exceptional prices are sometimes termed "key money" or "foot in value", the meaning of the expressions being self-evident. This value phenomenon suggests that the prevailing rents, which are accepted as market rents, may be falsely low due to the centre becoming an artificial market in the absence of actual market transaction evidence.

The valuer may also note prices being achieved considerably in excess of any valuation figure and in contradiction to all previous market evidence. The reasons may be many, such as the inexperience of the buyer, a mistake being made, the property having a special sentimental value and so on. The valuer needs to recognise such "fluke" transactions and to disregard or distinguish them when considering them as market evidence.

4. RICS valuation – professional standards, global and UK

This has already been referred to and is known as the Red Book. It was first published in 1995 and is now in its 2012 edition. Its use is mandatory on members of the RICS and the IRRV. RICS members undertaking Red Book valuations must join the RICS Valuer Registration Scheme. The standards are not concerned with valuation theory or methods of valuation. They are, however, directly concerned with the value definitions; processes, including advice on referencing; the assembly, interpretation and reporting of information; and the matters that valuers should include within their reports. The purpose of the standards is to provide a uniform level of service to be provided by valuers to their clients, and particular notice should be taken of the need to clarify the instructions on which the valuer is acting. In effect it sets out "best practice" for valuers to follow. No valuation governed by the Red Book should be prepared without a reference to the definitions and guidance notes. Some UK-specific standards are considered in more detail in Chapter 19.

5. The cost of building works

The valuer may have to estimate the cost of building works, including infrastructure, as part of a development appraisal, whether of new buildings, or in relation to the refurbishment of existing buildings, or for the repair or renewal of part of a building, or to assess the sum for which existing buildings should be insured against fire and third party risks.

The cost approach to be adopted will depend upon the accuracy desired and the information available.

In estimating the development value of large areas of land it is necessary to make assumptions as to the value which the land, or various parts of it, will command in the future. These estimates will be based on the evidence available at the time when the valuation is made, but will necessarily depend on factors that may vary considerably in the future.

In many cases, estimates have to be made of the cost of works to be carried out in the future. In view of the fluctuations in costs that may occur before the works are carried out, undue refinement or accuracy in the methods used might be out of place. However, in the case of, for example, major town centre developments where the planning stages are often prolonged, it may be necessary to project both income and costs into the future. Indices of past changes in rents and of building costs will provide some assistance, but some additional technique that enables various assumptions on future changes to be tested will also be applied.

In many instances, both in connection with building estates and with the development of individual sites, the particulars available of the type of building to be erected or what, in fact, can be erected on a particular site may be very scanty. In the absence of detailed plans and specifications, any estimate of cost can only be arrived at by means of an approximation.

It is not possible to formulate any general rule in this connection. The circumstances of each case must dictate the degree of accuracy to be attempted, and the usefulness of the result achieved will depend largely upon the experience of the valuers in applying their knowledge, derived from similar cases to the one under review.

(a) Bills of quantities

Where detailed drawings and a specification are available, it is possible for a bill of quantities to be prepared and for this bill to be priced by a quantity surveyor or cost estimator. This is the most accurate method of arriving at cost of works, but is often not practicable.

An approximate bill of quantities may be prepared whereby only the principal quantities in various trades are taken off, the prices to be applied being increased, beyond what is customary, to include for various labours that normally would be measured and costed separately.

In the case of works of alteration, approximate quantities may be the only way of arriving at a reasonably accurate estimate where the works are extensive or present unusual difficulties.

(b) Unit comparisons

The most common approximate method of comparing building costs is per square metre of floor area. It is important to check whether the unit costs are based on gross external or gross internal measurements.

The area of a proposed building can be arrived at with an acceptable degree of accuracy from sketch plans or from a study of the site on which the building is to be erected, taking into account the likely requirements of the local planning authority, the restrictions imposed by building regulations and any rights of light or other easements.

The gross internal area having been determined, the total cost can be estimated by applying a price per square metre derived from experience of the cost of similar buildings or from a publication such as *Spon's Architects and Builders Price Book*, which is revised annually. Costs are based on typical buildings and do not include external works. Adjustments will, therefore, be required if the building is not typical, if site conditions are abnormal and for external works. It must be remembered that costs taken from publications are necessarily based on tenders prior to publication and are therefore prone to inaccuracy, particularly if there has been a significant change in building costs since those tenders were prepared.

Gross internal floor area is usually determined by taking the total floor area of all storeys of the building measured inside external walls and without deduction for internal walls. The method is set out in detail in the *Code of Measuring Practice* published by the RICS (sixth edition, published in September 2007).

It should be noted that gross internal floor area will, in the case of offices, commonly differ from net lettable area or net internal area.

(c) Other methods of comparison

In some cases the gross external area of the building is utilised, but the valuer should be careful to exclude the reference to "gross external floor area". The unit of comparison being used is the gross building area and not the gross floor area. This method is used for insurance purposes of houses and flats, and for most mortgage purposes it is usual to use the online RICS *BCIS Guide to House Rebuilding Costs* service to determine the appropriate rebuilding cost.

In the case of building estates, the cost of road and sewer works is often estimated on a linear basis. This price should provide for the roads to be completed to the standard required by the local authority for adoption.

In the absence of a detailed development scheme, it is sensible to prepare a sketch plan of the proposed development in order to estimate the cost of the development works required. It may sometimes be possible, however, to make a comparison between one estate and another, whereby the total development cost per unit may be estimated by reference to the development costs of other land of similar character in the neighbourhood.

The usefulness of these methods depends upon the extent of the information available and upon the experience of the valuer, who must make full allowance for any differences between the comparable and the subject property.

(d) Alterations and repairs

The use of an approximate bill of quantities in the case of alterations has already been referred to. A similar degree of accuracy may often be arrived at by preparing a brief specification of the works required to be done and placing a "spot" amount against each item. Similar methods may be used in regard to repairs.

6. Claims for damages

(a) Dilapidations

Section 18(1) of the Landlord and Tenant Act 1927 limits the amount which landlords can recover from tenants by way of damages for breaches of the covenant to repair. This limit is known as the "diminution in the value of the landlord's reversion".

Section 18(1) applies only to the breach of the covenant to repair and not to breaches of other covenants, such as a breach of the covenant to decorate in the last year of the term or of the covenant to remove internal partitions or to reinstate the premises. In the latter cases regard may be had to the case of *Joyner* v *Weeks*[5] which held that damages were payable based on the cost of the work, irrespective of whether the landlord actually suffered any loss. However, this decision flies in the face of the normal rule of damages which is that the claimant should be put in the same position, so far as money can do it, as if the damages had not occurred.

---

5 [1891] 2 QB 31.

To calculate the diminution in value, the valuer needs two figures. The first is the value of the property on the basis that there had been no breach of the covenant to repair, and the second is the value of the building as it existed on the term date of the lease. Some items of disrepair will not have an effect on value because the item concerned will be superseded by works which the landlord proposes to carry out in order to modernise or upgrade the building. This concept of supersession was considered in the case of *Mather* v *Barclays Bank plc.*[6] In many cases the two valuations will have the appearance of two residual calculations. The calculation for the building in disrepair will show higher building costs, a possibly longer period of development, higher fees and higher interest than the valuation of the building as if in repair, but some of the items put into the calculations will be neutral to both valuations. They therefore require a lesser degree of accuracy in their ascertainment. This would apply to the rental value and yield in particular.

It sometimes happens that, by virtue of the nature of the landlord's interest, the value in repair may be negative but likewise the value in disrepair may be a greater negative. The difference between these two figures will still represent the diminution in the value of the landlord's reversion. This was the situation in the case of *Shortlands Investments Ltd* v *Cargill plc.*[7]

In some cases the value of the property in repair and the value of the property in disrepair may be the same, and therefore no damages will be payable to the landlord. An example of this is where a large multiple retailer leaves first-class premises with the interior in disrepair. Most large multiple retailers have their own internal style of fitting and will usually strip out the entire interior of a shop unit prior to a refit, and so the current state of interior repair will not lead to a reduction in value.

There will, of course, be no claim for dilapidations if the building is to be demolished. For a full consideration and understanding of dilapidations, reference should be made to the specialist textbooks in this area.[8]

(b) Other claims

The level of damages payable in many actions often relates to the difference in value between a property in a known state and one in an assumed hypothetical state. One example is in connection with the calculation of damages for negligence resulting from a defective structural survey. The valuer's duty in these cases is to advise on the loss in value to the plaintiff and not to consider other heads of damage which may, as a matter of law, also apply. Frequently the difference in value will relate to the cost of necessary repair, for example where structural movement or

6 [1987] 2 EGLR 254.
7 [1995] 1 EGLR 51.
8 *Handbook of Dilapidations* (Sweet & Maxwell); *West's Law of Dilapidations* (Estates Gazette, 11th edition, 2008); *Dilapidations: The Modern Law and Practice* (2008) by N. Dowding, K. Reynolds and A. Oakes (Sweet & Maxwell).

dry rot has been missed. It is usually impossible to find comparable evidence for a property in a given state of disrepair. Therefore, the valuer starts with the value in repair and then adjusts that figure to represent the cost of dealing with the defective items, financing them, and the profit to reflect the risk inherent in the purchase of an affected property as well as the reward to the persons carrying out the work in recompense for their inconvenience and, often, entrepreneurial input. It is very unusual for a difference in value to be represented solely by the cost of work without the other incidental costs being added to it, but regard must also be had to the possibility of betterment resulting from the work having been carried out, for example the benefit of a new roof.

An example of the betterment point is where the surveyor has failed to point out that a 40-year-old house needed a new roof. The comparable for the house in good repair would often be one where the roof is in an acceptable, but not new, condition. Once the house has been re-roofed, it would go up in value to reflect the absence of any cost relating to maintaining the roof over the foreseeable future. Similarly, in some areas, buyers welcome the opportunity to reconstruct the inside of a property and are not put off by the need to redecorate completely or to renew fixtures or fittings. Indeed, in some cases perfectly new and modern fixtures and fittings are replaced to reflect the buyer's own taste.

The approach to all of these valuations is the same. The valuer should produce two valuations in the same way as when considering diminution in the value of the landlord's reversion. The first valuation reflects the true actual position and the second valuation reflects the position before the event that has triggered the litigation was known. Loss in value is clearly the second value less the first value, having regard to all the relevant factors.

7. Management of investments

The valuer may be required to advise on the policy to be adopted in the management of investment properties.

No attempt is made here to deal with general estate management issues or with the type of problem that can arise where, for example, a major estate forms part of the wider investment portfolio of a large financial institution. However, an attempt is made to comment on issues with a valuation connotation which are common to most types of estate, large or small.

Speaking very generally, it may be said that good estate management should aim at securing the maintenance and maximisation of income and capital value.

To advise on a group of investment properties of varying types involves consideration of a number of factors, including:

(a) the possibilities of increase or decrease in future capital value;
(b) an examination of present rents with a view to possible increase or reductions; and
(c) the security of the income.

The latter point relates to both security in real terms and security in actual terms. In connection with the former, consideration of (a) above will apply. However, there is a second consideration which is the actual security of payment of the money, and this will depend on the quality of the covenant and the ability to seek payment from a predecessor in title. This latter point is bound up with a consideration of Privity of Contract.

Where the actual tenant is the first lessee, then the security of income will depend on the quality of that tenant's covenant and any guarantor who may have been a party to the lease. However, in many cases the lease will have been assigned and then it may be possible to go back to a previous tenant, or guarantor, to obtain any unpaid rent or to require them to pay damages for any breach of covenant. This is now governed by the Landlord & Tenant (Covenants) Act 1995 which came into effect on 1 January 1996.

There are two sets of rules depending upon whether the lease was entered into prior to 1st January 1996 or after that date. In the former case, Privity of Contract still applies, but for arrears of rent this only relates to a maximum of six months' rent which is due prior to the issue of a notice under the Act; any additional rent above six months' will not be recoverable. Where a lease is granted after 1 January 1996 there will be no Privity of Contract unless an Authorised Guarantee Agreement was entered into. The value of an investment will depend on the security of the income, and therefore different yields might be applicable to identical properties where the rules concerning Privity of Contract are different.

Other factors are:

(d)   consideration of outgoings;
(e)   the provision of reserves for future repairs or other capital expenditure, to enable the cost of repairs to be spread evenly over a period, and to provide sinking funds for wasting assets such as leasehold properties.

(a) Maintenance of capital values

From time to time it will be desirable to review the properties comprised in a portfolio in order to consider whether some should be sold or otherwise dealt with, and whether or not they are likely to increase or decrease in value.

As an obvious and desirable objective is to maintain or enhance the capital value of the estate, properties likely to depreciate should be sold and others sought to replace them of a character that may increase in value, or at least not depreciate. On the other hand, such properties may produce a high rate of return and, if a high level of income is desired, they may nonetheless be suitable for retention in the portfolio.

Opportunities may arise where other interests in the properties can be bought, or adjoining properties acquired, which will create added value. For example, a portfolio may include freehold interests in properties let on leases granted some years ago when rents were considerably lower than today's and where no provision was made for rent review. It will be advantageous to buy back such leases if they come up for sale, so that they can be extinguished and new ones granted on improved terms. This will not only increase income but probably increase the capital value of the freehold interest by more than the cost of the lease, as synergistic value will usually exist. Restructuring (or rearrangement of the terms, including rent) of leases is a further possibility. Alternatively, the freehold interest could be offered to the lessees on the basis of a share in the synergistic value, with benefits to both parties.

The possibilities of enhancement of capital value by redevelopment or refurbishment need to be kept in mind, subject to the availability of finance for such works. In the case of large-scale developments these might be carried out by developers under ground lease to encourage inward investment.

There is also the possibility of investment in properties of a type where increases in value may be expected and which are not dependent upon the grant of planning permission. For example, a block of shops offered for sale may be the subject of leases granted some time ago at rents which, at the end of the leases, may be expected to be substantially increased. An investment in property of this type offers possibilities of an inherent increase in capital value in the future.

Property in blocks should be considered for "break-up" value; this happens when the sum of the parts exceeds the value of the whole and where individual occupiers are likely to be interested in buying their freeholds. In these cases an owner/occupier premium may be paid by the existing tenants. It can occur in the case of small self-contained office buildings or shop parades.

(b) Rents

Periodic consideration of lease rents should be made to ensure that, as and when appropriate, the rent is increased to the MR. Such a review will also identify properties with falling rents or which are over-rented. Any excesses over MR need to be treated with caution as the only security for the enhanced rent is the tenant's covenant to pay, and in the event of the tenant's failure the rent will fall to the lower MR. Chapters 12 and 17 consider the valuation implications.

The general trend over many years has been for rental values to increase, but this pattern was broken by a significant fall in rental value from 1989 to 1995 and again in 2008, other than in the most prestigious London locations. Before 1989 it was normal for leases to contain upward-only rent review clauses operating at regular intervals, say every three, four or five years. The term was generally a multiple of the review periods, for example, 15 or 25 years' lease with five-year reviews. Since 1989 leases have tended to be shorter, with tenant's break clauses within the term. In the case of building leases, similar periodic rent review clauses are incorporated but the rents commonly remain as a percentage of the market rent so that they rise in parallel – they are "geared" to the MR or occupation rents.

(c) Outgoings

The current outgoings, particularly those liable to variation such as repairs and rates, need to be scrutinised. Where the landlord is responsible for outgoings, rating assessments should be checked to ensure they are correct, that short-term void rate allowances are claimed where properties are empty and that adequate provision is made for repairs and insurance.

(d) Reserves

Certain outgoings, in particular repairs, will vary in amount from year to year. For instance, external painting may be required perhaps every five years. It is desirable to maintain the income from an estate to avoid fluctuations and so, in those years when such expenditure is not incurred, an amount should be set aside for future use, so that if the cost of external painting every fifth year is estimated at say £50,000, a sum of £10,000 should be set aside each year to meet it, rather than have a £50,000 outgoing in one year swallowing up the whole rent.

Where property is held on lease with full repairing covenants, or where there is a covenant to reinstate after alteration, provision should be made for future expenditure either by setting aside a certain amount of income each year or on the lines indicated below.

Where income is derived from leasehold properties, consideration should be given to the replacement of capital expended. This may be done by means of a sinking fund policy (very unusual, despite the long established theory – see Chapter 9) or other amortisation plan. It may be possible to find a suitable investment – for example, ground rents with a reversion to rack rents which will come into hand at about the time when the leasehold interests run out. This portfolio mix can ensure that the estate overall maintains its rental and capital value. The usual situation with leaseholders is that the gross yield is such that an investor can fund other acquisitions through bank borrowing, and hence the new purchase is in effect subsidised by the leasehold investment so that effectively the new purchases comprise the amortisation plan.

The latter arrangement has certain advantages over the sinking fund method. The value of the reversion accumulates at a high rate of interest in most cases and a sinking fund, if established, has to be paid out of taxed income and reduces the net available revenue. It may be, however, that the capital resources of an estate do not permit the purchase of suitable investments, and provision for the future has to be made out of income.

A valuer may have to advise a particular investor in landed property on the type of investment best suited to the investor. In such a case, the valuer would have to take account of such a factor as the investor's liability to income tax. Value to a particular investor is known as "worth". This is considered more fully in Chapter 12.

A number of these practical points reflect a more normal world of property than the one valuers currently find themselves in. It is their responsibility to monitor and reflect the current market when offering value or investment advice; for this reason, whilst this chapter may help the valuation student, it should not be seen as suggesting any particular course of action under current market conditions.

Chapter 14

Principles of the law of Town and Country Planning

1. The Town and Country Planning Acts

(a) The 1947 Act

The modern system of planning control, that is to say control of land use and development, originated in the Town and Country Planning Act 1947. As is well known, there were several earlier planning statutes, but the system of control that they introduced was of limited scope and effect. The Act of 1947 is in a very real sense the starting point of planning as we know it in Britain. From that point on, no development of land was lawful without the grant of planning permission. Furthermore, development value was to be transferred to the State and a £300 million fund was set aside as compensation in respect of land which was ripe for development at the commencement of the Act. A "development charge" was to be levied in respect of the grant of planning permission, and there would be no compensation for the refusal of planning permission. However, the expropriation of development value discouraged development and it was abolished in 1952. The compensation fund was still available in respect of land ripe for development as at the commencement of the 1947 Act, but over time inflation depreciated the value of the fund and eventually it was not worth claiming, given the cost of recovery. So it was abolished by the Planning and Compensation Act 1991. Thus we have reached a stage where landowners who are refused planning permission receive no compensation, whereas landowners who receive planning permission obtain development value. The only planning compensation payable is for the removal of planning permission or lawful use rights by revocation, modification or discontinuance orders (and similar cases under the Planning (Listed Buildings and Conservation Areas) Act 1990) or the removal of permitted development rights under article 4 of the General Permitted Development Order 1995. There have been various failed attempts over the years to tax "betterment" or development value, but they have all been abolished. However, those landowners who receive the benefit of planning permission may, of course, be liable to capital gains tax and, as set out below, developers may have to make financial contributions for infrastructure and other community benefits by way of connected planning obligations or the Community Infrastructure Levy. In these respects, some of the value of development is returned to the State.

2. The planning system

(a) Planning authorities

National planning policy is determined by the Secretary of State for Communities and Local Government, but plan making and development control is carried out by local authorities. (The regional tier was abolished by the Localism Act 2011.) In some areas there is a two-tier local government system of counties and districts, in others there are unitary authorities. The unitary authorities are the local planning authorities and, in the case of two-tier authorities, the district councils are the local planning authorities, while the counties are planning authorities for those areas where there are no district councils. In the National Parks, the local planning authority is the National Park Authority. There are also various statutory bodies (e.g. the Environment Agency, English Heritage, Natural England and the Welsh and Scottish equivalents) that must be consulted in certain planning matters.

The Secretary of State exercises his authority by issuing policy guidance and determining planning appeals against the decisions of local planning authorities (LPAs). Plan preparation and development control must be in accordance with national policy. In March 2012 the coalition government introduced a much simplified system of policy guidance, replacing over 5000 pages of planning policy documents (the old Planning Policy Statements and Guidance Notes) with the new National Planning Policy Framework which consists of a mere 65 pages. LPAs must abide by the relevant policy in the NPPF when drawing up plans and when considering applications for planning permission. The NPPF creates a presumption in favour of sustainable development. It states that LPAs must grant planning permission for development that accords with the development plan. If development conflicts with the plan it must be refused unless material considerations indicate otherwise. Where the plan is silent, or its policies are out of date, the LPA must grant planning permission unless any adverse impacts of doing so would significantly and demonstrably outweigh the benefits when assessed against the policies in the NPPF, or when policies in the NPPF indicate that development should be restricted.

(b) The development plan system

Ever since the modern planning system was introduced, governments have tinkered with the development plan system in a series of attempts to make it quicker, flexible and adaptable to changes in circumstances. Originally a zoning system was employed, but this could not adapt to change quickly enough and was eventually replaced (in the 1968 Town Planning Act) by the strategic and flexible county-level "structure plan", coupled with more detailed local plans applying the strategic policies at the local level. (In a unitary authority these plans were comprised in a "unitary development plan", parts 1 and 2.) Unfortunately, many of these plans took a very long time to produce and update, resulting in official policies being out of date. So another attempt to improve the system was introduced under the Planning and Compulsory Purchase Act 2004. Under this system the "Regional Spatial Strategy", together with local development plan documents (called the Local Development Framework in the relevant circular), were to replace the old-style structure and local plans or unitary development plans. But then the regional tier of government (the Regional Development Agencies) was abolished and the Localism Act 2011 made provision for the abolition of all regional strategies, streamlined and simplified the local development plan process, and granted more powers at local level. In particular, there is a new right for communities to draw up a "neighbourhood development plan" through the auspices of a parish council or neighbourhood forum. However, this plan must be in accordance with national planning policy and the strategic vision of the local authority.

Development Plan Documents set out the local planning authority's policies relating to the use and development of land in its area and must be submitted to the Secretary of State for independent examination by an inspector appointed by him. However, the Localism Act and regulations remove the power of the inspector to impose changes on the local authority. Instead, the inspector will only be able to recommend changes to overcome any conflict between the plan and national policy. The purpose is to make the process more like a form of mediation rather than an imposition of central power.

It should be noted that London has its own development plan system. The Greater London Authority, comprising the Mayor and the Assembly for London, has strategic planning powers. The Mayor is required to produce a Spatial Development Strategy (the "London Plan") and can intervene in planning applications of strategic importance. Planning functions at the local level are carried out by the London boroughs. At the time of writing, the development plan for London consists of the London Plan alongside the approved borough Development Plan Documents (or the old unitary development plans).

(c) Effect of the development plan

The legal effect of a development plan is largely indirect. Its main function is to guide the authorities: their planning "determinations" must be made "in accordance" with it "unless material considerations indicate otherwise": Planning and Compulsory Purchase Act 2004, section 38(6). So obviously the strategies and policies set out in a development plan will affect land value as they will, in general, determine whether planning permission for development is likely to be forthcoming or not. Nevertheless, a planning decision may depart from the policies within the plan. For further discussion on this point, see "Determining planning applications" below.

There are a few positive consequences of a development plan: there is a require-ment to consider development plans when compulsory purchase compensation is claimed on the basis that it includes "development value", and when it is proposed to serve certain kinds of "blight notice". These matters are covered in Chapters 25 and 26.

Mention may be made here of registers which local planning authorities are required to keep for public inspection. In addition to registers of local land charges, which include various orders, agreements and notices relevant to planning and compulsory purchase, there are registers kept specifically for planning. Thus there are registers of planning applications, of applications for consent to display adver-tisements and of caravan site licences, together with the decision of the planning authority on the applications; and there are also lists of buildings of special architec-tural or historic interest. Prospective purchasers and their solicitors should always consult these registers and lists in appropriate circumstances, just as they normally apply for an official search of the local land charges register.

3. Development control

The Town and Country Planning Act 1990 (the Planning Act 1990) governs the control of development. Planning permission is required for any development of land (section 57), and section 55(1) defines "development" as "the carrying out of building, engineering, mining or other operations in, on, over or under land, or the making of any material change in the use of any buildings or other land". Thus there will be development either if an "operation" is carried out or if a "material change of use" is brought about. Often a project involves development because there will be one or more operations and a material change of use as well. Operations that are purely internal and do not affect the exterior of a building are excluded from the definition of "development"; but the Planning and Compulsory Purchase Act 2004, section 49, empowers the Secretary of State to prescribe maximum limits to the additional floor space which may be created by and within such operations, for the purposes of this definition.

An operation involves something that changes the physical characteristics of the land and has some degree of permanence.[1] Building operations are further defined as including demolition, rebuilding, structural alterations and additions to buildings, and operations normally undertaken by a person carrying on business as a builder. Building is defined in section 336 as any structure or erection and any part of a building, but does not include plant and machinery comprised in a building.

Section 55 of the Planning Act 1990 defines certain changes of use as material, e.g. use as two or more separate dwellings of a single dwellinghouse and the deposit of waste on land. It also defines certain other changes as not material, e.g. use of

1 Cheshire CC v Woodward [1962] 2 QB 126.

land for agriculture and forestry, use of land or buildings within the curtilage of a dwellinghouse for any purpose incidental to the use of a dwellinghouse as such, and uses within a "use class" (below). Outside of these categories, the question of whether a change is material or not is a matter of fact and degree.[2] The courts have developed some "tools of analysis" to assist the decision maker in deciding the question of materiality. One of the most important of these is to identify the "planning unit" on which the change of use is taking place. Judicial guidance can be found in *Burdle* v *Secretary of State for the Environment*.[3]

(a) The Use Classes Order 1987

Section 55 states that "in the case of buildings or other land which are used for a purpose of any class specified in an order made by the Secretary of State under this section", the use thereof "for any other purpose of the same class" is not development. Thus we have the Town and Country Planning (Use Classes) Order 1987, which lists various "use classes" in four groups A, B, C and D; and any change within a "use class" is not development at all, or in other words not "material". It is important for the valuer to know the use classes because the use to which land or buildings can be put without having to apply for planning permission has a significant effect on value. Furthermore, some changes from one use class to another are permitted by the General Permitted Development Order 1995 (below).

Class A1– Shops

This concerns the retail sale of goods where the sale, display or service is to visiting members of the public. This class also includes certain specific uses which, though not retail sale of goods (e.g. post office, hairdressing), are included with the class.

Class A2 – Financial and professional services

This includes financial services, professional services (other than health and medical) and any other services appropriate in a shopping area. The services must be provided principally to members of the public.

Class A3 – Restaurants and cafes

Use for the sale of food and drink for consumption on the premises.

---

2 East Barnet UDC v British Transport Commission [1961] 3 All ER 878 at p. 885.
3 [1972] 3 All ER 240.

Class A4 – Drinking establishments

Use as a public house, wine bar or other drinking establishment.

Class A5 – Hot food takeaways

Use for the sale of hot food for consumption off the premises.

Class B1 – Business

Use as an office (other than a use within A2), use for research and development, and use for any industrial process so long as, in all cases, the use can be carried out in any residential area without any detriment to local amenity (by reason of noise, vibration, smell, fumes, smoke, soot, ash, dust or grit).

Class B2 – General industrial

An industrial process other than one within Class B1.

(Classes B3–B7 abolished)

Class B8 – Storage or distribution

Use as a storage or distribution centre.

Class C1 – Hotels

A hotel, boarding house or guest house where no significant element of care is provided.

Class C2 – Residential institutions

Residential accommodation and care for people in need of care, hospital and nursing home, and a residential school, college or training centre.

Class C2A – Secure residential institutions

This includes a prison, young offenders' centre, secure hospital, military barracks, etc.

Class C3 – Dwellinghouses

Use as a dwellinghouse (whether or not as a sole or main residence) by:

(a) a single person or by people to be regarded as forming a single household;

(b) not more than six residents living together as a single household where care is provided for residents; or

(c) not more than six residents living together as a single household where no care is provided to residents (other than a use within Class C4).

Class C4 – HMOs

Use as a dwellinghouse by not more than six persons as a "house in multiple occupation".

Class D1 – Non-residential institutions

This includes uses for medical or health services, crèches, education, museums, art galleries, public libraries, exhibition halls, places of worship and law courts.

Class D2 – Assembly and leisure

Includes cinema, concert hall, bingo hall, dance hall, skating rink, gymnasium and indoor or outdoor sports or recreations (not involving motor vehicles or firearms).

Excluded uses

Certain uses are excluded from the Use Classes Order by article 3(6). These include the sale of cars, the sale of fuel, theatres, launderettes, amusement arcades, scrap yards, taxi and car hire, hostels, retail warehouse clubs, nightclubs and casinos. Normally changes to and from such uses will be material but in each case it is a question of fact and degree.

Sui generis uses

Unclassified uses are normally referred to as *sui generis* – literally, "of its own kind". Whether a change to or from a *sui generis* use is material is, as always, a question of fact and degree.

(b) Part 3, General Permitted Development Order (GPDO) – changes of use

Some changes from one class to another constitute permitted development under the GPDO. They include the following:

To A1 from A3, A4 or A5.
To A2 from A3, A4 or A5.
To A3 from A4 or A5.
To A1 from an A2 use with a ground floor display window.

A change from A1 to mixed use of A1 and a single flat above, and vice versa.
A change from A2 to mixed use of A2 and a single flat above, and vice versa.
To B1 from B2 or B8 (subject to maximum B8 use of 235 square metres floor space).
To B8 from B1 or B2 (subject to maximum B8 use of 235 square metres floor space).
To C3 from C4 and to C4 from C3.

At the time of writing the government has initiated a review of the Use Classes Order with the intention of expanding its scope so as to speed up the planning process.

4. Planning permissions

Section 57 of the Planning Act 1990 states that planning permission is "required" for carrying out development (subject to certain special exceptions).

(a) Development orders and local development orders

Section 59 empowers the Secretary of State to make "development orders" for the purpose, *inter alia*, of actually granting permission, on a general and automatic basis, for certain forms of development. Practitioners have long been familiar with the Town and Country Planning General Development Order, known for short as the "GDO", which gave such permission for specified classes of development. But with effect from 3 June 1995, the part of this order that permitted development was revised and became the General Permitted Development Order 1995 (GPDO). The other part governed planning applications. This is now governed by the Development Management Procedure Order 2010 (DMPO).

The GPDO (as amended) largely re-enacts the classes of permitted development previously specified, with some specialised additions, the whole being set out in Schedule 2 of the Order in 40 "Parts". The kinds of permitted development which apply most widely are to be found in Parts 1 (within the curtilage of a dwellinghouse), 2 (minor operations), 3 (changes of use – above) and 4 (temporary buildings and uses), also 6–8 (certain kinds of agricultural forestry and industrial development as specified). As regards Part 1, alterations, extensions and improvements within the curtilage of a dwellinghouse are permitted subject to restrictions relating to position, size and height. There are also restrictions on the materials to be used. A very useful "interactive house" illustrating what is permissible can be found on the government's Planning Portal website.

A direction under article 4 of the GPDO may withdraw some or all permitted development rights where evidence suggests that the exercise of such rights would harm local amenity or the proper planning of an area. The relevant circular (9/95) states that LPAs should only use an article 4 direction in exceptional circumstances.

The effect of such a direction is to require planning permission to be sought. If permission is sought within 12 months of the article 4 direction and it is refused (or granted subject to conditions different from those in the GPDO), a claim for compensation for depreciation in the value of land and abortive expenditure and any other loss or damage can be made (section 108, Planning Act 1990). In the case of development within the curtilage of a dwellinghouse, no compensation is due where notice of at least 12 months is given prior to revocation of permitted development rights. As the article 4 direction coupled with the refusal of planning permission effectively takes away planning permission, the compensation is on the same basis as that for a revocation order under section 107 of the Planning Act (below).

(b) Local development orders

Section 61A of the Planning Act 1990 provides for the creation of local development orders (LDOs). An LDO may grant planning permission for a specified development or for a specified class of development. The purpose of an LDO is to allow local planning authorities to extend permitted development rights, thus removing some of the barriers in the way of development in order to meet local policy objectives. They are used in association with "enterprise zones" (below). These are areas designated by the government where tax incentives and rate relief are provided in order to stimulate economic growth.

(c) Simplified planning zones and enterprise zones

Sections 82–87 and Schedule 7 of the Planning Act 1990 require local planning authorities to consider making simplified planning zone schemes, and the Secretary of State may direct them to do so. A scheme resembles a development order, in that it automatically grants planning permission for specified classes of development, with or without conditions, in the whole or part of a zone.

Enterprise zones were introduced under section 179 and schedule 32 of the Local Government and Land Act 1980. The Secretary of State is empowered to approve schemes designating these zones, which are prepared by local authorities or new town or urban development corporations at his invitation. Planning permission is automatically granted for various kinds of development specified in each scheme by an associated local development order; it may in some cases be "outline" permission.

(d) Public authority development

Public authorities, including the Crown (Planning Act and Compulsory Purchase Act 2004, sections 79 and 90), are subject to planning control with certain practical reservations. For local authorities, when any project that involves expenditure requires the approval of a government department, such approval may also be expressed to confer "deemed" planning permission, with or without conditions, if needed (1990 Act, sections 58 and 90). This rule also applies to "statutory undertakers", that is the public utility authorities and similar public bodies; but with them there is also another factor, the difference between their "operational" and non-operational land (the latter being offices, houses, investment property and any other land that is not the site of their operating functions). "Operational" land has the benefit of special rules in planning law, for example in regard to compensation for restrictions on development. Local planning authorities are subject to special rules when they propose to carry out development and require the Secretary of State's approval if the development is not in accordance with the development plan (Planning Act 1990, Part XV, section 316; Town and Country Planning General Regulations 1992).

Planning proposals by public bodies which "blight" land of nearby owners may entitle the latter to serve a "blight notice" on the public body in question under Part VI of the 1990 Act (see Chapters 22 and 26).

5. Applications for planning permission

Section 62 of the Planning Act 1990 (as amended) provides that a development order may make provision as to the form and content of applications for planning permission. The local planning authority may also require such particulars as they think necessary, including evidence relating to the application. The relevant development order is the Town and Country Planning (Development Management Procedure) Order 2010 (DMPO).

(a) Outline planning permission

The DMPO provides for the grant of planning permission for the erection of a building, subject to a condition requiring the subsequent approval of the local planning authority (LPA) with respect to one or more "reserved matters", i.e. access, appearance, landscaping, layout and scale. The original idea was that the developer can obtain approval in principle without wasting time and expense on detailed plans which might be refused. However, the changes made to the requirements for outline applications (some of which were precipitated by the need for compliance with the EU Directive on Environmental Impact Assessment) have resulted in a significant amount of detail being required by the DMPO. So, in respect of layout, approximate locations are required for buildings, routes and open spaces; in respect of scale, the upper and lower limits for all the dimensions of the buildings must be stated; and in respect of access, the situation of the access points must be stated. Furthermore, "Design and Access Statements" are required to be supplied with a planning application for certain types of building operation. The purpose is to ensure that LPAs have sufficient information to apply local policies and encourage high-quality design. All these requirements (plus the others that are required by each individual LPA to accompany an application) have resulted in many applicants seeking a full permission instead of an outline, unless they simply wish to sell the land for development.

(b) Applicants and applications

Any person may apply for planning permission; but an applicant must notify all owners and farm tenants, either directly or, if that is not possible, by local press publicity. There are also certain classes of controversial development that must be publicised. The persons notified by these methods may "make representations", which the authority must take into account.

A prospective developer who is not certain whether a project amounts to "development" at all may apply to the authority for a "certificate of lawfulness of proposed use or development" (Planning Act 1990, section 192) or a "certificate of lawfulness of existing use or development" (section 191), as the case may be.

Any "person interested in land" can agree with the authority concerning its detailed development, including payments by way of planning obligations for "planning gain" (infrastructure and other community benefits) under section 106 of the 1990 Act. But planning obligations are being superseded by the "Community Infrastructure Levy" (described later in this chapter).

On receiving an application, the local planning authority must consult other authorities, government departments and public bodies as appropriate, and have regard to the provisions of the development plan. The local planning authority may grant permission "unconditionally or subject to such conditions as they think fit", or refuse it (1990 Act, section 70). Obviously a developer who acts on a planning permission cannot continue any previous use of the land which is inconsistent with it. The possibility that a project could be regulated under some other statutory procedure does not preclude a refusal of planning permission, even if that other procedure might carry with it a right to compensation.

(c) Environmental Impact Assessment

Certain development projects are subject to Environment Impact Assessment (EIA) before planning permission can be granted. Where EIA is required, the developer must compile an Environmental Statement containing detailed information about the likely environmental effects of the development. This statement must be publicised so that the general public as well as public bodies can make representations. The relevant planning authority must take the statement into account, together with any representations made, in deciding whether or not to grant planning permission. The process and requirements are governed by the Environmental Impact Assessment Regulations.

Schedule 1 of the regulations sets out the categories of development (EIA development) for which environmental impact assessment is required. It includes such projects as power stations, oil refineries, airports, railways, motorways, certain chemical installations, incineration, landfill, extraction of petroleum and gas, etc.

Schedule 2 sets out categories of development which require EIA if they are "likely to have significant effects in the environment by virtue of factors such as its nature, size and location". Such development will therefore require a "screening opinion" from the LPA or the Secretary of State to assess whether EIA is required. Schedule 2 development includes industrial estates and urban development projects (including the construction of shopping centres and car parks, sports stadiums, leisure centres and multiplex cinemas) where the area of the development exceeds half a hectare.

(d) Nationally significant infrastructure projects

The Planning Act 2008 set up a new procedure to deal with nationally significant infrastructure projects (NSIPs). This procedure requires a "development consent" for any development which is a "nationally significant infrastructure project" in the fields of energy, transport, water resources, waste water or hazardous waste. One of the purposes of the legislation was to set up an independent Infrastructure Planning Commission (IPC) to deal with such applications, but the Localism Act 2011 abolished the IPC and handed back its powers to the Secretary of State. There is a stringent pre-application process required for NSIPs which includes a statement as to how the developer will consult the local community.

(e) Planning fees

The provisions for charging fees for planning applications are set out in section 303 of the Planning Act 1990 (as amended by the Planning Act 2008). Details are set out in the Fees for Applications and Deemed Applications regulations. A charge may also be levied under the Community Infrastructure Levy, below.

(f) Determining planning applications

Planning law requires that all applications are determined in accordance with whatever policies were in force *at the time the decisions were taken* (per Lord Brown of Eaton-under-Heywood (The *Times*, 22 May 2009)). Section 38(6) of the Planning and Compulsory Purchase Act 2004 provides that the determination must be made in accordance with the development plan unless material considerations indicate otherwise. However, the old policy presumption in favour of development has been restored (in the National Planning Policy Framework (NPPF), above), subject to the condition that the development must be sustainable. As plans should be based on a policy of sustainable development and be in accordance with the NPPF, any potential conflict between the presumption in favour of sustainable development and an up-to-date development plan policy may be limited. Nevertheless, it provides an opportunity for developers to attempt to justify a departure from the development plan – especially if the plan is ambiguous. It must also be observed that it has always been the case that the development plan can be departed from if "material considerations indicate otherwise". A consideration that relates to the use and development of land is capable of being a planning consideration[4] and therefore material. The weight to attach to any material consideration is a planning, not legal, matter, so it is within the discretion of the decision-maker, i.e. the local planning authority or, on appeal, the planning inspector or Secretary of State. However, a decision that is manifestly unreasonable (so unreasonable that no reasonable person would come to such a decision) can be overturned by the courts. Such a high degree of unreasonableness is usually known as *Wednesbury unreasonableness* from the case *Associated Provincial Picture Houses Ltd v Wednesbury Corporation*[5] (a case concerning cinema licences). The courts will only step in where the degree of unreasonableness is very high, otherwise they would be involved in assessing the planning merits of the decision. See "Judicial control of planning decisions", below.

(g) "Calling in" procedure

Section 77 of the Planning Act 1990 provides that, if he so wishes, the Secretary of State may direct that a planning application be "called in" (as it is usually termed), i.e. referred to him instead of being decided by the local planning authority.

The call-in power is generally only exercised in respect of applications of more than local significance which conflict with national policy, or may give rise to substantial controversy, or raise issues of national security, or raise novel planning issues and the like.

In respect of "Major infrastructure projects", section 76A provides that the Secretary of State may direct that an application must be referred to him if he thinks that the development to which it relates is of national or regional significance.

(h) Revocation, modification and discontinuance orders

Planning permission can be revoked or modified (1990 Act, sections 97–100). The authorities which do this must pay compensation for any abortive expenditure and for any depreciation in relation to development value which, having come into existence by virtue of the permission, disappears because of the revocation or modification. Revocation or modification orders must be confirmed by the Secretary of State, except in unopposed cases. If permission is given automatically by the GPDO, referred to earlier, it may in effect be revoked or modified. If permission given by an "article 4 direction" under that order is partly or wholly withdrawn, and a specific application is then made which is refused or only granted subject to conditions, compensation is paid in these cases also.

4 Stringer v Minister of Housing and Local Government [1971] 1 All ER 65.
5 [1948] 1 KB 223

In so far as authorised development has actually taken place, even if only in part, revocation or modification orders and "article 4 directions" are ineffective. For a planning authority to put an end to any *actual* development or "established use" of land (except of course where it is the necessary consequence of acting on a planning permission that this should happen) requires a discontinuance order, which must also be confirmed by the Secretary of State (1990 Act, section 102). Compensation must be paid for loss of development value and abortive expenditure and also the cost of removal or demolition; but as compliance involves physical action there is also an enforcement procedure in cases of recalcitrance, similar in essentials to ordinary enforcement of planning control discussed below. Discontinuance of mineral working, or its temporary suspension, is governed by a specially modified code of regulation and compensation (1990 Act, Schedule 9).

6. Effect of planning permission

Section 75 of the Planning Act 1990 provides that any grant of permission "shall enure for the benefit of the land and of all the persons for the time being interested in it". To enure is to take effect or to come into operation. So once a planning permission has been implemented it cannot be abandoned unless it becomes physically impossible because of an inconsistent development.[6] (However, in appropriate circumstances a condition can be attached to a planning permission requiring a new use to cease after a certain time, thereby creating a temporary planning permission.)

7. Duration of planning permission

The statutory time limits for the commencement of development are set out in sections 91 and 92 of the Planning Act 1990. Development must be commenced within three years of the grant of full planning permission; otherwise the permission lapses. In the case of outline planning permission, application for full permission must be made within three years and commencement must be within two years of final approval (section 92). In both cases the LPA can impose different time limits if they consider it appropriate.

Whether development has commenced depends on whether a "material operation", as defined by section 56, has been carried out. "Material operation" is widely defined and includes "any operation in the course of laying out or constructing a road or part of a road" as well as "the digging of a trench which is to contain the foundations, or part of the foundations of a building". So developers wishing to preserve planning permission may undertake such activities although at the time they may have neither the intention nor the finance to complete the development. However, if the LPA is of the opinion that the development will not be completed within a reasonable period, it has the option of serving a "completion

6 Pioneer Aggregates (UK) Ltd v Secretary of State for the Environment [1985] AC 132.

notice" specifying a time, not less than 12 months, by which development must be complete or else the permission "will cease to have effect" (section 94). Note that although a developer can apply for permission to develop without compliance with any condition attached to a planning permission (section 73), this does not include an application to extend the time limits.

8. Planning conditions and planning obligations

Apart from meeting the policy requirements of the Secretary of State, a planning condition must meet certain legal tests of validity. These tests are usually known as the "Newbury tests" from *Newbury DC* v *Secretary of State for the Environment*.[7] The tests are that a planning condition must be for a planning purpose, be fairly and reasonably related to the permitted development and not be *Wednesbury* unreasonable. A further test of certainty must also be met. If a condition cannot be given any sensible or ascertainable meaning, it will be void.[8]

It was held in *Hall & Co Ltd* v *Shoreham UDC*[9] that a condition requiring the developer to provide an ancillary road with public access along the entire frontage of the development was *Wednesbury* unreasonable because it required the provision of what was, in effect, a public highway at private expense. The purpose of the condition was to relieve congestion on a busy main road. The consequence of this landmark decision is that a developer cannot be required by way of a planning condition to provide public infrastructure and other community benefits at his own expense, even if the need for the infrastructure is a consequence of the development itself. This led to planning authorities using section 52 agreements (Town and Country Planning Act 1971) and their replacement, section 106 planning obligations in the Planning Act 1990, in order to achieve by way of agreement or obligation what could not be achieved by condition. Section 106 enables a planning authority to require, inter alia, financial contributions from developers. The purpose was to make developers pay for the 'externalities' imposed by the development itself. For example, a new retail development may require new roads, road junctions, access roads and bridges to be built. New housing development will necessitate the provision of local services and facilities including primary schools, recreational space, and other local services (as well as a contribution to affordable housing). These could be achieved by way of a planning obligation so long as, according to the policy in the relevant circular, the obligation was necessary to make the development acceptable in planning terms, directly related to the development, fairly and reasonably related to the development in scale and in kind, and reasonable in all other respects. Unfortunately, planning obligations could be used by big developers to "buy" planning permission and could also be

7 [1981] AC 578.
8 Fawcett Properties Ltd v Buckingham County Council [1961] AC 636.
9 [1964] 1 All ER 1.

exploited by planning authorities to extract more in the form of planning gain than the circular allowed. So long as the obligation was within the terms of section 106 and there was some connection (not "*de minimis*") between the obligation and the related planning permission, both the obligation and the permission were lawful.[10] This was because the law only required that the obligation was a material consideration. It did not require that it was necessary – the latter was a matter of policy, not law. This has now been changed by the Community Infrastructure Levy (CIL) regulations, which turn the policy of necessity into law.[11]

9. The Community Infrastructure Levy (CIL)

The Community Infrastructure Levy (CIL) was introduced by the Planning Act 2008 and will largely supersede section 106 obligations by permitting local authorities to impose a levy on development instead of negotiating a financial contribution. The levy is to be based on a costed assessment of local and sub-regional infrastructure (such as roads, schools, medical facilities, recreational and sporting facilities, flood defences, etc.) to support development contemplated by the development plan. The levy will be a sum of money per square metre of gross internal floor space, charged on new dwellings and on any development or change of use of 100 m^2 or more of gross internal floor area. Section 106 obligations are to be retained for affordable housing (at least for the time being) and to mitigate the direct impact of new development on site.

In order for a CIL to be levied, there are certain requirements: an up-to-date development plan; the carrying out of satisfactory infrastructure planning; the creation and approval of a charging schedule; the grant of a planning permission for specified development; and commencement of that development.

10. Appeals – section 78

Appeal lies to the Secretary of State against a refusal of planning permission, or a grant made subject to conditions, or a failure to give any decision within the appropriate time limit, and must be on a completed appeal form in compliance with the requirements of the DMPO. The standard time limit is six months from the date of notice of the adverse decision or expiry of the time limit, but is 12 weeks for householder applications (i.e. existing dwellinghouse development). Where the application is substantially the same as a development in respect of which an enforcement notice has been served, the time limit is 28 days from the relevant date.

The appeal may be by written representations or take the form of a hearing or a local inquiry. The Secretary of State through the Planning Inspectorate (PINS) determines which form of appeal is appropriate according to criteria published in

10 Tesco Stores Ltd v Secretary of State for the Environment [1995] 2 All ER 636.
11 Community Infrastructure Regulations 2010, r.122.

PINS Procedural Guidance Note 01/2009. The great majority of appeals (around four-fifths) will be dealt with by written representations. The actual procedure is governed by a variety of regulations explained in the PINS Procedural Guidance Note. Guidance on costs can be found in Circular 02/2009. Practitioners should advise caution and careful consideration before an appeal is launched, as unreasonable behaviour can result in costs being awarded against the appellant. The six-month appeal period should be utilised for consultation and negotiation with the planning authority.

The Secretary of State has wide powers in determining an appeal under section 78 as he may allow or dismiss the appeal or reverse or vary any part of it. He can deal with the appeal as if it were an application for planning permission made to him in the first instance, so his hands are not tied in any way by the decision of the planning authority. The Secretary of State's decision, or that made by a planning inspector on his behalf, is "final" (section 79(5)) and cannot be challenged in a court except in the circumstances set out in Part XII of the Planning Act 1990 in relation to the judicial control of planning decisions (section 284(3)).

(a) Appeals and overlapping and repeated applications

When an application for planning permission is made to a local planning authority, a period is prescribed within which the authority must issue notice of its decision. That period is normally eight weeks (DMPO art. 29), but it is increased to 13 weeks for major development and 16 weeks in cases involving an Environmental Impact Assessment; and it may in any event be extended by written agreement with the applicant. If this prescribed or agreed period expires without the authority having given notice of their decision, the applicant is entitled to appeal to Secretary of State under section 78(1) as if the application had been refused (the "deemed refusal").

The applicant may prefer, instead of lodging an appeal, to continue in negotiations with the local planning authority, but then risks missing out on a permission and having to lodge an appeal in any event. A practice therefore developed of applicants lodging two identical or closely similar applications (i.e. "twin-tracking") and then, if the period expires without a decision having been given, appealing on one application while continuing to negotiate on the other. So section 70B was inserted by the Planning and Compulsory Purchase Act 2004 to grant the planning authority the power to decline an application which is substantially the same as one already under consideration.

Note that the developer and planning authority may agree in writing on a longer period for a decision to be made. Prior consultation with the authority to obtain a realistic idea as to how long the decision may take, and an appropriate agreement in writing, will save the developer time in the long run if this avoids an unhelpful appeal against a deemed refusal.

Another difficulty for busy planning authorities was having to deal with repeated applications designed to wear down the authority and objectors. So section 70A was inserted by the Planning and Compulsory Purchase Act 2004. The effect, broadly speaking, is that if a planning application is substantially the same as one that has been dismissed on appeal (or no appeal was pursued), and there has been no change in any relevant considerations, the LPA may decline to determine it.

II. Judicial control of planning decisions

No one holding public office has unrestricted freedom in decision-making, i.e. arbitrary power. Subject to alteration by Act of Parliament, the "discretion" of all public authorities has express or implied limits, enforced by the courts. Activities outside those limits are *ultra vires* (beyond the powers) and the courts can invalidate them. But there are stringent restrictions on recourse to the courts in planning, as in other matters within the scope of the *ultra vires* principle.

Many, but not all, decisions of the Secretary of State (as distinct from those of local planning authorities) can only be challenged by a "person aggrieved" within six weeks, to the High Court, under the relevant section in Part XII of the 1990 Act (section 288 in respect of a Secretary of State's decision on a planning appeal). The Court can only quash such decisions on the ground that they are "not within the powers of [the relevant Act], or that the interests of the applicant have been substantially prejudiced" by some procedural default. This represents a tightened definition of the *ultra vires* principle itself. The courts have adopted an approach broadly similar to that taken in judicial review (below). In the *Seddon* case[12] five principles were set out as follows: the Secretary of State must not act perversely (this resembles the *Wednesbury* principle set out above); he must take account of relevant material and ignore irrelevant material; he must abide by the statutory procedures and his reasons must be proper, adequate, clear and intelligible; he must not depart from the principles of natural justice; if he differs from his inspector on a finding of fact or takes account of new evidence not canvassed in the appeal, he must notify the parties and give them the opportunity of making further representations. ("Natural justice" comprises two basic rules, namely that the decision maker must not be biased and that both sides are given a proper hearing on the points at issue.)

In contrast, local planning authorities and other public bodies are liable to have their *ultra vires* decisions challenged by a more general procedure in the High Court, "judicial review", in accordance with the Senior Courts Act 1981, section 31, subject to a three-month limitation period (Civil Procedure Rules 1998, Part 54, rule 5(1)). It has already been noted above that the courts' approach to statutory challenge to a Secretary of State is largely based on the principles of judicial review. In order to seek judicial review, a claimant must have a "sufficient interest" in the decision.

The courts will not intervene on the merits of a planning decision: that would be to usurp the administrative function of the decision maker. So even if the

---

12 Seddon Properties Ltd v Secretary of State for the Environment [1978] JPL 835.

Secretary of State, planning inspector or local planning authority may have come to the wrong conclusion, it will not be overturned so long as it is a conclusion that was within the jurisdiction of the decision maker and that it was entitled to make (*Anisminic* v *Foreign Compensation Commission*,[13] *per* Lord Pearce.)

12. Enforcement of planning control

(a) Enforcement notices

Part VII of the 1990 Act, as amended (sections 171A–196C), sets out procedures for the enforcement of planning control. It is not however a criminal offence to develop land without planning permission. If this happens, the local planning authority should first consider whether it would be "expedient" to impose sanctions, "having regard" to the development plan and to any other material considerations. If it would, the local planning authority may issue an "enforcement notice" and serve a copy on the owner and occupier of the land to which it relates and on any other person having a material interest in the land. The notice must specify the "breach of planning control" complained of, the steps required to remedy it, the date when it is to take effect, and the time allowed for compliance. (The Enforcement Notices and Appeal Regulations 1991 also require some additional information, such as the reasons for the notice and the boundaries of the land to which the notice relates.) "Breach of planning control" occurs when development takes place either (a) without the necessary permission or (b) in breach of conditions or limitations contained in a permission (section 171A(1)).

Enforcement action cannot be taken after certain time limits which are set out in section 171B. In the case of operational development, the time limit is four years. For change of use of a building to a single dwellinghouse, it is also four years. In respect of any other breach (i.e. breach of condition and changes of use other than to a dwellinghouse) the time limit is 10 years. For the time limit to operate, enforcement action must have been possible at any time during the relevant period. So a period of cessation of a change of use (not a temporary cessation such as a holiday period) will start the time period running from the beginning again.[14] In respect of unlawful building operations, the time period only starts to run when the building is substantially complete. So even though the only remaining work is to the interior, the time period may not have begun and the unfinished building can be enforced against.[15]

If enforcement action has not been taken within the time limits, the operation, change of use or breach of condition becomes lawful and the owner may apply for a certificate of lawfulness of the existing use or development. This is particularly

13 [1969] 1 All ER 208, at 234.
14 Thurrock BC v Secretary of State for the Environment, Transport and the Regions [2002] 2 PLR 43.
15 Sage v Secretary of State for the Environment, Transport and the Regions et al. [2003] 2 All ER 689.

useful when selling land subject to lawful development rights and will obviously have an important effect on value.

(b) Planning contravention notices

A local planning authority to whom "it appears ... that there may have been a breach of planning control" may also serve a "planning contravention notice" on the owner or occupier of the land in question, or on the person carrying out oper-ations thereon, requiring the recipient to furnish specified information which will enable the authority to decide what action (if any) to take by way of enforcement (section 171C). Failure to comply within 21 days is a criminal offence, punishable by a fine up to level 3 on the standard scale; the giving of false information either knowingly or recklessly is punishable by a fine up to level 5 on the standard scale.

(c) Stop notices

The period specified in an enforcement notice before it takes effect (minimum 28 days) is intended to allow for making an appeal. Further, the notice is "of no effect" while any appeal is going forward. This may encourage a recalcitrant developer to press on with their activities in the meantime, in the hope of creating a *fait accompli*. Such action may also be damaging to local amenity. Local planning authorities therefore have the additional power, during this period, to serve a "stop notice" prohibiting any activity which is "specified in the enforcement notice as an activity which the local planning authority requires to cease, and any activity carried out as part of that activity or associated with that activity" (section 183). The LPA may take such action where they consider the activity should cease before the expiry of the period for compliance with the enforcement notice. Stop notices cannot be used in respect of dwellinghouses or activities that commenced more than four years ago.

If the enforcement notice is subsequently quashed or varied on any of the appeal grounds except ground (a), or if it is withdrawn, compensation will be due for loss or damage accruing from the unnecessary cessation resulting from the stop notice. No compensation is due for a successful appeal on ground (a) as this is an appeal on the ground that planning permission should be granted rather than on the ground of a legal or factual error in the notice. (See para (f) below for the grounds of appeal against enforcement notices.)

Sections 171E–171H of the 1990 Act empower a planning authority to serve a "temporary stop notice" where the authority considers it necessary to stop the breach of planning control immediately in order to safeguard amenity. This differs from the ordinary stop notice in that the LPA does not have to wait before an enforcement notice is issued. It prohibits specified actions for a maximum of 28 days; but it may involve payment of compensation if the prohibited action is in fact lawful, or the notice is withdrawn. In addition to the normal restrictions on the use of a stop notice (above), a temporary stop notice cannot be used for the stationing of a residential caravan unless the LPA consider that the risk of harm to a compelling public interest is so serious so as to outweigh any benefit to the occupier of the caravan. During the 28-day period, the planning authority must consider whether it is appropriate to take enforcement action coupled with a normal stop notice. (A second temporary stop notice cannot be served.)

(d) Breach of condition notices

The Planning and Compensation Act 1991 inserted the breach of condition notice into the Planning Act 1990 in order to deal with the problem of developers flouting planning conditions and then appealing against the resulting enforcement notice. As the planning condition could have been challenged by appeal to the Secretary of State when the permission was granted (and under section 73 an application seeking release from a planning condition can also be made), it was thought inappropriate for the developer to effectively challenge a planning condition by way of an enforcement appeal. This practice made it difficult and time consuming for planning authorities to uphold planning conditions and provided opportunities for developers to sidestep them, particularly if the land was sold on. So section 187A provides that the LPA, instead of serving an enforcement notice, may serve a breach of condition notice on "any person who is carrying out or who has carried out the development; or any person having control of the land", requiring him to secure compliance in not less than 28 days with such of the conditions as are specified in the notice. For the reasons stated above, there is no right of appeal against a breach of condition notice. Failure to comply with the notice is an offence punishable, on summary conviction, with a fine not exceeding level 3 on the standard scale.

(e) Injunctions

A powerful weapon added to the armoury of the LPA by the Planning and Compensation Act 1991 is the injunction. Under section 187B of the Planning Act 1990, if the LPA considers it necessary or expedient for any actual or apprehended breach to be restrained by injunction, they may apply to the court for an injunction, whether or not they have exercised or are proposing to exercise any of their other enforcement powers. Breach of a court injunction is contempt of court and the offender can be imprisoned. But in accordance with the European Convention for Human Rights, any enforcement action taken must be proportionate. This is a particular issue in the eviction of travellers and gypsies from their homes as article 8 of the Convention (right to respect for the home) is engaged (*South Buckinghamshire DC v Porter*).[16]

---

16 [2004] 4 All ER 775 (HL).

(f) Appeals against enforcement notices

An appeal may be made against an enforcement notice by the recipient "or any other person having an interest in the land" within the time specified *before* it is to take effect. It must be made in writing (which includes electronic transmission) to the Secretary of State, and may be on one or more of seven grounds set out in section 174:

(a) Permission ought to be granted or a condition or limitation ought to be discharged.

(b) The alleged breach of planning control has not occurred.

(c) The matters alleged (if they occurred) do not amount to a breach of planning control.

(d) The alleged breach occurred more than four years ago in respect of operations or a material change of use to a single dwelling, or more than 10 years ago for other changes of use or breaches of condition.

(e) Copies of the notice were not served as required.

(f) The specified steps for compliance exceed what is necessary to remedy the breach or any injury to amenity caused by the breach.

(g) The specified time for compliance is too short.

The Secretary of State may uphold, vary or quash the enforcement notice and also grant planning permission if appropriate. He may "correct any defect, error or misdescription" in the notice if "satisfied" that the correction will not cause injustice to the appellant or the local planning authority, and he may disregard a failure to serve it on a proper party if neither that party nor the appellant has been "substantially prejudiced". The Court of Appeal has stated that "an enforcement notice is no longer to be defeated on technical grounds. The Minister ... can correct errors so long as, having regard to the merits of the case, the correction can be made without injustice". That was said in a case in which it was held to be at most an immaterial misrecital for an enforcement notice to allege development "without permission" when in fact a brief temporary permission had existed under the General Development Order. "The notice was plain enough and nobody was deceived by it" (*Miller-Mead* v *Minister of Housing and Local Government* (1963) *per* Lord Denning).[17]

Further appeal from the Secretary of State's decision on an enforcement notice lies to the High Court on a point of law (1990 Act, section 289). Apart from these appeals, as a general rule a person with sufficient interest may challenge the validity of an enforcement notice by way of judicial review, but no such challenge can be made on any of the seven grounds in section 174 otherwise than by appeal to the Secretary of State (section 285). So a person who does not appeal on one of

---

17 [1963] 1 All ER 459.

the grounds in section 174 and sleeps on his rights will lose the opportunity of challenging the enforcement notice on that ground altogether.[18] One possibility of challenge is in defence to a prosecution for non-compliance. However, a prosecution cannot be challenged if the enforcement notice is valid on its face and not quashed by way of appeal or judicial review.[19] The only exception is where the defendant did not have a copy of the notice served upon him, did not know and could not reasonably be expected to know the notice had been issued, and is substantially prejudiced thereby (section 285(2)).

(g) Non-compliance with an enforcement notice

It is a criminal offence to fail to comply with an enforcement notice (section 279). A person guilty of the offence is liable, on summary conviction, to a fine not exceeding £20,000 and on indictment to an unlimited fine. In determining the fine, the court must take account of any financial benefit accruing in consequence of the offence.

In addition to prosecution after failure to comply with an effective enforcement notice within the time specified in it, the authority also has the power to enter the land and carry out the steps prescribed by the notice, other than discontinuance of any use, and recover from the owner the net cost reasonably so incurred, who may in turn recover from the true culprit, if different, such reasonable expenditure on compliance (section 178).

Anyone wishing to take precautions against being subjected to enforcement procedures can, in suitable cases, cover themselves by applying for a certificate of lawfulness of existing or prospective use or development (sections 191–2). This is particularly useful when selling land which has no planning permission but benefits from existing use rights.

13. Amenity and safety

The other major concerns of planning law apart from development are amenity and safety. Amenity "appears to mean pleasant circumstances, features, advantages"; but it is not statutorily defined, nor is safety. The provisions of planning law governing amenity cover trees, buildings of special interest, advertisements, caravan sites and unsightly land; those governing safety.

(a) Protection of trees and special buildings

To grow or cut trees is not of itself development. But local planning authorities are specifically empowered, under the 1990 Act, Part VIII, "in the interests of

---

18 R v Challinor [2007] EWCA Crim 2527.
19 R v Wicks [1997] 2 All ER 801.

amenity", to make "tree preservation orders" (TPOs) for specified "trees, groups of trees or woodlands", restricting interference with the trees except with the consent of the local planning authority. Dangerous trees, however, may be cut if necessary. There are also provisions governing compensation and replanting. Unauthorised interference with any protected tree calculated to harm it is a criminal offence (1990 Act, Part VIII, section 210).

In *Perrin* v *Northampton Borough Council*,[20] the Court of Appeal held that, if a protected tree is causing a nuisance, such as subsidence next door, the possibility of engineering works as a viable alternative to "cutting down, uprooting, topping or lopping" the tree in question can justify refusing to permit any such physical interference with the tree as a cure for the "nuisance".

A TPO is made and confirmed by the local planning authority or the Secretary of State after considering any objections from owners and occupiers of the relevant land, though, if necessary, a provisional TPO taking immediate effect can be made for up to six months. Regulations are prescribed governing the procedure for making TPOs, and the content. Standard provisions in TPOs lay down essentially the same procedure for applying for consents to interfere with protected trees as exists for making planning applications.

"Amenity" is not expressly mentioned in relation to buildings of special interest, which are now governed by the Planning (Listed Buildings and Conservation Areas) Act 1990. Part II of the Act refers to "areas of special architectural or historic interest, and character or appearance of which it is desirable to preserve or enhance", and requires local planning authorities to determine where such areas exist and designate them as "conservation areas". When one of these areas has been designated, "special attention shall be paid to the desirability of preserving or enhancing the character or appearance of that area" by exercising appropriate powers to preserve amenities under planning legislation, and also by publicising planning applications for development which in the authority's opinion would affect that character or appearance. On this, see *South Lakeland District Council* v *Secretary of State for the Environment* (1992).[21] All trees in conservation areas are protected in the same way as if subject to a TPO, and no buildings in a conservation area can be demolished without conservation area consent.

The phrase "special architectural or historic interest" applies chiefly to buildings, although trees and other objects may affect their character and appearance. The Secretary of State has the duty, under Part I of the Planning (Listed Buildings and Conservation Areas) Act 1990, of compiling or approving lists of such buildings, after suitable consultations, and supplying local authorities with copies of the lists relating to their areas. Such authorities must notify owners and occupiers

20 [2008] 4 All ER 673.
21 [1992] 1 PLR 143.

of buildings included in (or removed from) these lists. The Secretary of State may, when considering any building for inclusion in a list, take into account the relationship of its exterior with any group of buildings to which it belongs and also "the desirability of preserving ... a man-made object or structure fixed to the building or forming part of the land and comprised within the curtilage of the building". If a building is not "listed", the local planning authority may give it temporary protection by a "building preservation notice" while they try to persuade the Secretary of State to list it.

It is a criminal offence to cause such a building to be demolished or altered "in any manner which would affect its character as a building of special architectural or historic interest", without first obtaining and complying with a "listed building consent" from the local planning authority or the Secretary of State, unless works have to be done as a matter of urgency. A consent may be granted subject to conditions, contravention of which is also a criminal offence; and it is normally effective for three years, although the LPA may impose a shorter or longer period if appropriate.

The procedure for applying for listed building consents, and for appeals and revocations, is laid down on lines very similar to the procedure in ordinary cases of planning permission for development; and so is the procedure for listed building enforcement notices and purchase notices. Compensation is payable for depreciation or loss caused by revocation or modification of listed building consent, and for the service of a building preservation notice that is not upheld by the Secretary of State. If an owner fails to keep a listed building in proper repair, a local authority or the Secretary of State may first serve a "repairs notice" and, if this is not complied with after two months, may then compulsorily purchase the property. Local authorities can, on seven days' notice to the owner, carry out urgent works at the owner's expense to preserve any unoccupied building, or part of a building, which is listed.

(b) Regulation of advertisement displays and caravan sites

Control of the display of advertisements is provided for, under section 220 of the Planning Act 1990, in the interests of amenity and safety, but not censorship. The details are laid down in the Town and Country Planning (Control of Advertisements) Regulations. The use of any land for the display of advertisements requires in general an application to the local planning authority for consent, which in normal cases is for a period of five years. Appeal lies to the Secretary of State. There are several categories of display in which consent is "deemed" to be given, including the majority of advertisements of a routine nature and purpose; but "areas of special control" may be declared where restrictions are greater. If the authority "consider it expedient to do so in the interests of amenity or public safety", they may serve a "discontinuance notice" to terminate the "deemed" consent of most kinds of advertisement enjoying such consent; but there is a right to appeal to the Secretary of State. Contravention of the regulations is a criminal offence. Consent under the regulations is also "deemed" to convey planning permission should any development be involved.

The control of caravan sites may also be regarded as a question of amenity. Such control involves questions of public health, and there is authority for the view that control for purposes of public health must not be exercised for purposes of amenity. But there can be little doubt in practice that although control is concerned with health and safety on the caravan site itself, it preserves amenity for the neighbourhood of the site.

Planning permission must be sought for caravan sites, but the detailed control of the use of each site is governed by a system of "site licences", obtainable from the local authority (Caravan Sites and Control of Development Act 1960).

There are two authorities which have power to control caravan sites. On the one hand, there is the planning authority, and on the other hand, there is the site authority. The planning authority ought to direct their attention to matters in outline, leaving the site authority to deal with all matters of detail. Thus the planning authority should ask themselves this broad question: Ought this field to be used as a caravan site at all? If Yes, they should grant planning permission for it, without going into details as to number of caravans and the like, or imposing any conditions in that regard.

Many considerations relate both to planning and to site. In all matters there is a large overlap, where a condition can properly be based both on planning considerations and on site considerations (per Lord Denning in *Esdell Caravan Parks Ltd* v *Hemel Hempstead Rural DC*).[22]

It is the "occupier" of land who must apply for a site licence, which must be *granted* if the applicant has the benefit of a specific planning permission, and *withheld* if the applicant has not; and it must last as long as that permission lasts, perpetually in a normal case. The practical question, therefore, is what conditions a site licence should contain. These are "such conditions as the authority may think if necessary or desirable to impose", with particular reference to six main kinds of purpose. Appeal may be made to a magistrate's court against the imposition of any conditions, and the court may vary or cancel any condition which is "unduly burdensome".

There are several categories of use of land for caravans which are exempted from control, and also additional powers conferred on local authorities in special cases.

(c) Unsightly land and hazardous substances

Next comes the question of unsightly land: neglected sites, rubbish dumps and the like. Local planning authorities are empowered, when "the amenity of any part of their area, or of any adjoining area, is adversely affected by the condition of land

22 [1965] 3 All ER 737.

in their area", to serve a notice on the owner and occupier, specifying steps to be taken to remedy the condition of the land (section 215 of the Town and Country Planning Act 1990). As with enforcement notices, two time limits must also be specified: a period (of 28 days or more) before the notice takes effect, and the time of compliance. Failure to comply is a summary offence (1990 Act, section 216 as amended).

Appeal lies, at any time before the notice takes effect, to a magistrate's court on any of the following grounds:

- the condition of the land is not injurious to amenity;
- the condition of the land reasonably results from a use or operation not contravening planning control;
- the specified steps for compliance are excessive; and
- the specified time for compliance is too short.

The magistrates may uphold, quash or vary the notice and "correct any informality, defect or error" if it is not material (1990 Act, section 217). A further appeal lies to the Crown Court (1990 Act, section 218), during which the notice is in suspense.

An additional set of controls over the use of land has been introduced into planning law by the Planning (Hazardous Substances) Act 1990 whereby the presence of a hazardous substance on, over or under land requires the consent of the hazardous substances authority unless "the aggregate quantity of the substance ... is less than the controlled quantity". The Secretary of State is empowered to define "hazardous substances" by specifying them in regulations, together with "the controlled quantity of any such substance".

Control of land, the use of which involves hazardous substances, is to be exercised whenever such substances are present in an appreciable amount ("controlled quantity"). The Act requires applications for "hazardous substances consents" to be made to "hazardous substances authorities" which are, by and large, the local planning authorities, including county councils where sites used for mineral workings or waste disposal are involved, and in most National Parks, as well as certain urban development corporations and housing trusts. Central government is also involved because the "appropriate ministers" are the authorities for the "operational land" of "statutory undertakers". The system of consents (with or without conditions), plus revocations, appeals, enforcement, etc., is broadly similar to planning control, and in fact was previously integrated with it; the purpose of the separation is to free this system of control of land use from being tied to the concept of "development" as against safety, which is the true consideration.

There is a requirement for "appropriate consultations" to take place with the Health and Safety Executive of the Health and Safety Commission.

Chapter 15

Development properties

1. Introduction

Development properties are properties whose value can be increased by capital expenditure, or by a change in the permitted use to a more valuable use, or by a combination of capital expenditure and change of use. These properties may be areas of undeveloped land likely to be in demand for future building purposes; they may be individual undeveloped sites in towns; or other properties where the buildings have become obsolete and are now ripe for refurbishment, change of use or redevelopment for the same or another use. The value, which in these cases is latent in the property, can only be released by development or refurbishment if the necessary planning consent is given.

2. Valuation approach

The residual method of valuation is the one most used for the valuation of development properties (Chapter 11). Direct sales comparison can be used where there is evidence of actual sales of comparable development properties.

The RICS Information paper 12 *Valuation of Development Land* provides a summary of the information needed for a full development valuation. In brief, the valuer will need to:

- Establish the most suitable scheme of development (which includes redevelopment or refurbishment) for the site for which planning permission is available or might reasonably be assumed; if assumed, all assumptions and conditions must be clearly stated as the valuation will be conditional on obtaining planning permission.
- Estimate the market value of the developed building or land when put to the proposed use.
- Consider the total development period, the construction period and the letting period if the building is to be let.
- Estimate the cost of carrying out the works required to put the property to the proposed use, together with such other items involved as legal costs and agent's commission on sales and purchases and fees for planning applications.

- Assess the cost of financing the development. Where the development is forecast to take a considerable time, or where the development is to be phased, the valuer must also consider the effect that the future may have on the market value of the developed building(s) and on the costs, and hence on today's land value.
- Establish the developer's profit margin to cover entrepreneurial risk.

Where data of recent similar transactions exist, the valuer may be able to use the comparative method of valuation. The valuer may do so even if it requires consideration of values outside the area in which the property is situated, for example sales of office sites in other towns with the same level of office rental values or residential sites with closely similar house prices. These indirect comparisons will affect the valuer's considerations. The variability in development sites is often greater than that for existing buildings and the use of the comparative method in these circumstances requires extra care and skill in adjusting for the differences, apart from the obvious one of site area. Even when the value can be established by comparison, it is sensible to check the value using the residual method. Since this is essentially a forecast of sales and expenditure, it is possible to set out the figures as a viability or feasibility report or study. Examples are given below, but these contain the same figures as used in a residual valuation.

In the residual method, errors made in any of the estimates will impact on the valuation, and so considerable skill is required when using this method if consistent and accurate results are to be obtained. Moreover, as the estimates of completed value and cost are very large compared to the existing value of the property, a small error in the estimates will create a large error in the residual.

At one time the residual method was the only method of valuation used for valuing development land, but direct comparison with other similar properties may be a more reliable method and is certainly preferred by the Lands Tribunal.[1] The Lands Tribunal is now the Lands Chamber of the Upper Tribunal, but colloquially it is still generally known as the Lands Tribunal. The residual method has been accepted by the tribunal in some cases,[2] particularly where no comparables are available, and it was accepted by the Court of Appeal when six residual valuations were submitted.[3] This was a case claiming negligence concerning a valuation of a development site, in which the valuations of six experts were submitted. As to the method of valuation to be adopted, it was said that "All are agreed that a residual valuation is the appropriate method for a site on which development is to be carried out by the purchaser".

---

1 *Fairbairn Lawson Ltd* v *Leeds County Borough Council* [1972] 222 EG 566; *South Coast Furnishing Co Ltd* v *Fareham Borough Council* [1977] 1 EGLR 167 and *Essex Congregational Union* v *Colchester Borough Council* [1982] 2 EGLR 178.
2 *Baylis's Trustees* v *Droitwich Borough Council* [1966] RVR 158; *Clinker & Ash Ltd* v *Southern Gas Board* [1967] RVR 477; *Trocette Property Co Ltd* v *Greater London Council* [1974] 28 P&CR 256.
3 *Nykredit Mortgage Bank plc* v *Edward Erdman Group Ltd* [1996] 1 EGLR 119.

The tribunal's criticism of the residual method is essentially that of the inherent weakness referred to above, coupled with the fact that valuations for the tribunal are not "tested" in the market. The quality of the inputs is probed by both sides in a willing buyer/willing seller scenario, but in the end both want a completed transaction; this probing is not possible in the non-market scenario of a reference to the Lands Tribunal.[4] In *Wood Investments*,[5] Mr John Watson, a member at that time of the Lands Tribunal said, with regard to a witness's residual valuation, that it:

> provides a telling illustration of its [the residual method's] uncertainties. The key figures are (a) the value of the completed buildings estimated at £265,537 and (b) the cost of providing it estimated at £230,210. £35,327, which is (a) less (b) is the land, but (a) and (b) are necessarily rough estimates and there must be some margin of error. If (a) turned out to be only 5% too high and (b) 5% too low the residual value of the land would be approximately £10,500 instead of approximately £30,000 and if the errors happen to be the other way round it would be over £60,000.

Where an optimistic view of values and costs is adopted (high values and low costs), a high land value will emerge, whilst a pessimistic view will produce a low value. If such opposing but genuinely held views are adopted by the parties to a dispute, then the arbitrator will be faced with markedly different values. This is the common experience of the tribunal and it results from the fact that no actual transaction is contemplated; it is only hypothesised. The residual method is thus mistrusted in such cases. Concerns were also raised in *Ridgeland Properties Ltd (Claimant* v *Bristol City Council (Acquiring Authority*[6]) where a residual approach was used by both parties. The Lands Tribunal, as it still was, noted that:

> This tribunal has repeatedly stressed its reluctance to use this method. Its enforced use in this reference does not mean that its faults are any the less; it remains a valuation method of last resort which is inherently very sensitive to even small changes in the input variables.

However, in commercial situations, where there is a willing buyer and a willing seller, a realistic view must be adopted and the residual method is accepted as a proper approach. In effect, the various participating parts of the valuation are "tested" by both sides and a compromise price or value for the land is usually reached.

---

4 *Liverpool and Birkenhead House Property Investment Co* v *Liverpool Corporation* [1962] RVR 162.
5 *Wood Investments Ltd* v *Birkenhead Corporation* [1969] RVR 137.
6 [2009] UKUT 102 (LC)

3. Types of development

At any given time there is a general demand for land for building purposes which is dependent upon international, national and local economic, social and political factors.

General factors affecting the demand for land include the national prosperity and demographic factors. In prosperous times there will be an increase in demand for sites for offices, shops, manufacture and other commercial and leisure purposes, and for an improved standard of housing, both in quality and quantity. When the population is increasing there will be a larger demand for homes.

Unless a valid planning consent is in existence, it will be necessary to obtain an indication of the sort of planning consent likely to be granted by informal discussion with the local planning authority before carrying out even a tentative valuation. If the existing consent is approaching the end of its life (it will only normally be valid for five years and sometimes less), then discussions with the local planners will be necessary in order to assess whether the consent will be renewed. In the course of such discussion any requirements for "planning gain" (sometimes more correctly known as section 106 payments) and section 278 works under the Highways Act 1980 will become apparent. For example, the planning authority may require the provision by the developer of children's play areas in residential developments or a community centre in a district shopping centre. The financial implications of such arrangements must be established and allowance made for any additional costs that may fall on the developer, including planning-related costs such as the Community Infrastructure Levy.

But simply because a certain type of development is provided for in the development plan or planning proposals for the area, it does not mean that land can be developed profitably. For instance, land may be allocated for manufacturing but unless manufacturers are willing to set up businesses in these areas the development of the land will not occur. Profitable development of any parcel of land will depend largely on user demand in the neighbourhood. The valuer must also consider general trends affecting development in the country as a whole.

There is a difference between a valuation for the purposes of a sale and a valuation for mortgage purposes. The former can be made conditionally upon the assumed planning consent being granted, whereas the latter can only be based on a planning consent which exists, or is so certain to be granted that there is no doubt that it will exist, unless of course the loan or mortgage valuation is conditional upon a consent being granted, when it can of course be given.

4. Viability statements (studies)

These are forecasts of the profits which will be earned from a development. They are based on forecasts of sales and costs of development. They can be presented in a number of different forms including that of profit and loss accounts. The form used in Example 15-1 is for a situation where the land cost is known and the profitability of the scheme is being tested.

Example 15–1
A developer is negotiating to buy a freehold site for a block of four flats and garages. It can be bought for £250,000 subject to contract and to outline planning consent. The flats will have a gross floor area of 70 m^2 each (including common parts). Comparable sales evidence suggests that selling prices will be £160,000 each, including a garage, for a 99-year lease at a ground rent of £100 per annum. A viability statement might be prepared along the following lines.

Viability statement

Sales:				
4 flats at £160,000 each. Ground rents: 4 × £100 p.a. at 7 Years' Purchase (YP)			£640,000	
			£2,800	£642,800
Costs:				
(a) Land costs:				
Land	£250,000			
Stamp duty on land at 1%	£2,500			
Legal costs on purchase of the site at 1%	£2,500	£255,000		
(b) Costs of development:				
Cost of building:				
280m^2 at £750 per m^2	£210,000			
4 garages at £4,000 each	£16,000			
Site preparation, approach road and gardens	£10,000	£236,000		
Architect's fee for plans (agreed)				
		£15,000	£506,000	
(c) Finance costs:				
Land costs £255,000 for, say, 12 months at 9%	£22,950			
Building costs £506,000 for, say, 6 months at 9%	£22,770	£45,270		
(d) Sale fees:				
Agent's commission on sale of flats (agreed)		£6,000		
Legal costs on sale of flats (agreed)	£3,000	£9,000	£54,270	£560,720
Estimated profit				£82,080

Developers can judge whether the profit level is sufficient or not by expressing the estimated profit as an initial yield on cost or on the total estimated realisation price (i.e. the Gross Development Value [GDV]). In this case it is 14.64% (£82,080/£560,270) × 100 = 14.64%). This figure can be used to compare one development with another of the same type. Alternatively, the profit can be expressed as a percentage of the total sales, £82,080 on £642,800 producing a figure of 12.77%. The profit per flat of £20,520 can be used as a further measure of viability.

Developers usually look for a profit of 12.5–15% of total sales or 15–20% on total costs. The profit on this scheme is probably just acceptable, depending on market activity and whether prices are rising faster than costs, or costs faster than values. If a developer was looking for a profit of 15% on sales, then a residual valuation could be completed as follows:

Sales – as before		£642,800
Costs:		
Cost of building – as before	£251,000	
Finance on building cost	£22,770	
Agent's commission	£6,000	
Legal costs	£3,000	
Profit – say 15.0% of sales	£96,420	£379,190
		£263,610
Less interest for 12 months at 9%		
(£263,610 – (£263,610/1.09))		£22,179
		£241,431
Less		
Stamp duty (SDLT) and legal costs on purchase		£4,734
at 2% (£241,431/1.02)		
Value of land		£236,697

The adjustment for SDLT is at 1% because at this stage in the calculation the expected residual looks as if it lies in the lowest SDLT rate. If the end result were above £250,000 or above £500,000, then the appropriate rates of 3% and 4% respectively would need to be used. If a 15% profit is to be achieved, then the developer will need to buy the site for £237,000 or less.

5. Factors affecting value

In dealing with a particular area of land the supply and demand for different uses of land in the area must be carefully considered.

Proper access to the land is essential. The proximity and availability of public services such as sewers, gas, electricity, water and telephone supplies, is also of major importance. The title must be checked for any restrictive covenants, and easements must be noted. The factors affecting development potential and value will vary in each case.

In the case of residential land these factors will include the proximity and quality of travel facilities, and the existence and proximity of shops, schools, places of worship and the like, or the possibility of the provision thereof in the future. Local employment conditions also affect demand. Regard must be had to the presence of open spaces, parks, golf courses and other leisure facilities, and the reservation of land under planning control for similar uses. The character of the neighbourhood must also be considered in determining the most suitable type or types of development.

The principal factor in the case of retail development is location. Where the land is in an established prime shopping location, it is important to determine whether there are proposals such as road schemes or major retail developments nearby which might draw shoppers away from that location. Edge-of-town developments, normally large retail warehouse schemes or new shopping centres, depend on access to major roads in the area and on the site being large enough to provide extensive customer car parking and of a scale to compete with existing centres.

Offices depend on several factors, such as access by road and public transport generally and an adequate supply of suitable labour. Many occupiers need to be close to other companies in the same area of business, as witnessed by the grouping of professional firms or those engaged in financial services. The success of an office scheme also depends on proximity to the established town centre. Office workers usually like to visit shops, banks, etc. and they like to travel on foot to these from their place of work. Many office workers have to fit their home requirements into their work regime at minimal cost and inconvenience. Non-centralised office developments can fail because they have not taken these requirements into consideration, although offices set in attractive landscaped areas have met with success.

In the case of business parks, warehouses and distribution centres and industrial developments, access by road and an adequate supply of suitable labour are the key factors. Proximity to ports and markets and to sources of power and materials are now less important than they were, due to the improvement of road and air transport facilities and national supplies of energy. But they can be very significant, as is the case with data storage facilities which require substantial and reliable sources of energy.

The physical state of the land, availability of services, etc. are also important factors to be taken into account. Until 1947, these were the only factors to be considered, but since then planning has become the overriding factor which the valuer has to consider. Reference has already been made to the RICS VIP 12, which is a useful source of checks and prompts for students and practitioners.

6. Residential development schemes

(a) Generally

When the site is small it may be comparatively simple to assess the best use for the land, having regard to general trends of development in the area, the factors affecting the particular piece of land under consideration and the planning position.

Example 15–2
What is the value of land fronting a residential road, made up and adopted by the local highways authority, and enjoying outline consent for residential

development? There are foul and surface water sewers in the road and all the utilities are available. The property has a frontage of 146 metres and a depth of 36 metres. This is the remaining vacant land in this road; the rest of the road frontage has been developed for houses currently selling at about £275,000 with plot frontages of about nine metres each. Similar plots in the area have sold at prices ranging from £100,000 to £120,000 each. There is a demand for houses of the character already erected and it is expected that all the plots could be built on and the houses disposed of within a year.

Valuation

The construction of 16 semi-detached houses appears to be the obvious best scheme of development. After inspecting the land and the area and taking into account the prices realised and the upward trend in values, it might be considered that a fair value per plot would be £120,000, giving a total value of £1,920,000 on the assumption that full permission for this form of development would be granted subject only to usual conditions.

In this instance the value has been estimated by comparison with the sale prices of other similar land in the area. A builder, before buying, would be wise to prepare a viability report, and a bank lending on the scheme will usually insist on such a report in order to check the probable rate of profit. A valuer advising on the sale or purchase price of the land would be wise to do the same as this may bring to light some factor that might otherwise be overlooked or given too little weight.

Viability statement

Sales: 16 houses at £275,000[1] each			£4,400,000
Costs: (a) Land	£1,920,000		
Stamp duty on land at 5%[2]	£96,000		
Legal costs on purchase of land	£19,200	£2,035,200	
(b) Costs of development:	£1,440,000		
Building costs [3]—16 houses of 90 m^2 each at £1,000 per m^2 including garages, say			
Plus plans[4]	£30,000	£1,470,000	
(c) Finance costs:			
Land costs £2,035,200 for 12 months at, say, 9%	£183,168		
Building costs £1,470,000			
For, say, 6 months at 9%[5]	£66,150		
	£249,318		
Say		£249,318	
(d) Sale fees:			
Agent's commission at £3,000[6] per house	£48,000		
Legal costs on sale, say	£16,000	£64,000	
		£3,818,518	
Say			£3,819,000
Estimated profit			£581,000

£581,000 represents 15.22% of the total costs of £3,819,000, or £36,313 per house, or 13.20% of the sale price. If this is a reasonable level of profit, then it would indicate to the developer that the land price is fair for a scheme of this scale in this location.

Notes
1.   The sale price of £275,000 is supported by reference to sales of similar houses in the same area. If prices are rising the developer might deliberately increase this estimate to allow for a continuation of increase, or reduce the profit requirement, which would be made up from the rising prices. This is the classic developers' predicament; if they adhere to house prices current at the date of the purchase of the land they might find that they are constantly losing opportunities due to higher bids from other developers. The sale price might also be increased to reflect the fact that these will be new houses in an area where new houses command a premium. The developer would probably do a number of such calculations within a price bracket. Valuers generally feel it is preferable to work on prices current at the date of purchase and with a lower profit margin and to regard any increase in prices obtained in the event as "super-profits". If the calculations are based on the possibility of rising sale prices, then consideration needs to be given to the possibility of rising costs.
2.   SDLT is at 5% on residential land and property sales between £1m and £2m.
3.   There is an obvious interdependence between the figures of selling prices – £275,000 – and building costs – £1,000 per m^2. The better the quality of the house offered, the higher will be the price realised and the higher the cost of building. The developers' object is to maximise the difference between the two figures through their expertise and skill in design and efficiency in sales organisation and building cost control.
4.   Most developers use stock plans or adapt plans they have used before, hence the reduced architect's fee. Where fresh plans have to be prepared, a larger figure would have to be allowed for. If the developer offers the designer too small a fee, the design and hence the saleability of the houses is likely to suffer. Many developers have their own in-house architects/designers so that the costs would be reflected in their profit targets with no specific allowance for architects' fees or plans. In deciding on the level of architects' fees, regard should be had to the number of times the design will be repeated on the site.
5.   Since proceeds of sales will start to be received in, say, six months because of a phased building scheme, it will not actually be necessary for the developer to borrow the whole of the estimated land and building costs for the whole period of the development (12 months). Under half the total figure for just over half the total period will provide a good approximation of the finance cost in this case.
6.   There is a similar interdependence between the selling agents' commission and the prices obtained. If selling agents are paid too little they will not use maximum efforts to sell. On the other hand, a developer is able to

negotiate better terms for multiple instructions, particularly if the market is buoyant.

In order to support this figure, a "cash flow" should be prepared. The cash flow in Example 15–2C (below) assumes the sale of two houses per month, starting in the sixth month, and three houses per month in the last two months. Such a cash flow would be useful to (and probably required by) any bank providing the money for the development.

Most valuations and appraisals of this kind are now computer based and the programs provide for the generation of cash flows. These programs will also allow for increases in selling prices and costs to be incorporated; and they can cater for a more accurate spread of building expenditure which typically takes the shape of an S-curve.

It may be necessary when valuing larger sites to prepare sketch plans of a suitable scheme of development, allocating specific areas for shops, flats and houses, probably with varying densities. Although the problem here is more complicated, the principles are the same.

A residual valuation for a simple scheme is relatively straightforward, but the greater the complexity, the greater is the need to create a site-specific spreadsheet or to make use of a development software package. There are several development programs available to valuers. Two popular versions are those provided by Kel and by Argus, previously known as Circle. The Academic version of Argus Developer 5.00.001 has been used to generate appraisal summaries and Excel spreadshhets for this chapter. Software packages offer the potential for itemising costs in more detail; each data screen allows the valuer to operate at a macro level or to "open up" or "drill down" to cover a much wider range of variables than would normally be included in a simple spreadsheet or paper-based residual. Reports, when generated, show all the variable factors used by the valuer. These packages can accommodate phased schemes and can generate sensitivity analysis. The latter allows the valuer to explore the "what if" question for any of the variables, e.g. what happens to the residual land value if costs go up by 10%. The use here of Argus Developer is not an endorsement per se, merely an illustration of some aspects of the flexibility and additional resource that such packages offer the valuer.

In Example 15-2A Argus uses a cash flow to assess the build-up of interest on construction on a month-by-month basis and so reflects here, and in the cash flows, the front weighting of the land purchase and the S-curve build-up of construction costs. This cash flow assumes no sales until completion, and so interest is payable on land purchase for a year.

The cash flow version of this, without phasing of house sales, is shown in Example 15-2B. Argus provides a variety of measures to give profit on cost of 16.56%; profit on GDV and NDV of 14.21%. In addition, the internal rate of return is given at 33.86% and, with interest running at 9%, all profit would be eroded in one year and nine months if there were no sales. The differences in profit and profit measures arise mainly from differences in the treatment of interest in Argus.

The message is that valuers must fully appreciate what a software package is doing when they use it to support opinions of value or viability.

Example 15-2C illustrates how simple it is to present a cash flow on any assumption as to phased sale of units. The impact on finance and on the profit and profit measures is significant. The Argus figures of profit rise from £625,051 to £714,452. Profit on cost is now 19.39% and on GDV it is 16.24%. This is due to the reduction in interest charges arising from the revenue flows starting in month six. A sensitivity analysis has not been shown for this example.

These examples illustrate some of the potential of such programs and the added information that can be generated to support the valuer's opinion and client advices.

(b) User of land

When preparing an outline scheme the extent of the various uses to which the land is to be put must be determined, and a layout prepared of the site showing the various plots, allowing for conditions relating to the density of development and

---

APPRAISAL SUMMARY   **ACADEMIC COPY**

Example 15-2A ARGUS Developer Summary Appraisal
of Example 15-2

Summary Appraisal for Phase 1

REVENUE

Sales Valuation	Units	Unit Price	Gross Sales	
Houses	16	£275,000	4,400,000	

NET REALISATION				4,400,000

OUTLAY

ACQUISITION COSTS

Fixed Price			1,920,000	
Stamp Duty		5.00%	96,000	
Legal Fee			19,200	
				2,035,200

CONSTRUCTION COSTS

Construction	m²	Rate m²	Cost	
Houses	1,440.00	£1,000.00	1,440,000	1,440,000

PROFESSIONAL FEES

Plans			30,000	
				30,000

DISPOSAL FEES

Sales Agent Fee	16.00 un	3,000.00 /un	48,000	
Sales Legal Fee	16.00 un	1,000.00 /un	16,000	
				64,000

FINANCE
Debit Rate 9.000% Credit Rate 0.000% (Nominal)

Land			156,928	
Construction			48,821	
Total Finance Cost				205,749

TOTAL COSTS				3,774,949

PROFIT				625,051

Performance Measures

Profit on Cost%	16.56%	
Profit on GDV%	14.21%	
Profit on NDV%	14.21%	
IRR	33.86%	
Profit Erosion (finance rate 9.000%)	1 yr 9 mths	

Example 15-2A Argus Developer Summary Appraisal of Example 15-2.

Heading	Category	Total	Jan 12	Feb-12	Mar-12	Apr-12	May-12	Jun-12	Jul-12	Aug-12	Sep-12	Oct-12	Nov-12	Dec-12
			1	2	3	4	5	6	7	8	9	10	11	12
Revenue														
Sale - Houses	2	4,400,000	0	0	0	0	0	0	0	0	0	0	0	4,400,000
Disposal Costs														
Sales Agent Fee	9	-48000	0	0	0	0	0	0	0	0	0	0	0	-48000
Sales Legal Fee	10	-16000	0	0	0	0	0	0	0	0	0	0	0	-16000
Acquisition Costs														
Fixed Price	86	-1920000	-1920000	0	0	0	0	0	0	0	0	0	0	0
Stamp Duty	14	-96000	-96000	0	0	0	0	0	0	0	0	0	0	0
Legal Fee	17	-19200	-19200	0	0	0	0	0	0	0	0	0	0	0
Construction Costs														
Con. - Houses	24	-1440000	-31131.58	-73605.26	-108184.21	-134868.42	-153657.89	-164552.63	-167552.63	-162657.89	-149868.42	-129184.21	-100605.26	-64131.58
Professional Fees														
Plans	38	-30000	-30000	0	0	0	0	0	0	0	0	0	0	0
Finance Details														
Total VAT paid		0	0	0	0	0	0	0	0	0	0	0	0	0
VAT recovered on cycle date		0	0	0	0	0	0	0	0	0	0	0	0	0
Net period total		830,800	-2096331.58	-73605.26	-108184.21	-134868.42	-153657.89	-164552.63	-167552.63	-162657.89	-149868.42	-129184.21	-100605.26	4,271,868
Period Total for Interest			-2096331.58	-2096331.58	-2169936.84	-2310118.07	-2444986.49	-2598644.38	-2818350.13	-2985902.76	-3148560.66	-3365575.18	-3494759.39	-3595364.65
Inflation Set 1. Rate pa = 0.000%			0	0	0	0	0	0	0	0	0	0	0	0
Interest Set 1. Debit Rate pa = 9.000% var.			9	9	9	9	9	9	9	9	9	9	9	9
Interest Set 1. Credit Rate pa = 0.000%			0	0	0	0	0	0	0	0	0	0	0	0
Total for Interest Set 1		-205748.74	0	-15722.49	-16274.53	-17325.89	-18337.4	-19489.83	-21137.63	-22394.27	-23614.2	-25241.81	-26210.7	0
Total Interest (All Sets)		-205748.74	0	-15722.49	-16274.53	-17325.89	-18337.4	-19489.83	-21137.63	-22394.27	-23614.2	-25241.81	-26210.7	0
Period Total For IRR		830,800	-2096331.58	-73605.26	-108184.21	-134868.42	-153657.89	-164552.63	-167552.63	-162657.89	-149868.42	-129184.21	-100605.26	4,271,868
Cumulative Total C/F		625,051	-2096331.58	-2185659.33	-2310118.07	-2462312.37	-2634307.67	-2818350.13	-3007040.39	-3192092.55	-3365575.18	-3520001.2	-3646817.16	625,051

Example 15-2B Argus Developer cash flow of Example 15-2.

Heading	Category	Total	Jan 12 / 1	Feb-12 / 2	Mar-12 / 3	Apr-12 / 4	May-12 / 5	Jun-12 / 6	Jul-12 / 7	Aug-12 / 8	Sep-12 / 9	Oct-12 / 10	Nov-12 / 11	Dec-12 / 12
Revenue														
Sale - Houses	2	4,400,000	0	0	0	0	0	550,000	550,000	550,000	550,000	550,000	825,000	825,000
Disposal Costs														
Sales Agent Fee	9	-48000	0	0	0	0	0	0	0	0	0	0	0	-48000
Sales Legal Fee	10	-16000	0	0	0	0	0	0	0	0	0	0	0	-16000
Acquisition Costs														
Fixed Price	86	-1920000	-1920000	0	0	0	0	0	0	0	0	0	0	0
Stamp Duty	14	-96000	-96000	0	0	0	0	0	0	0	0	0	0	0
Legal Fee	17	-19200	-19200	0	0	0	0	0	0	0	0	0	0	0
Construction Costs														
Con. - Houses	24	-1440000	-31131.58	-73605.26	-108184.21	-134868.42	-153657.89	-164552.63	-167552.63	-162657.89	-149868.42	-129184.21	-100605.26	-64131.58
Professional Fees														
Plans	38	-30000	-30000	0	0	0	0	0	0	0	0	0	0	0
Finance Details														
Total VAT paid		0	0	0	0	0	0	0	0	0	0	0	0	0
VAT recovered on cycle date		0	0	0	0	0	0	0	0	0	0	0	0	0
Net period total		830,800	-2096331.58	-73605.26	-108184.21	-134868.42	-153657.89	385,447	382,447	387,342	400,132	420,816	724,395	696,868
Period Total for Interest			-2096331.58	-2169936.84	-2310118.07	-2444986.49	-2634307.67	-2048644.38	-1714225.13	-1331777.76	-944435.66	-574232.37	121,583	845,978
Inflation Set 1. Rate pa = 0.000%			0	0	0	0	0	0	0	0	0	0	0	0
Interest Set 1. Debit Rate pa = 9.000% var.			9	9	9	9	9	9	9	9	9	9	9	9
Interest Set 1. Credit Rate pa = 9.000% var.			9	9	9	9	9	9	9	9	9	9	9	9
Total for Interest Set 1		-116348.29	0	-15722.49	-16274.53	-17325.89	-18337.4	-15364.83	-12856.69	-9988.33	-7083.27	-4306.74	912	0
Total Interest (All Sets)		-116348.29	0	-15722.49	-16274.53	-17325.89	-18337.4	-15364.83	-12856.69	-9988.33	-7083.27	-4306.74	912	0
Period Total For IRR		830,800	-2096331.58	-73605.26	-108184.21	-134868.42	-153657.89	385,447	382,447	387,342	400,132	420,816	724,395	696,868
Cumulative Total C/F		714,452	-2096331.58	-2185659.33	-2310118.07	-2462312.37	-2634307.67	-2264225.13	-1894634.45	-1517280.68	-1124232.37	-707723.32	17,583	714,452

Example 15-2C Argus Developer Cash Flow of Example 15-2 assuming phased sale of units.

site coverage. In the case of large residential estates it may be necessary to reserve land for: open spaces, a school or a church or community centre, and for similar local amenities. The scheme will need to be phased over a number of years on the basis of a realistic sales programme (see section 8 below).

It may be necessary, in relation to those areas allocated for commercial purposes, to allocate land for a petrol filling station, publichouse or neighbourhood shopping centre.

Even though a scheme is prepared primarily for the purpose of arriving at the value of the land, it is usual to discuss the scheme informally with the local planning authority. Such discussions will identify whether different densities should be allowed for in different parts of the land and what provision should be made for roads, open spaces, etc. Indeed, as the scheme is worked up, further valuations will be required to monitor the value implications.

(c) Infrastructure

Close attention must be given to the provision of infrastructure such as roads and services since these will represent a significant cost even for a modest scheme. This applies not only to the area of the scheme but also to off-site works.

The internal road layout should be such as to provide good and easy access to all parts of the estate. If there are any changes of level, regard must be had to the provision of easy gradients. So far as is practicable, plots of regular shape and of suitable size must be produced by the road layout, at the same time avoiding undue monotony and lack of amenity. Roads producing no building frontage should be kept to a minimum. The layout of soil and surface water sewers must be determined in relation to the available outfall. Roads and sewers will need to be constructed to the requirements of the local authorities prior to them being taken over (or "adopted") by the highway authority or statutory undertaker.

In addition to agreeing the internal road and sewer provisions with the appropriate authorities, it may well be necessary to agree off-site works. For example, the local road serving the site may need to be widened or a roundabout or traffic lights provided; the existing sewerage system might be inadequate and may need to be upgraded with larger pipes or by enlarging the sewage works. Such works will be carried out by the authorities with the developer bearing all or part of the costs.

In some cases it will not be possible to connect to a public sewer by gravity and allowance will have to be made for the cost of a pumping station. Surface water disposal might require the provision of a balancing pond, which could provide an attractive amenity feature.

Enquiries must be made as to the terms upon which supplies of gas and electricity can be obtained, and the costs of installing telephone cables must be discovered; the nearest mains or cables may be at some distance and supplies inadequate, all of which will increase infrastructure costs.

The nature of the soil and subsoils must be checked as these could affect building costs if, for example, piling is necessary to overcome poor land-bearing capability.

The possibility of ground contamination and potential areas of archaeological interest must be checked as these can add significantly to costs. Valuers must make adequate enquiries before making assumptions as to these matters and then they must state their assumptions.

(d) Environmental and other factors

Climate change and the growing awareness of issues such as ground contamination, flooding, built-up ground and environmental legislation simply mean that developers and their advisors have to spend more time assessing the potential of each site in the light of their investigations. Costs of ground clearance and cleaning can be very high and need to be included. Construction on flood plains or areas which have experienced flooding in recent years may require cooperation with environmental agencies over river control and water runoff requirements such as on-site storage of storm water. The need to create energy-efficient homes and to reduce the carbon footprint of buildings generally means that construction costs are now higher; developers and their advisors in all markets will be conscious of the potential additional development costs and benefits that may arise in the light of the Climate Change Act 2008 and the Energy Act 2008. They have to be fully aware of the changing mood of buyers in respect of the same issues; most buyers have been alerted to these problems by the suffering of owners whose homes have been flooded and who may spend a year or longer in temporary accommodation whilst insurers and builders deal with the task of repairs. Solicitors during the conveyancing process now request an environmental report, so it is as well for the valuer undertaking the development valuation to make similar enquiries.

These do not affect the form of a residual valuation or a viability study, but they do impact on costs and sale prices and in some cases may render land unsuitable for development.

7. Commercial development schemes

The same principles apply to commercial schemes as they do to residential schemes, save that the unit of supply is different; it may be an office block or a light industrial estate or a mixed scheme of shops, offices and multiscreen cinema.

The development period may be much longer than a year, so the consideration given to finance and interest charges becomes more significant. The effects of changing rents/yields and rising costs over a longer time frame take on greater importance. Some developments will be phased and others will be single phase.

Most development calculations are prepared on computers. The programs used are very sophisticated and allow for interlocking phasing as well as overlapping phases. A consideration of the range of programs available is beyond the scope of this book, but for illustration purposes the Argus-generated examples from the 10th edition are retained. The individual nature of some schemes is such that occasionally it may be preferable to develop a scheme-specific spreadsheet.

8. Period of development

There are two main factors that determine the speed of development: the time taken to construct the building or buildings and the rate at which the completed buildings can be sold or let.

In the case of a small housing estate, it may be reasonable to assume that each house will be sold prior to completion so that no deferment of development costs and sale proceeds is necessary. However, in the case of an estate of several hundred houses, it will be necessary to estimate the rate at which the market will take up the houses and to assume that the development will be appropriately phased over a period of years. This will depend on a number of factors, including the strength of demand in the area and any competing developments being carried out. The easiest way to value a large area of land of this kind is to consider first the value of the land for an average phase, say, of 50 houses. If the total scheme envisaged 300 houses there would be six phases, but, because of the holding cost of the land, its current value could not be six times the first phase value. The value (in the absence of any assumptions as to increasing sales prices and costs) would have to be reduced by the compound interest cost of each phase, as shown below.

Assume that the land value per phase is £3m, each phase lasts one year, and the holding cost is 10% p.a.:

Phase 1 land		£3,000,000
Phase 2 land	£3m	
× PV of £1 in 1 yr at 10%	0.91	£2,730,000
Phase 3 land	£3m	
× PV of £1 in 2 yrs at 10%	0.83	£2,490,000
Phase 4 land	£3m	
× PV of £1 in 3 yrs at 10%	0.75	£2,250,000
Phase 5 land	£3m	
× PV of £1 in 4 yrs at 10%	0.68	£2,040,000
Phase 6 land	£3m	
× PV of £1 in 5 yrs at 10%	0.62	£1,860,000
		£14,370,000

It may not be possible to buy at this price because it assumes that land values will remain constant for six years. It may therefore be necessary to build in an inflation factor.

Sometimes the buyer of a large site will sell off parcels of serviced land to other developers in order to recoup some of the initial costs of providing the basic infrastructure. However, the introduction of different developers with their own design standards can add variety to the house types and layouts on offer and so give an impetus to interest in the whole scheme with benefits to the developers through increased take-up of houses. Developers might even join together from the start and form a joint venture consortium to carry out the scheme.

9. Incidental costs

In many schemes additional expenses will be incurred beyond the bare constructional costs of roads and sewers. One typical cost is that of landscaping; such costs must be incorporated.

Professional services are required for the preparation of detailed layouts, drawings and specifications for constructional works; fees for this may vary from 4% to 12.5% on the cost of the works, depending on the amount of professional input required. Even where a developer has an in-house design department, some independent professional advice may still be required, particularly from highway and civil engineers. In the case of large and contentious schemes where the planning permission must be pursued through the appeals procedure with a public hearing, the fees of planning lawyers and consultants can amount to very large sums.

Legal charges may be incurred on the purchase of the land (and stamp duty) and on the sales. The amount of these charges will depend on whether the land is registered or unregistered. Further legal charges may be incurred where planning, highway and other legal agreements are required, including financing agreements.

An estimate of the cost of advertisements and commission on sales payable to agents must be made, including the costs, where appropriate, of furnishing and staffing a show house.

10. Site assembly

The acquisition of land for large schemes almost invariably involves the purchase of several land holdings. The problem for developers is that they are usually buying land without planning permission. They will be unwilling, and probably unable, to commit large sums of money without the certainty of being able to carry out development and so justify the expenditure.

The normal solution to this problem is for the parties to enter into option agreements. These may be call options, whereby the developer can require the owner to sell the land on the grant of planning permission, or put and call options, whereby either party can require the other to carry out the sale/purchase of the land.

The option agreement must cover many matters, including the means of determining the price payable. Sometimes this will be at a stated figure, although when land values are rising this is unfavourable to the landowner. Alternatively, the price will be the market value, to be determined at the time of the exercise of the option. Commonly, the price will be a percentage of market value, say 70–90%, to reflect the costs of assembly and of obtaining planning permission, which will be borne by the developer.

It is essential that the provisions for determining the price should be set out to enable the valuers for the parties to have a clear understanding of the approach to the valuation. For example, where the price is to be assessed at an agreed price per hectare, it must be clear whether this is the gross area or some lesser area, such as the net developable area, which needs to be defined. Where it is price per plot, how the valuer should deal with plots which straddle the boundaries of land ownership needs to be stated. If a landholding controls the access to other land, the valuer needs to know whether the special value of this strategic land – "ransom value" – must be reflected. In options where a developer may exercise the option without planning permission having been granted, the planning assumptions for the valuation should be present. If, as is usual, the developer has to bear the cost of extensive and expensive infrastructure, the valuer needs to know how this factor should be reflected in the valuation.

Considerable care is required in the preparation of an option if extra complications, in what is in any event a difficult and contentious area of valuation, are to be avoided.

In urban areas many sites are assembled by a developer buying properties at existing-use value and then letting them on short-term leases with rebuilding clauses. In these cases the holding costs are met wholly or partly out of the rents.

11. Hope value

The development value of land can be identified by the existence of a planning permission. However, planning permissions do not appear out of the blue so that, on one day, a planning permission for residential development is given and the land acquires a residential development value. What happens in practice is that a view is taken as to the likelihood of planning permission being given, either now or at some time in the future. If the likelihood is nil, such as prime agricultural land in the Green Belt in an Area of Outstanding Natural Beauty and with rare wild flowers found nowhere else, it is unlikely in the extreme that anyone would pay more than agricultural value.

In other cases, whilst it is agricultural land with no immediate prospect of development, it may be felt that in a few years, when, say, a proposed new road has been built nearby, some form of development might be allowed. In such cases buyers might be found who will pay above agricultural value in the hope that, after a few years, they will realise development value and make a large profit. This price, above a value that reflects only the existing use but below full development value, is termed hope value.

A valuation to determine hope value is often impossible other than by adopting an instinctive approach, particularly in the stages when the hope of permission is remote; it can only be a guesstimate of the money a speculator would be prepared to pay. As the hope crystallises into reasonable certainty of a permission being granted, a valuation can be attempted, based on the potential development value deferred for the anticipated period until permission will be forthcoming, but with some end deduction to reflect the lack of certainty. Indeed, since most developers will buy only when permission is certain (preferring an option to buy or a contract conditional on the grant of permission before certainty has been reached), any sale in the period of uncertainty will probably require a significant discount on the full hope value.

In some cases there may be little risk that planning consent will be refused, e.g. with an infill site for one house in a whole road of houses and where there is no view that will be obstructed by the development. In such a case the market may assume that consent will undoubtedly be granted and there will be little if any discount from the development value.

12. Urban sites

(a) New development

The previous sections of this chapter have been concerned mainly with land, not built upon, in the vicinity of existing development, commonly referred to as greenfield sites.

In the centre of towns it is often necessary to consider the value of a site that has become vacant following the demolition of buildings, sometimes termed a brownfield site, or to value a site occupied by obsolete buildings which do not utilise the site fully.

The general method of approach suggested in section 2 can be used, i.e. to value by comparison with sales of other sites by reference to an appropriate unit, e.g. per square metre or metre frontage, and to check the result by drawing up a viability statement or carrying out a valuation by the residual method. For example, with office sites, there is a correlation between market rent, building cost and site value. In areas with closely similar market rents and site conditions, an office site might be valued at £x per gross square metre of offices to be built. Thus, in an area where offices are letting at £160 per m^2, analysis of site sales might show that prices represent £1,000 per m^2 of the gross floor area of the offices to be built. So a site with permission to build 5,000 m^2 gross of offices would be valued at £5 million.

As in the case of residential estates, the type of property to be erected and the use to which it can be put when completed is entirely dependent upon the planning permission being granted. An indication will probably be obtained by an inspection of the development plan and discussion with the local planning authority.

Sites are often restricted as to user, height of buildings, and the percentage of the site area that may be covered at ground floor and above, and by conditions imposed when planning permission is granted. Car parking standards can have a significant effect on the size of building that can be erected. There may be further restrictions on use, owing to the existence of easements of light and air, or rights of way.

A site may be bare but may contain old foundations, or may be contaminated, necessitating high clean-up costs, whilst on the other hand it may be possible to take advantage of existing drains.

Costs of development may be significantly increased by the presence of underground water or the difficulty of access for building materials in a crowded and busy thoroughfare.

Factors of this kind must be carefully considered both in relation to any suggested scheme of development and also when comparing two sites which apparently are very similar but which are subject to different restrictions or conditions.

The existence of restrictive covenants may reduce the value of a site, although the possibility of an application for their modification or removal under section 84 of the Law of Property Act 1925 (as amended by the Landlord and Tenant Act 1954, section 52, and by the Law of Property Act 1969) must not be lost sight of. It is frequently possible to insure against restrictions being enforced where these are contained in old documents and where there is considerable doubt as to whether they are any longer extant. Where applications are made to the Lands Tribunal it must be remembered that the tribunal can award compensation to the person having the benefit of the covenant. In the case of restrictions preventing the conversion of houses, it may be possible to take similar action to that under section 84 of the Law of Property Act 1925, under the Housing Act 1985 (section 610). The alternative to a section 84 application is to buy out the restriction by agreement with those who have the benefit of the covenants or rights.

Where possession of business premises is obtained against tenants in occupation at the termination of their leases for purposes of redevelopment under section 30 of the Landlord and Tenant Act 1954, the amount of compensation to be paid must be deducted as part of the costs of development (see Chapter 17).

As in the case of a residential estate, the value of urban sites is best found by direct comparison using an appropriate unit; but where comparables are few, or are complicated by a mix of uses, then the residual method is usually adopted.

(b) Refurbishment

In urban areas, development can take the form of refurbishment rather than redevelopment. In conservation areas, or with listed buildings, this may be the only possible form of development. Even in other cases it may be found that it is more profitable to modernise a property rather than to demolish and replace it.

When considering refurbishment, it is necessary to consider the need for planning permission. This may be needed because alterations will be made to the external appearance, such as inserting new windows or making additions such as plant rooms on the roof, or because the building is listed as being of architectural importance or is in a conservation area. Planning permission will be needed for a change of use, e.g. from office to residential.

The other important issue is to estimate the cost of the refurbishment works. A typical refurbishment scheme of an office building would include installation of suspended ceilings and raised floors where possible, upgrading of all services including heating, toilets and lifts, re-design of the entrance hall and common parts, and even the re-cladding of external parts. Similar considerations apply to blocks of flats, shopping centres and other buildings. Unlike new development, where building costs may be made available by comparison with other new buildings, each refurbishment is unique and needs the services of a building cost surveyor to estimate the costs involved.

A common practice is for planning authorities to allow new buildings subject to the retention of existing façades. This film-set architecture produces a hybrid new development/refurbishment scheme which draws upon the elements of both.

Since refurbishment is a form of development, the residual method of valuation is appropriate in valuing such property. Indeed, the valuer will often need to prepare a residual valuation assuming refurbishment and also one assuming redevelopment. A comparison of the valuations will provide a strong indication of which approach is to be favoured.

Example 15–3

A freehold office building erected 30 years ago is on five floors with 400 m^2 NIA on each floor. There is surface car parking for 40 cars. The site area is 1,200 m^2. The property is occupied by a tenant whose lease has two years to run at a current rent of £100,000 p.a.; the building lacks modern amenities. The rateable value is £100,000.

It would cost £2,000,000 to bring the property up to current standards The planning authority has indicated that a new office building would be permitted at a plot ratio of 25:1, producing 3,000 m^2 gross, 2,300 m^2 NIA, with 15 car spaces at basement level. The market rent would be £150 per m^2.

What is the value of the freehold interest, assuming the refurbishment would take eight months and the new build 18 months?

Valuations
Here it is necessary to establish the value assuming refurbishment and assuming redevelopment.

A. *Refurbishment*
GDV:

Offices, 2,000m^2 at £150 per m^2		£300,000 p.a.	
Car spaces, 40 spaces at £500 per space p.a.		£20,000	
		£320,000 p.a.	
YP or PV£1p.a. in perp. at 6.75%		14.8148	say £4,740,740
Less			
Building costs			
Estimated costs	£2,000,000		
Add			
Fees at say 10%	£200,000	£2,200,000	
Finance:			
Assume 8-month scheme £2,200,000 for 4 months at 9%, say		£66,000	
Letting costs, say 15% of £320,000		£48,000	
		£2,314,000	
Developer's profit at 15% of £4,740,740		£711,111	£3,025,111
Land balance including acquisition costs and interest			£1,715,629

Less		
Interest for 8 months at 9% (PV £1)	0.9441674	
Land balance including acquisition costs (5.75%)	£1,619,841	(a)
Less		
Acquisition costs at 5.75%	£88,076	
Land value	£1,531,765	(b)
Value assuming refurbishment, say	£1,530,000	(c)

Notes

(a) This figure can be found by multiplying the land value, including interest, by the PV of £1 for the development period, which will automatically allow for compound interest.

(b) This figure is calculated direct by dividing the land balance including acquisition costs by 1.0575. Note that this figure includes 4% for SDLT and is dependent on the probable rate of SDLT. The figure of 5.75% has been retained for these examples, although technically some would argue that these should now be 5.80% now that VAT on fees is at 20%. But a counter argument is that fee competition during the recession may have reduced the 1.5% plus VAT to 1% plus VAT, which emphasises just how important, for a MV calculation, it is to use market-comparable and supportable evidence for all variables.

(c) If the developer's profit is calculated at a percentage of costs, then a further allowance for profit may be made against the residual; if at 15%, then the land balance would be divided by 1.15. If this adjustment is not made, then the developer is only working on a margin equivalent to a contractor's profit margin (£1,790,000/1.15 = £1,556,521). Most descriptions of the method show a single adjustment for profit as a percentage of GDV.

B. New development

GDV:		£575,000		
Offices, 2,300 m² at £250 per m²				
Car spaces, 15 spaces at £600 per space p.a.		£9,000	p.a.	
Say		£584,000	p.a.	
YP or PV£1p.a. in perp. at 6.25%		16.0		£9,344,000
Less Building costs, 3,000 m² at £1,500 per m²	£4,500,000			
Basement, say 400 m² at £800 per m²	£320,000			
	£4,820,000			
Add				
Fees at 12%	£578,400	£5,398,400		
Finance		£372,472		
Assume 18-month scheme, £5,398,000 for 9 months at 9%				

Letting costs, say 15% of £584,000	£87,600	
Developer's profit at 15% of £9,344,000	£1,401,600	£7,260,072
Land balance including acquisition costs and interest		£2,083,928
Less		£252,693
Interest for 18 months at 9% p.a.		
Land balance including acquisition costs (5.75%)		£1,831,235
Less		£99,570
Acquisition costs (divide by 1.0575)		
Land value		£1,731,664
Say		£1,725,000

Value

Redevelopment and refurbishment are of similar value. There is a difference of about £200,000 but, as redevelopment is often simpler and easier to cost than refurbishment and shows the higher land value, it is reasonable to suggest that the choice will be to redevelop. The final decision will be influenced by the resources available and the state of the market at the end of the lease, and will have regard to the degree of confidence the valuer has in the estimates of each variable.

Valuation

Term – rent reserved	£100,000 p.a.	
YP or PV£1p.a. for 2 yrs at 10%	1.74	£180,000
Reversion – to development value, say	£1,700,000	
Less Compensation to tenants, say 2 × RV £100,000[1]	£200,000	
	£1,500,000	
PV £1 in 2 yrs at 6.75%	0.88	£1,320,000
		£1,500,000
Value, say		£1,500,0002

Notes

1. The requirement to pay compensation to business tenants only applies where the tenancy has not been contracted out of the Landlord and Tenant Act 1954 as amended.
2. In practice, valuations should also be made on the basis that the building would be redeveloped, or be modernised to lesser standards at reduced costs of improvement, and a check made to assess the value, if any, if no action is taken. Here the future continuation of any tenancy at any rental for any significant period of time is doubtful. If left unmodernised, the office rent might continue at £100,000 for a few more years, but even if perpetuity is assumed the property will still have a greater value if sold for modernisation or redevelopment. The property is at the end of its economic life in its present condition.

If the refurbishment work could be carried out with the tenants remaining in occupation, it could bring refurbishment back into consideration.

The differences between the Argus Developer solutions – Examples 15-3A and 15-3B – and the paper-based calculations arise from the more correct assessment of interest and the use of 5.80% for acquisition costs.

The advantage of the cash flow (see Example 15-3C) is that it allows an opportunity to view the expected expenditure on a month-by-month or quarter-by-quarter

APPRAISAL SUMMARY	ACADEMIC COPY

Example 15-3A ARGUS Developer Summary Appraisal for Ex 15-3 showing the value for refurbishment as £1,574,751

Summary Appraisal for Phase 1

REVENUE

Rental Area Summary

	Units	m²	Rate m²	Initial MRV/Unit	Net Rent at Sale	Initial MRV
Offices	1	2,000.00	£150.00	£300,000	300,000	300,000
Car Spaces	40			£500	20,000	20,000
Totals	41	2,000.00			320,000	320,000

Investment Valuation
Offices

Current Rent	300,000	YP @	6.7500%	14.8148	4,444,444	
Car Spaces						
Current Rent	20,000	YP @	6.7500%	14.8148	296,296	
					4,740,741	

NET REALISATION   4,740,741

OUTLAY

ACQUISITION COSTS

Residualised Price		1,574,751	
Stamp Duty	4.00%	62,990	
Agent Fee	1.02%	16,062	
Legal Fee	0.78%	12,283	
			1,666,087

CONSTRUCTION COSTS

Construction	m²	Rate m²	Cost	
Offices	2,000.00	£1,000.00	2,000,000	**2,000,000**

PROFESSIONAL FEES

Other Professionals	10.00%	200,000	
			200,000

MARKETING & LETTING

Letting Agent Fee	15.00%	48,000	
			48,000

FINANCE
Debit Rate 9.000% Credit Rate 0.000% (Nominal)

Land		76,009	
Construction		39,534	
Total Finance Cost			115,543

TOTAL COSTS   4,029,630

PROFIT   711,111

Performance Measures

Profit on Cost%	17.65%
Profit on GDV%	15.00%
Profit on NDV%	15.00%
Development Yield% (on Rent)	7.94%
Equivalent Yield% (Nominal)	6.75%
Equivalent Yield% (True)	7.04%
IRR	59.22%
Rent Cover	2 yrs 3 mths
Profit Erosion (finance rate 9.000%)	1 yr 10 mths

Example 15-3A Argus Developer summary appraisal of Example 15-3 showing the value for refurbishment as £1,575,496.

APPRAISAL SUMMARY

Example 15-3B ARGUS Developer Summary Appraisal for
Example 15-3 New Build Alternative value £1,814,742

Summary Appraisal for Phase 1

REVENUE

Rental Area Summary

	Units	m²	Rate m²	Initial MRV/Unit	Net Rent at Sale	Initial MRV
Offices	1	2,300.00	£250.00	£575,000	575,000	575,000
Car Spaces	15			£600	9,000	9,000
Totals	16	2,300.00			584,000	584,000

Investment Valuation
Offices

Current Rent	575,000	YP @	6.2500%	16.0000	9,200,000
Car Spaces					
Current Rent	9,000	YP @	6.2500%	16.0000	144,000
					9,344,000

NET REALISATION   9,344,000

OUTLAY

ACQUISITION COSTS

Residualised Price		1,814,742	
Stamp Duty	4.00%	72,590	
Agent Fee	1.02%	18,510	
Legal Fee	0.78%	14,155	
			1,919,997

CONSTRUCTION COSTS

Construction	m²	Rate m²	Cost	
Offices	3,000.00	£1,500.00	4,500,000	
Car Spaces	400.00	£800.00	320,000	
Totals	3,400.00		4,820,000	4,820,000

PROFESSIONAL FEES

Other Professionals	12.00%	578,400	
			578,400

MARKETING & LETTING

Letting Agent Fee	15.00%	87,600	
			87,600

FINANCE
Debit Rate 9.000% Credit Rate 0.000% (Nominal)

Land	242,153	
Construction	294,249	
Total Finance Cost		536,403

TOTAL COSTS   7,942,400

PROFIT
                                                        1,401,600

Performance Measures

Profit on Cost%	17.65%
Profit on GDV%	15.00%
Profit on NDV%	15.00%
Development Yield% (on Rent)	7.35%
Equivalent Yield% (Nominal)	6.25%
Equivalent Yield% (True)	6.50%
IRR	30.10%
Rent Cover	2 yrs 5 mths
Profit Erosion (finance rate 9.000%)	1 yr 10 mths

Example 15-3B Argus Developer summary of new build alternative. Value for this purpose is £1,815,600.

basis. When interest rates are high or the development period is long, it can be important to try and structure payments to reduce interest charges.

Two versions of a sensitivity analysis are shown as Example 15-3D and Example 15-3E. These are self-explanatory and illustrate the sensitivity of a residual to a small 5% change to the specified variables. This is a useful tool for the valuer and, generally, an expectation when presenting a valuation of land as security for a loan or as part of an application to a bank or financial institution for funding of a development project.

Example 15-3C Argus Developer cash flow of new build.

Heading	Category	Total	Jan-12 (1)	Feb-12 (2)	Mar-12 (3)	Apr-12 (4)	May-12 (5)	Jun-12 (6)	Jul-12 (7)	Aug-12 (8)	Sep-12 (9)	Oct-12 (10)	Nov-12 (11)	Dec-12 (12)	Jan-13 (13)	Feb-13 (14)	Mar-13 (15)	Apr-13 (16)	May-13 (17)	Jun-13 (18)
Revenue																				
Cap - Car Spaces	4	144,000	0	0	0	0	0	0	0	0	0	0	0	0	0	0	0	0	0	144,000
Cap - Offices	4	9,200,000	0	0	0	0	0	0	0	0	0	0	0	0	0	0	0	0	0	9,200,000
Acquisition Costs																				
Residualised Price	12	-1814742.13	-1814742.13	0	0	0	0	0	0	0	0	0	0	0	0	0	0	0	0	0
Stamp Duty	14	-72589.69	-72589.69	0	0	0	0	0	0	0	0	0	0	0	0	0	0	0	0	0
Agent Fee	16	-18510.37	-18510.37	0	0	0	0	0	0	0	0	0	0	0	0	0	0	0	0	0
Legal Fee	17	-14154.99	-14154.99	0	0	0	0	0	0	0	0	0	0	0	0	0	0	0	0	0
Construction Costs																				
Con - Offices	24	-4500000	-48891.33	-111537.52	-166873.78	-214600.1	-255616.47	-280022.9	-315119.4	-333905.95	-345382.55	-349549.22	-346405.95	-335952.73	-318189.57	-293116.47	-260733.43	-221040.45	-174037.52	-119724.66
Con - Car Spaces	24	-320000	-3476.72	-7931.56	-11866.56	-15281.78	-18177.17	-20552.74	-22408.49	-23744.42	-24560.54	-24856.83	-24633.31	-23889.97	-22626.81	-20843.84	-18541.04	-15718.43	-12376	-8513.75
Professional Fees																				
Other Professionals	38	-578400	-8284.17	-14336.29	-21448.84	-27621.83	-32855.24	-37149.08	-40503.35	-42918.04	-44393.17	-44928.73	-44524.71	-43181.12	-40897.97	-37675.24	-33512.94	-28411.07	-22369.62	-15388.61
Marketing/Letting																				
Letting Agent Fee	41	-87600	0	0	0	0	0	0	0	0	0	0	0	0	0	0	0	0	0	-87600
Finance Details																				
Total VAT paid		0	0	0	0	0	0	0	0	0	0	0	0	0	0	0	0	0	0	0
VAT recovered on cycle date		0	0	0	0	0	0	0	0	0	0	0	0	0	0	0	0	0	0	0
Net period total	1,938,003	-1978649.38	-133805.37	-200189.21	-257803.71	-306648.88	-346724.72	-378031.23	-400568.41	-414336.26	-419334.78	-415563.97	-403023.82	-381714.35	-351635.55	-312787.41	-265169.95	-208783.15	9,112,773	
Period Total for Interest		-1978649.38	-1978649.38	-2112454.76	-2343327.24	-2601130.95	-2907779.83	-3313396.34	-3691427.57	-4091995.98	-4589558.39	-5008863.17	-5424457.14	-5940152.78	-6321667.13	-6673502.68	-7128306.51	-7393476.46	1,741,740	
Inflation Set 1. Rate pa = 0.000%			0	0	0	0	0	0	0	0	0	0	0	0	0	0	0	0	0	0
Interest Set 1. Debit Rate pa = 9.000% var.			9	9	9	9	9	9	9	9	9	9	9	9	9	9	9	9	9	9
Interest Set 1. Credit Rate pa = 0.000%			0	0	0	0	0	0	0	0	0	0	0	0	0	0	0	0	0	0
Total for Interest Set 1		-536402.82	0	-14839.87	-15843.41	-17574.95	-19508.48	-21808.35	-24850.47	-27685.71	-30089.97	-34421.69	-37566.7	-40683.43	-44561.15	-47414	-50051.27	-53462.3	-55451.07	0
Total Interest (All Sets)		-536402.82	0	-14839.87	-15843.41	-17574.95	-19508.48	-21808.35	-24850.47	-27685.71	-30089.97	-34421.69	-37566.7	-40683.43	-44561.15	-47414	-50051.27	-53462.3	-55451.07	0
Period Total For IRR	1,938,003	-1978649.38	-133805.37	-200189.21	-257803.71	-306648.88	-346724.72	-378031.23	-400568.41	-414336.26	-419334.78	-415563.97	-403023.82	-381714.35	-351635.55	-312787.41	-265169.95	-208783.15	9,112,773	
Cumulative Total C/F	1,401,600	-1978649.38	-2127294.63	-2343327.24	-2618705.9	-2944863.27	-3313396.34	-3716278.04	-4144532.16	-4589558.39	-5043314.86	-5496445.53	-5940152.78	-6360418.28	-6765467.83	-7128306.51	-7448938.76	-7711172.98	1,401,600	

**Example 15-3D ARGUS Developer Summary Appraisal for
Ex 15-3 showing the value for refurbishment as £1,574,751**

Table of Land Cost and IRR%

Rent: Rate pm²	Construction: Rate pm²				
	-10.000%	-5.000%	0.000%	+5.000%	+10.000%
	900.00 pm²	950.00 pm²	1,000.00 pm²	1,050.00 pm²	1,100.00 pm²
-20.000%	(£1,102,350)	(£1,001,130)	(£899,909)	(£798,689)	(£697,469)
120.00 pm²	61.9619%	63.6746%	65.5070%	67.4723%	69.5863%
-10.000%	(£1,439,771)	(£1,338,550)	(£1,237,330)	(£1,136,110)	(£1,034,890)
135.00 pm²	59.0524%	60.4107%	61.8478%	63.3711%	64.9888%
0.000%	(£1,777,192)	(£1,675,971)	(£1,574,751)	(£1,473,531)	(£1,372,311)
150.00 pm²	56.9215%	58.0420%	59.2181%	60.4544%	61.7556%
+10.000%	(£2,114,613)	(£2,013,392)	(£1,912,172)	(£1,810,952)	(£1,709,732)
165.00 pm²	55.2929%	56.2439%	57.2362%	58.2726%	59.3563%
+20.000%	(£2,452,034)	(£2,350,813)	(£2,249,593)	(£2,148,373)	(£2,047,152)
180.00 pm²	54.0075%	54.8321%	55.6884%	56.5785%	57.5045%

Sensitivity Analysis : Assumptions for Calculation

Construction: Rate pm²
Original Values are varied by Steps of 5.000%.

Heading	Phase	Rate	No. of Steps
Offices	1	£1,000.00	2 Up & Down

Rent: Rate pm²
Original Values are varied by Steps of 10.000%.

Heading	Phase	Rate	No. of Steps
Offices	1	£150.00	2 Up & Down

Example 15-3D Illustrating the effect on land price of 5% steps in construction cost, 10% steps in rents, no change in yield.

13. Ground rents

A common feature in the development process is the release of land to developers by landowners who wish to retain an interest in, and some control over the future use of, the land to be developed. This is achieved by the grant of a ground lease to the developer.

Ground leases have been a common feature of the development process for several centuries. The best known examples from the past are the large estates controlled by families or charities whereby the overall estate remained in the ownership of the estate owner who controlled the development of the estate in accordance with a general estate plan. The Grosvenor and Cadogan Estates in Central London, Alleyn's Estate in Dulwich, the Calthorpe Estate in Birmingham and the Crown Estate generally are typical examples. Control was, and still is, exercised by the granting of ground leases which impose restraints on the manner in which any parcel of land is to be developed for the benefit of the estate at large.

More recently, the ground lease has been used by local authorities and new town corporations whereby areas of land have been developed by the private sector whilst control of such development has been exercised through the landlord's powers under the ground lease, notwithstanding the powers they may already have in their capacity as planning authorities.

Ground leases, however, are not confined to large-scale estate development. Many individual sites may be offered on a ground lease basis since such an approach may be attractive to others who are taking a long-term view. This is particularly so with the advent of rent review clauses. A further significant area

SENSITIVITY ANALYSIS REPORT | ACADEMIC COPY

Example 15-3E ARGUS Developer Summary Appraisal for
Ex 15-3 showing the value for refurbishment as £1,574,751

Table of Land Cost and IRR%

Rent: Yield 5.7500%

Rent: Rate pm²	Construction: Rate pm²				
	-10.000% 900.00 pm²	-5.000% 950.00 pm²	0.000% 1,000.00 pm²	+5.000% 1,050.00 pm²	+10.000% 1,100.00 pm²
-20.000% 120.00 pm²	(£1,617,057) 57.7588%	(£1,515,837) 58.9787%	(£1,414,617) 60.2637%	(£1,313,397) 61.6192%	(£1,212,176) 63.0514%
-10.000% 135.00 pm²	(£2,013,868) 55.6441%	(£1,912,647) 56.6381%	(£1,811,427) 57.6769%	(£1,710,207) 58.7639%	(£1,608,987) 59.9024%
0.000% 150.00 pm²	(£2,410,678) 54.0604%	(£2,309,458) 54.8968%	(£2,208,237) 55.7660%	(£2,107,017) 56.6698%	(£2,005,797) 57.6106%
+10.000% 165.00 pm²	(£2,807,488) 52.8298%	(£2,706,268) 53.5505%	(£2,605,048) 54.2961%	(£2,503,827) 55.0680%	(£2,402,607) 55.8675%
+20.000% 180.00 pm²	(£3,204,299) 51.8461%	(£3,103,078) 52.4784%	(£3,001,858) 53.1303%	(£2,900,638) 53.8027%	(£2,799,417) 54.4966%

Rent: Yield 6.2500%

Rent: Rate pm²	Construction: Rate pm²				
	-10.000% 900.00 pm²	-5.000% 950.00 pm²	0.000% 1,000.00 pm²	+5.000% 1,050.00 pm²	+10.000% 1,100.00 pm²
-20.000% 120.00 pm²	(£1,339,115) 59.7704%	(£1,237,895) 61.2174%	(£1,136,675) 62.7528%	(£1,035,454) 64.3852%	(£934,234) 66.1246%
-10.000% 135.00 pm²	(£1,703,855) 57.2862%	(£1,602,635) 58.4501%	(£1,501,415) 59.6737%	(£1,400,194) 60.9618%	(£1,298,974) 62.3199%
0.000% 150.00 pm²	(£2,068,595) 55.4455%	(£1,967,375) 56.4156%	(£1,866,155) 57.4286%	(£1,764,935) 58.4874%	(£1,663,714) 59.5953%
+10.000% 165.00 pm²	(£2,433,335) 54.0267%	(£2,332,115) 54.8565%	(£2,230,895) 55.7185%	(£2,129,675) 56.6147%	(£2,028,454) 57.5472%
+20.000% 180.00 pm²	(£2,798,076) 52.8995%	(£2,696,855) 53.6233%	(£2,595,635) 54.3723%	(£2,494,415) 55.1477%	(£2,393,194) 55.9509%

Rent: Yield 6.7500%

Rent: Rate pm²	Construction: Rate pm²				
	-10.000% 900.00 pm²	-5.000% 950.00 pm²	0.000% 1,000.00 pm²	+5.000% 1,050.00 pm²	+10.000% 1,100.00 pm²
-20.000% 120.00 pm²	(£1,102,350) 61.9619%	(£1,001,130) 63.6746%	(£899,909) 65.5070%	(£798,689) 67.4723%	(£697,469) 69.5863%
-10.000% 135.00 pm²	(£1,439,771) 59.0524%	(£1,338,550) 60.4107%	(£1,237,330) 61.8478%	(£1,136,110) 63.3711%	(£1,034,890) 64.9888%
0.000% 150.00 pm²	(£1,777,192) 56.9215%	(£1,675,971) 58.0420%	(£1,574,751) 59.2181%	(£1,473,531) 60.4544%	(£1,372,311) 61.7556%
+10.000% 165.00 pm²	(£2,114,613) 55.2929%	(£2,013,392) 56.2439%	(£1,912,172) 57.2362%	(£1,810,952) 58.2726%	(£1,709,732) 59.3563%
+20.000% 180.00 pm²	(£2,452,034) 54.0075%	(£2,350,813) 54.8321%	(£2,249,593) 55.6884%	(£2,148,373) 56.5785%	(£2,047,152) 57.5045%

Rent: Yield 7.2500%

Rent: Rate pm²	Construction: Rate pm²				
	-10.000% 900.00 pm²	-5.000% 950.00 pm²	0.000% 1,000.00 pm²	+5.000% 1,050.00 pm²	+10.000% 1,100.00 pm²
-20.000% 120.00 pm²	(£898,242) 64.3593%	(£797,021) 66.3853%	(£695,801) 68.5726%	(£594,581) 70.9420%	(£493,360) 73.5183%
-10.000% 135.00 pm²	(£1,212,112) 60.9579%	(£1,110,891) 62.5397%	(£1,009,671) 64.2251%	(£908,451) 66.0250%	(£807,230) 67.9519%
0.000% 150.00 pm²	(£1,525,982) 58.4977%	(£1,424,761) 59.7881%	(£1,323,541) 61.1503%	(£1,222,321) 62.5907%	(£1,121,101) 64.1165%
+10.000% 165.00 pm²	(£1,839,852) 56.6347%	(£1,738,631) 57.7208%	(£1,637,411) 58.8594%	(£1,536,191) 60.0546%	(£1,434,971) 61.3108%
+20.000% 180.00 pm²	(£2,153,722) 55.1748%	(£2,052,501) 56.1103%	(£1,951,281) 57.0858%	(£1,850,061) 58.1042%	(£1,748,841) 59.1682%

Rent: Yield 7.7500%

Rent: Rate pm²	Construction: Rate pm²				
	-10.000% 900.00 pm²	-5.000% 950.00 pm²	0.000% 1,000.00 pm²	+5.000% 1,050.00 pm²	+10.000% 1,100.00 pm²
-20.000% 120.00 pm²	(£720,470) 66.9943%	(£619,250) 69.3922%	(£518,029) 72.0077%	(£416,809) 74.8732%	(£315,589) 78.0278%
-10.000% 135.00 pm²	(£1,013,828) 63.0202%	(£912,608) 64.8603%	(£811,387) 66.8362%	(£710,167) 68.9641%	(£608,947) 71.2627%
0.000% 150.00 pm²	(£1,307,186) 60.1850%	(£1,205,965) 61.6680%	(£1,104,745) 63.2433%	(£1,003,525) 64.9200%	(£902,305) 66.7085%
+10.000% 165.00 pm²	(£1,600,544) 58.0594%	(£1,499,323) 59.2964%	(£1,398,103) 60.5998%	(£1,296,883) 61.9754%	(£1,195,663) 63.4296%
+20.000% 180.00 pm²	(£1,893,902) 56.4062%	(£1,792,681) 57.4643%	(£1,691,461) 58.5725%	(£1,590,241) 59.7345%	(£1,489,020) 60.9546%

Sensitivity Analysis : Assumptions for Calculation

Construction: Rate pm²
Original Values are varied by Steps of 5.000%.

Heading	Phase	Rate	No. of Steps

Example 15-3E Illustrating the effect on land price of 5% steps in construction cost, 10% steps in rents, 0.5% increase in yield.

SENSITIVITY ANALYSIS REPORT   **ACADEMIC COPY**

Example 15-3E ARGUS Developer Summary Appraisal for
Ex 15-3 showing the value for refurbishment as £1,574,751

Offices	1	£1,000.00	2 Up & Down

Rent: Rate pm²
Original Values are varied by Steps of 10.000%.

Heading	Phase	Rate	No. of Steps
Offices	1	£150.00	2 Up & Down

Rent: Yield
Original Values are varied in Fixed Steps of 0.50%

Heading	Phase	Cap. Rate	No. of Steps
Offices	1	6.7500%	2 Up & Down
Car Spaces	1	6.7500%	2 Up & Down

Example 15-3E (Continued)

where ground leases have become common is that of developments of flats and maisonettes. The legal system in the UK, apart from Scotland, is so structured that a freehold interest in a flat which is only part of a property raises considerable problems. This remains the case even after the introduction of commonhold title. These problems are more easily solved by the grant of a ground lease of the flat.

At one time ground leases were usually granted for 99 years at a fixed rent, although other terms were sometimes adopted. For example, there are in existence leases granted for 999 years at a peppercornground rent, which are effectively freehold interests.

Recently, two important changes have taken place. First, there has been a movement towards leases of 125 years' duration. The pressure for this appears to be twofold. One is that the pace of change is such that it is felt that the life of buildings is shortened. Hence, if a building will last around 60 years, then a 99-year lease does not allow sufficient time to justify the redevelopment of the building after 60 years. The other is that the major funding institutions, the pension funds and insurance companies, have argued forcefully that a 125-year lease is the minimum period to justify their investing in such interests. Hence, building leases for 125 years have become more common. However, the power of local authorities to grant leases of more than 99 years was for some time tightly controlled by central government, so that they normally granted leases for the traditional term of 99 years.

The other change that has occurred is far more significant: this is the adoption of rent review clauses. At first, rent reviews were introduced into 99-year leases after 33 years and 66 years, but over time these intervals have shortened and it is now common for a rent review to operate every five years, as in occupation leases.

A further issue in residential schemes is the creation of management companies within whom the freehold is vested and where the leaseholders are the sole shareholders.

The valuer, therefore, is likely to be faced with a variety of ground leases. The current forms of ground lease for residential and commercial schemes are considered here since these are the ones to be found in most instances.

(a) Residential ground leases

As has been explained, developers of flats have overcome the problems of divided freehold interests by the grant of long leases. Typically, such flats will be offered for sale at a stated price plus a ground rent. The ground rents are commonly quite small, with few reviews. Technically, what is offered is a lease at a lower rent and a premium, but such terminology is rarely, if ever, used when they are marketed.

The market continues to change following the introduction of the Leasehold Reform, Housing and Urban Development Act 1993, where qualifying lessees can require their landlords to take a surrender of the existing lease and grant a new lease for the original term plus 90 years, or can collectively buy the freehold (see Chapter 16). This is likely to lead to leases being sold once again on the basis of 999 years at a peppercorn rent.

The determination of the original ground rent bears no relationship to the rental value of the land but tends to be derived from prevailing levels of such rents charged in the area. For this reason they are not truly ground rents, though so described. This may not be the case on review.

Many ground rents have fixed rent reviews; thus a typical ground rent may be £30, doubling every 25 years. Where the ground rent is subject to a market review the critical factor is whether regard should be had to the premium payable for the lease; if not, the ground rent payable on review will be a true rent, being the reciprocal of the MV in exactly the same way as a section 15 rent under the Leasehold Reform Act 1967 is a true ground rent (see Chapter 16).

In valuing such ground rents, the principal characteristics to note are that the sums tend to be small, with a consequently disproportionately high management cost, and that the income, though secure, is fixed for a long period.

Example 15–4

Value the freehold ground rents derived from a block of 24 flats erected two years ago. Each flat owner pays a ground rent of £100 p.a. rising to £200 p.a. in 32 years' time and £300 p.a. in 65 years' time.

Income				
24 flats at £100 p.a.		£2,400	p.a.	
PV£1p.a. or YP 32 years at 9%		10.41		
				£24,980
Reversion to		£4,800	p.a.	
PV£1p.a. or YP in, say, perp. at 9%	11.11			
× PV£1 32 years at 9%	0.7	0.77		£3,696
				£28,676

Where the unexpired period of the ground lease is shortening to the point where purchasers of the ground lease have difficulties in raising mortgages, the ground lessee may be prepared to offer a sum significantly in excess of the normal investment value to the freeholder in order to overcome the problem. Alternatively, the ground lessee may be prepared to pay a premium for a new lease or an extended lease. This extra payment is generated because of the marriage value which will exist. A lower yield may also be achieved because some purchasers will acquire the "estate" in order to manage it and derive management fees and insurance premiums. Thus, where these are available, the value in the above example may easily rise to £30,000 or more. Valuation here is likely to be based on analysis of auction sale results where very high multipliers are frequently paid, suggesting very low yields.

(b) Commercial ground leases

Where commercial sites are offered on ground leases, the ground rent will tend to represent the annual market rent of the site. The market rent clearly depends on the lease conditions.

A typical approach to the matter is that the freeholder offers the site on ground lease. Initially, the developer or tenant will be granted a building agreement which requires a proposed development to be carried out. The agreement provides that, on satisfactory completion of the development, a building lease will be granted in the form of the draft lease attached to the agreement. The lease will require the tenant to be responsible for the property under the normal repairing and insuring obligations. The initial ground rent will be agreed at the prevailing ground rental value. Current practice is to provide for the ground rent to be reviewed at frequent intervals, commonly five years, and for the basis of review to be a geared rent. This means that, at the grant of the lease, the parties will agree on market rent of the property to be built. This establishes a relationship between initial ground rent and occupation market rent, whiich will normally be expressed as a percentage. Thereafter, at each review, the ground rent will rise to that same percentage of the then prevailing market rent.

For example, on the grant of a ground lease of an office site, the ground rent may be fixed at £200,000 p.a., and it is agreed that the market rent of the offices, if they were already built, would be £800,000 p.a. It is clear that the initial ground rent is 25% of the office rents. At the first review, it may be agreed that the office rents are £1,000,000 p.a. If so, the ground rent will become 25% of £1,000,000 = £250,000 p.a.

In this way the ground rent will follow the rental pattern of the finished building rather than movements in land values. This tends to be preferred since it brings greater certainty in the rental value pattern, links movements in income to larger and more acceptable types of investment, and obviates arguments over land values at each review which tend to be less easily determined and thus less certain of agreement.

The valuer's task is therefore to determine the initial ground rent and, in due course, to estimate the MV of the freehold ground rent.

(c) Ground rent

Since the ground rent is the annual equivalent of the site's market value, it may be derived from the market value of the site, as previously described.

From the freeholder's point of view, the ground rent should be at a level which, when capitalised, will produce the MV of the site prior to the grant of the lease. This level is therefore the level derived from applying the prevailing yield for such ground rents. Hence, if a site has a capital value of £3,000,000, and ground rents are valued at 7% of market rent, then the freeholder will require a ground rent of 7% of £3,000,000 = £210,000 p.a.

Such a result might not emerge in practice. Developers who have been offered a ground lease might prefer such an arrangement, as they have to raise less building finance, and might therefore go above this level. On the other hand, they might find it more difficult to raise development funds if they can offer only a lease as security, and so might reduce their offer to encourage an outright sale of the freehold. These factors depend on the state of the market and the nature of development, but the general rental value can be derived from such a straight-forward approach, which at least provides a starting point. A developer's/lessee's approach to ground rents is different from a landlord's. To them a ground rent is an outgoing, and they may choose to determine the ground rent by a residual approach. The same principles will apply as to the residual method adopted to determine MV, as described earlier, save that the annual cost of the development coupled with an annual profit will be deducted from the prospective annual income, any difference representing the "surplus" they can offer on a ground rent.

Example 15–5

X has been offered a ground lease of a site for which planning permission exists to erect a warehouse of 4,500 m² gross internal area. The lease is to be for 99 years from completion of the building. What rent can X afford to pay for the ground lease?

MR of completed building 4,500 m² gross internal at £80 per m² =	£360,000	p.a.
Development costs	£1,500,000	
Building costs 4,500 m² at £330 per m², say		
Professional fees at 10%	£150,000	
Finance, say 6 months at 9%	£74,250	
Acquisition costs		
Legal and agent's fees and stamp duty, say	£40,000	
Letting costs		
Legal and agent's fees, say	£30,000	
	£1,794,250	
Say	£1,800,000	

Annual equivalent of costs			
Interest at 8%	0.080		
Annual sinking fund (ASF), 50 yrs at 3% adj. for tax at 35%	0.014		
Profits at 1.5%	0.015	0.109	£196,200 p.a.
Surplus for ground rent			£163,800 p.a.

A comparison can be made with a market value approach, as shown below:

GDV:			
Rent	£360,000	p.a.	
YP or PV£1p.a. in perp. at 8%	12.5		£4,500,000
Less Sale costs at say 3%			£135,000
			£4,365,000
Less Costs of development	£1,794,250		
Deduct Acquisition costs	£40,000		£1,754,250
Land balance			£2,610,750
Land value	1.000 ×		
Stamp duty 4%	0.040 ×		
	1.040 ×		
Funding 1 year at 9%	0.092 ×		
	1.132 ×		
Profits at 20%	0.200 ×	1.332 × =	£2,610750
Land value (x) =			£1,960,022
Say			£1,950,000

A ground rent of £163,800 p.a. represents a return of close to 8.4%, which can be regarded as realistic.

(d) Market value of ground rents

Once a building lease is in operation, the valuation of the ground rents produced follows the general principles in determining the MV of any investment.

However, a ground rent is generally less than the market rent of the property so that, in the case of a ground rent with geared reviews, apart from having the same qualities as the investment from which it derives, the rent is that much more certain of receipt and has no risk of income interruption because of voids. For these reasons a ground rent will generally be valued at yields below the prevailing yields for the type of property from which it derives.

On the other hand, it must be stressed that the general covenants in the ground lease will affect the yield. If rent reviews are widely spaced, so that the rent is fixed for longer than an acceptable period, then the yield will rise to reflect this. Similarly, at one time developers commonly entered into leasing arrangements whereby the landlord received a certain minimum proportion of the rent of the buildings, leaving the developer with a share of the marginal rent (a "top slice" arrangement). Clearly, the valuation of the landlord's interest would reflect the landlord's added security, whereas the valuation of the ground lessee's interest would need to reflect exposure to the changing fortunes of the market place.

The above comments have concentrated on the traditional situation where the freeholder grants a ground lease to the developer. Recent development schemes have seen the use of ground leases where several parties have come together to carry out large-scale developments, typically town centre redevelopments or the creation of out-of-town retail centres. The parties will include a developer, who carries out the development and manages it thereafter, a funding institution, which puts up the development funds, and the local authority, which provides planning and other support including the use of compulsory purchase powers to ensure site assembly. In these cases the agreement might be that the freehold interest will vest in the local authority who will grant a head lease to the funding institution who in turn will grant an under-ground lease to the developer, i.e. a sub-lease but of the ground only. The terms of the ground leases will reflect the agreement between the parties and their financial involvement. In these cases the ground lease terms will be arrived at in a different manner from that previously described. For example, the under-ground lease might well provide for a rent that is a percentage of money given to the developer and, as such, the agreement is more of a funding document than a traditional ground lease. Even so, the agreements must be related to the development, and "side by side" agreements are common whereby all the parties share in the growth in market rent whilst sharing the downside risks of a fall.

Ground rents and ground leases have seen a fairly rapid development in their nature and make-up over recent years. At one time the valuer was concerned with whether they were "well secured" or not and, if so, they attracted minimum yields. Today the ground lease has evolved into the expression of a commercial arrangement and the valuation of ground lease investments has changed accordingly. They are now seen as an investment to be judged critically along with the other investment opportunities available.

Chapter 16

Residential properties

I. Generally

This is the area of property that touches on most people in their everyday lives. The range of properties is vast, with tenements and cottages at one extreme and country estates at the other. It is also the area of property which is most affected by legislation because of the perceived need to protect the individual against exploitation by landlords.

Notwithstanding the development of the buy-to-let market, the residential property market is still dominated by the individual who purchases for owner occupation, as compared to the commercial property market which is dominated by the investor. The great majority of rented property that offers security of tenure is owned by the public sector (local authorities and housing associations) with only a comparatively small percentage being privately owned. Approximately 65% of all UK residential properties are owner-occupied and only about 14% are privately rented without offering any security of tenure. The difference between residential property and commercial properties is important because the direct correlation between rent and capital value which exists in the commercial market does not usually apply in the residential market.

The main statutory areas of law which will be looked at are those concerned with:

- the protection of residential tenants in terms of security of tenure and rent control; and
- the protection of owner-occupier tenants in connection with the management of their property, and the rights of these residential tenants to extend their leases or acquire the freehold of the property that they occupy.

It is necessary to distinguish between properties which are let on tenancies enjoying security of tenure or which are only suitable for lettings (with these being bought and sold for investment), and the great bulk of properties which are usually on the market with actual or potential vacant possession. The two markets are, however, not independent, as the former category is linked to the latter category because of the potential for capital gain if and when vacant possession is achieved.

Properties offered on the market with vacant possession are valued by direct comparison with achieved sales of similar property with the same accommodation in comparable locations. Where no direct comparable exists, then recourse should be had to the best comparable available, suitably adjusted for differences in accommodation and location. Except in the most unusual of situations, the value of a let property cannot be higher than the value of a vacant property since the latter represents the highest value achievable. This is because every vacant property can be let, but vacant possession is rarely immediately available from a let property.

There are a number of factors that influence the value of any residential property. The following list is not exhaustive but encompasses the principal factors, not necessarily in order of importance.

- Location.
- House or flat type. In the case of houses, whether they are detached, semi-detached or terraced. In the case of flats, whether they are standard, penthouses, duplexes or maisonettes (i.e. flats with their own independent street access).
- Accommodation; including the number of bathrooms and the extent of ancillary areas such as utility rooms, play rooms, swimming pools, etc.
- Design and layout.
- Energy performance.
- Number of storeys.
- Extent of the grounds, gardens, etc.
- Topography of the site.
- State of repair.
- Standard of finish.
- Historical associations.

Most of the above are self-explanatory and all could be the subject of significant amplification. It will be appreciated that many of these points are subjective and what one owner might think of as an improvement could be regarded as the opposite by a prospective purchaser; in some areas the proliferation of gilt and marble might detract from rather than add to the value of a property.

Of all the items on the list, the most important is location, by which is meant the reputation of the area as well as its locational advantages and disadvantages. The latter relates to proximity to work, schools and shopping centres, as well as to communication, places of worship, golf courses, etc. The factors that determine the quality of a location are not always susceptible to definition since there may be historical reasons why a given area is regarded as "upmarket" whilst other areas are considered to be "downmarket". Historically, much had to do with the wind direction (i.e. being on the fresher side of the town and that in many towns west tends to be more fashionable than east) or due to the decision of a particular person to settle in an area which then attracted friends to move close by.

The quality of an area can also change, and in valuing a property regard must be had to changes in fashion which might be caused by a change in infrastructure or even by a change in the political complexion of the local authority.

To summarise the effect of position and the other factors, it can be said that the general level of values of a neighbourhood is determined by locational factors, with differences in value between individual properties being determined by the nature and extent of the accommodation offered.

Of increasing importance is energy performance. Before any house or flat is let or can be sold or let the buyer/tenant must be presented with an Energy Performance Certificate (EPC). With the present cost of heating, and increasing awareness of green issues, these will become of more significance to buyers/tenants when given a choice between similar properties but with different fuel consumption levels. Obviously the one with the lower running cost will be the one preferred and thus the more valuable. However, this is not the only issue facing property owners; it is the Government's stated policy to ban the letting of all British property whose EPC level is F and G from 2018. This means that all these properties will need to be improved if they are to be let. A similar policy must soon come for property for sale.

2. The Rent Acts

Until the passing of the Housing Act 1988, the main body of legislation concerned with rent control of dwellinghouses was contained in the Rent Act 1977. This control is effectively being phased out in consequence of the 1988 Act as no new such tenancies can be created, but will, though diminishing, continue to exist for some years to come.

The Rent Act 1977, where it continues to apply, contains the law enacted in the Rent Act 1965 which first brought in the concept of "fair rent". It applies to all tenancies granted before 15 January 1989 where the rent passing was more than two-thirds of the rateable value and where the rateable value was less than £1,500 in London and £750 elsewhere. It will be referred to in the remainder of this account of the law, together with later amendments, as the 1977 Act. It should be noted that certain kinds of tenancy are excluded from Rent Act protection even if they are not at "low rents" and do not exceed the rateable value limits. Examples are holiday lettings, lettings by certain educational institutions to their students, lettings where the rent includes payment for board or attendance, lettings by the Crown or other public bodies, and lettings of parsonage houses, public houses or farms.

Tenancies at low rents (less than two-thirds of rateable value or less than £25,000 p.a.) granted for terms in excess of 21 years come within the provisions of Schedule 10 of The Local Government and Housing Act 1989 and also within the Leasehold Reform Act 1967 and the Leasehold Reform, Housing and Urban Development Act 1993 (but where the lease was granted for more than 35 years the low rent test does not apply).

Tenancies within the ambit of the 1977 Act are known as "regulated tenancies". The tenants have security of tenure and the right to have their rents fixed by an independent local official known as a "rent officer", although there is a right of appeal to a Rent Assessment Committee. Where a rent is fixed it is known as a "registered rent" and in the absence of a major change in circumstances the rent is fixed for two years. At the end of two years when the fair rent is varied it is known as a re-registration. Under the Rent Acts (Maximum Fair Rent) Order 1999, except where there is a major change in circumstances, from 1 February 1999 the maximum increase on the first re-registration is 7.5% above inflation and, on subsequent re-registrations, 5% above inflation. Inflation is measured by the change in the Retail Price Index between the month before the date of the earlier registration and the month before the re-registration.

(a) Rents under regulated tenancies (1977 Act, Part III)

Under the regulated tenancy system, the landlord or tenant or both can apply to the rent officer for the registration of a fair rent (subject to a right of appeal to the local Rent Assessment Committee). Once such a rent has been determined and registered it is the maximum rent that can be charged for the property during the continuance of the regulated tenancy. Mixed business and residential premises are protected by Part II of the Landlord and Tenant Act 1954 (see Chapter 18).

A fair rent is an open market rent subject to certain special rules laid down in section 70 of the 1977 Act, which states that in determining a fair rent, regard must be had to all circumstances (other than personal circumstances) and in particular to the age, character and locality of the dwellinghouse and to the state of repair; it must be assumed that the number of persons seeking to become tenants of similar dwellinghouses in the locality on the terms (other than those relating to rent) of the tenancy is not substantially greater than the number of such dwellinghouses in the locality available for letting on such terms. To be disregarded are disrepair or other defects attributable to the tenant and any improvements carried out, otherwise than in pursuance of the terms of the tenancy, by the tenant or the tenant's predecessors. Where a landlord can show that there is no scarcity of housing for rental in the locality where the tenant can obtain security of tenure, then no discount from the open market rental will be made.[1]

(b) Security of tenure (1977 Act, Part VII)

Limitation of rents would be of little use to protected tenants if they did not also have security of tenure, because otherwise their landlords could at common law terminate their tenancies by notice to quit (in the case of periodic tenancies) or refuse to renew them (in the case of fixed-term tenancies). The 1977 Act therefore

---

1 BTE Ltd v Merseyside and Cheshire Rent Assessment Committee [1992] 1 EGLR 116.

also provides that when the contractual ("protected") tenancy comes to an end in this way it is prolonged indefinitely in the form of a "statutory tenancy", which is transmissible on the tenant's death (though not otherwise, except by agreement) to a surviving spouse or other relative (if resident), and similarly a second time on that person's death (there can only be two successions). However, where the tenancy is transmitted to a non-spouse the rent ceases to be a Fair Rent but becomes an Assured Rent.

This security of tenure, however, can be terminated if the County Court grants possession to the landlord; though the court can only do this in certain specified cases, some of which arise because of the requirements of the landlord and others because of some default by the tenant. These various grounds will be found in Schedule 15 to the 1977 Act. Some of these grounds are discretionary and some are mandatory. In the case of the discretionary grounds the landlord will succeed only if he/she can satisfy the court that:

(a) it would be reasonable for the landlord to be awarded possession;
(b) there is suitable alternative accommodation; and
(c) there exists no special reason for withholding the award of possession.

Alternative accommodation is not a requirement of the Schedule for obtaining possession except where this is the sole ground, but in practice it will be required. The Act defines what is "suitable" as being suitable to the means as well as the needs of the tenants, with particular reference to their place of work. The alternative accommodation need not be the same size as the existing accommodation and it could, for example, comprise part of the existing premises.

As stated above, security may only be transmitted twice following the death of the original tenant. However, the rent will only be a regulated or fair rent only if the transferee is the spouse of the original tenant. Where the transferee is not the spouse, the successor only acquires an Assured Tenancy (see later) where there is security of tenure, but the rent is the open market rent. County courts also have jurisdiction over "rental purchase agreements", which were previously outside the scope of statutory protection (Housing Act 1980, Part IV).

3. The Housing Act 1988 – assured and assured shorthold tenancies

Under this Act no new regulated tenancies can come into existence after 15 January 1989 except by Order of the County Court (e.g. on the grant of a tenancy of suitable alternative accommodation). Two new forms of tenancy were created: (i) assured; and (ii) assured shorthold.

Where an assured tenancy is granted, the tenant has security of tenure but pays an open market rent. The tenancy may be either periodic or for a fixed term. At the end of the contractual period the parties can agree to a revised rent by the landlord serving a notice with a proposal which the tenant does not oppose. If the parties do not agree a revised rent, the tenant must make an application to the Rent Assessment Committee, which fixes a market rent. Possession can only be obtained on specified grounds which generally follow Schedule 15 to the Rent Act 1977 but also include the ground of redevelopment. The letting can provide for rent reviews which can be fixed to an actual assessment or to a formula, although the formula must be fair and not contrived so as to force the tenant to leave for financial reasons.

Where an assured shorthold tenancy is granted, the tenant does not have security of tenure but at the commencement of the tenancy the tenant can challenge the rent as being higher than comparable rents in the locality; the Rent Assessment Committee has powers to lower the rent. If the tenancy was created prior to the coming into effect of the Housing Act 1996, great care had to be exercised in complying with the formalities pertaining to the grant of the tenancy, as otherwise the tenancy would become an assured tenancy. The term had to be a term certain of not less than six months, determinable on not less than two months' notice, and the tenant had to be notified in advance of signing the tenancy of its nature and the fact that he or she would not have security of tenure. Best practice dictated that this notice was given more than 24 hours before the tenancy was signed and that the tenant acknowledged receipt of the notice by signing and returning a duplicate copy stating the time and date received.

Under the Housing Act 1996, all tenancies created after the coming into effect of the Act (i.e. after 1 March 1997) are automatically assured shorthold unless an assured tenancy is specifically created by agreement.

4. Schedule 10 to the Local Government Act 1989, the Leasehold Reform Act 1967 and the Leasehold Reform, Housing and Urban Development Act 1993 (Part 1)

The protection of tenants occupying dwellinghouses on ground leases had caused concern for a number of years, and these three pieces of legislation represent the steps taken to ensure that such tenants are not automatically dispossessed when their contractual right to remain in occupation has expired.

Schedule 10 to the Government Act 1989 (replacing Part I of the Landlord and Tenant Act 1954) applies to houses let on "long tenancies" (i.e. for more than 21 years) which, on account of the rent being a "low rent" (i.e. less than two-thirds of the rateable value on 23 March 1965, or when first rated thereafter), are outside the protection of the Rent Acts. The limits of rateable value within which Part I of the 1954 Act applies are those applicable under the Rent Act 1977, referred to above. A tenant is not protected if the landlord is the Crown or a local authority, the Development Corporation of a new town or certain housing associations and trusts. Where a tenancy is entered into after 1 April 1990 it is a tenancy at a low rent if it is £1,000 or less in Greater London or £250 or less elsewhere.

The effect of the Act is to continue the tenancy automatically where the tenant is in occupation after the date when it would normally expire, on the same terms as before, until either the landlord or the tenant terminates it by one of the notices prescribed by the Act. However, the landlord and the tenant can agree on the terms of a new tenancy to take the place of the long tenancy.

A long tenancy can be terminated by the landlord by giving one of two types of notice, each of which must be in prescribed form. If the landlord is content for the tenant to stay in the house, the landlord must serve a landlord's notice proposing a statutory tenancy. If the landlord wishes the tenant to leave, the landlord must serve a landlord's notice to resume possession.

Should a tenant wish to terminate a long tenancy, the tenant must give not less than one month's notice in writing.

A landlord's notice proposing a statutory tenancy must set out the proposed terms as to rent and repairs, including "initial repairs". The landlord and tenant can negotiate on these terms and come to an agreement in writing. If they cannot do so, the landlord can apply to the County Court to decide those items which are in dispute.

Long tenancies expiring on or after 15 January 1999 become assured tenancies with open market rents (see section 186 of the Local Government and Housing Act 1989, Schedule 10).

Where tenants remain in possession after the end of the long tenancy, they are relieved of any outstanding liability in respect of repairs arising under that tenancy. The terms proposed by the landlord may provide for the carrying out of repairs when the new terms come into force. These repairs are known as "initial repairs" and the tenant may have to bear some or all of the cost of them. Where tenants leave at the end of the long tenancy, their liability under the tenancy is not affected by the Act. "Initial repairs" may be carried out either by the landlord or by the tenant or partly by one and partly by the other; neither need do any unless they so wish. If the landlord carries out the repairs, the landlord is entitled to recover from the tenant the reasonable cost of the repairs in so far as they are necessary because the tenant did not meet his/her obligations under the long tenancy. Payment can be made by the tenant either by a lump sum or by instalments, as agreed between the parties or as determined by the County Court.

Where a landlord serves notice to resume possession, the tenant, if he/she wishes to remain in the house, should so inform the landlord. If the tenant does not agree to give up the house, the landlord can apply to the County Court for a possession order. The grounds upon which a landlord can apply for possession are those set out in Part II of Schedule 2 to the Housing Act 1988, which include the landlord's wish to redevelop.

The Leasehold Reform Act 1967 came into force on 27 October 1967, and represented a very radical departure from previous property law. The White Paper[2]

2 Cmnd. 2916 of 1966 "Leasehold Reform in England and Wales".

which preceded the legislation stated the Government's view that "the basic principle of a reform which will do justice between the parties should be that the freeholder owns the land and the occupying leaseholder is morally entitled to the ownership of the building which has been put on and maintained on the land". This principle has, however, only been applied to the limited range of properties to which the 1967 Act applies; and this excludes flats and maisonettes, similar units produced by subdivisions of buildings where the dividing line is not substantially vertical.[3]

These exclusions are covered by the 1993 Act, which is much less advantageous to the tenant in terms of price for lower value properties.

The 1967 Act enables qualified leasehold owner-occupiers either to purchase the freehold reversion from the ground landlord or to obtain an extension of the term of the lease.

The following requirements must be met before a leaseholder is qualified and entitled to the benefits conferred by the Act:

- The original term of the existing lease must be more than 21 years and the rent reserved must be less than two-thirds of the rateable value for the valuation to be calculated in accordance with S.9(1) of the Act. Where the lease was for more than 35 years, the low rent test will not apply.[4]
- The leaseholder must have owned the house for at least two years. Use of part of the premises for another purpose, for example as a shop, does not disqualify.[5] If the property was designed either in full or in part as a dwelling then, unless residential use is prohibited by the terms of the lease or it extends to a small part only of the building (probably less than 15%), the Act will apply.
- The rateable value of the house on 23 March 1965, or when first rated thereafter, must not exceed £400 in Greater London and £200 elsewhere. The Housing Act 1974, section 118(1), increased these limits, with effect from 1 April 1973, to £1,500 and £750 respectively for tenancies created on or after 18 February 1966, and to £1,000 and £500 respectively for tenancies created before that date. (For tenancies granted after 1 April 1993 the Leasehold Reform, Housing and Urban Development Act 1993 applies.) Where the RV exceeds these limits, the leaseholder can still acquire the freehold but the valuation must be calculated in accordance with S.9(1A) of the Act, which is based on the leaseholder paying the market value of the freeholder's present interest plus half the marriage value.

3 Malekshad v Howard de Walden Estates.
4 Housing Act 1996.
5 Boss Holdings Ltd v Grosvenor West End Properties Ltd [2008] 05 EG 167 (CS) and Prospect Estates Ltd v Grosvenor Estates Ltd [2008] EWCA Civ 1281.

"Shared ownership leases" granted by various public authorities and housing associations are excluded from the operation of the 1967 Act (Schedule 4A, added by the Housing and Planning Act 1986).

(a) Enfranchisement by way of purchase of the freehold

Where leaseholders are qualified under the 1967 Act and give the landlord written notice of their desire to purchase the freehold interest, then, except as provided by the Act, the landlord is bound to make to the leaseholder and the leaseholder is bound to accept, if he wishes to proceed (at a price and on the conditions to be agreed), a grant of the house and premises for an estate in fee simple absolute, subject to the tenancy and to the leaseholder's encumbrances but otherwise free from encumbrances. The tenant is not bound to proceed if the price proves too much for him but is then barred from serving another notice for 12 months.

There are therefore two methods of valuation. The first is where the lessee qualifies under the Act as originally enacted in 1967. The price payable is effectively confiscatory as it assumes that there will be a reversion equivalent only to site value for fifty years and only then will possession of the house be available to the landlord. Purchasers of these investments in the open market must be very cautious and should never pay more than what a qualifying tenant would pay under the Act, noting that there is no requirement for the qualifying tenant to be either a resident or even an individual. The second is for all other houses where the price payable, as defined in section 9(1A) of the 1967 Act as amended by section 23 of the Housing and Planning Act 1986, is the amount which the landlord's reversion to the house and premises might be expected to realise on the assumption that it is to be sold in the open market by a willing seller (the tenants and members of their family having no right, for the purpose of this assumption, to buy the freehold or an extended lease but they are assumed to be in the market). It must be assumed also that the freehold interest is subject only to the existing lease, ending on the original date of termination even if extended under the 1967 Act; i.e. marriage value is assumed to be payable.

Within one month of the ascertainment of the price payable, the tenants may give written notice to the landlord that they are unable or unwilling to acquire at that price, in which case the notice of their desire to have the freehold ceases to have effect. In such circumstances, the tenants must pay compensation to the landlord for the costs. The tenant can start the procedure again after one year.

(b) Valuations to determine enfranchisement price – "original method" (i.e. lower value properties only)

The valuation approach to determine the price payable for lower value properties (i.e. below rental value (RV), £1000/£500) under what is now section 9(1) of the Act (as amended) proved in practice to be highly contentious. One of the earliest cases turned on whether the valuation approach should reflect the tenant's position as a special purchaser and so allow for marriage value.

This led to an amendment which provided that the tenant's bid should be disregarded but only in relation to properties within the RV limits. In relation to higher value properties first introduced by the Housing Act 1974 and then by the Leasehold Reform, Housing and Urban Development (LRHUD) Act 1993, the tenant's bid is not excluded and the properties are valued on the section 9(1A) basis.

For lower value properties the original cases turned chiefly on two aspects, the method of determining the modern ground rent and the choice of the capitalisation rates within the valuation of the freehold interest.

The typical situation is where the freehold interest is subject to a ground lease granted several years ago at an annual ground rent which is now considered to be a nominal sum. At the end of the lease it must be assumed (whether or not the tenant acquires the freehold or takes an extended lease) that the tenant will extend the lease for a further period of 50 years at the current ground rental value (a "modern ground rent, known as a S.15 rent") subject to review after 25 years. Thereafter the house and land will revert to the freeholder. This involves a three-stage valuation:

1. Term of existing lease

This will be the valuation of a ground rent for the outstanding period of the lease. From an investment viewpoint this is an unattractive proposition since the income is fixed and it is usually small, so that management costs are disproportionately high. This has led to disputes as to the appropriate capitalisation rates to be adopted.

As a general rule, the Lands Tribunal decisions tended to adopt rates between 6 and 8%,[6] but since 1980 most leasehold valuation tribunal decisions have adopted 7%. Higher rates have been employed where the outstanding term is relatively long or where there is market evidence.[7]

The most recent case at the date of writing this book,[8] heard before the Upper Tribunal (Lands Chamber), previously the Lands Tribunal, determined an interest rate of 5.5% for (1) the capitalisation of the ground rent, and (2) to arrive at the S.15 rent from the site value and to discount the final reversion. However, in connection with the latter two regard was had to the location of the house, in Birmingham, and the Tribunal had regard to its decision in Zuckerman v Calthorpe Estates[9] where it had departed from the standard Sportelli[10] rate by raising the

---

6 See, for example, Carthew v Estates Governors of Alleyn's College of God's Gift (1974) 231 EG; Nash v Castell-y-Mynach Estate (1974) 234 EG 293.

7 See, for example, Leeds v J & L Estates Ltd (1974) 236 EG 819.

8 Re: Clarise Properties Ltd [2012] UKUT 4 (LC217).

9 [2010] 1 EGLR 187.

10 Earl Cadogan v Sportelli [2007] 1EGLR 153.

discount rate by 0.5% to reflect location. Therefore it is realistic to assume that both the interest rate to arrive at, and to capitalise, the S.15 ground rent and to discount the final reversion would have 0.5% lower had the house been located in a standard Sportelli location. This rate would therefore have been 5% and not the standard 4.75% standard house rate, to reflect the extra volatility of land values as opposed to house values. The Tribunal rejected the creation of an "adverse differential", although this deserves further debate. This case also removed the argument that the final reversion (known as the Haresign addition) was an exception rather than the rule, but it accepted the argument that as the tenant, under current law, has the right to occupy the house at the end of the lease as an assured tenant, there should be a discount to the final reversion value. The writer finds this latter point difficult to accept since no tribunal (either LVT or Lands Tribunal) has ever discounted the freeholder's reversion in all 1993 Act cases except where the reversion was extremely short and not, as here, 78.5 years away. There will no doubt be further argument on this point. These yields are not fixed in stone but will depend on long-term rates of interest applicable in the investment market generally but adjusted to reflect property's illiquidity and underlying risks.

2. Reversion to modern ground rent

At the end of the existing lease the rent will rise to the "modern ground rent" which will be receivable for 50 years, subject to one review after 25 years.

The "modern ground rent" is in essence the current ground rental value, being the annual equivalent of the cleared site value. However, in many built-up areas evidence of site values may be non-existent and therefore one approach has emerged which is commonly adopted, whereby the value of the freehold interest in the house and land is apportioned between land and buildings and the modern ground rent is then derived from the value of the land as apportioned. This is known as the "standing house approach".[11]

Alternative methods may be adopted, involving a valuation direct to site value or to site rental value, and are generally to be preferred where evidence exists.[12]

The standing house approach starts therefore with the valuation of the freehold interest in the house and land developed to its full potential. The valuation assumes vacant possession, ignoring for example the fact that all or part is let to tenants, or any disrepair.[13] The reason for this is that what is sought is the full unencumbered value of the property. For the same reason it will reflect any potential for conversion into separate flats to be sold with vacant possession.[14] Once this full value has been

---

11 See, for example, Hall v Davies (1970) 215 EG 175; Kemp v Josephine Trust (1971) 217 EG 351; Nash v Castell-y-Mynach Estate, supra.
12 Farr v Millerson (1971) 218 EG 1177; Miller v St John the Baptist's College, Oxford (1976) 243 EG 535; Embling v Wells and Campden Charity's Trustee [1978] 2 EGLR 208.
13 Official Custodian for Charities v Goldridge (1973) 227 EG 146713.
14 See, for example, Graingers v Gunter Estate Trustees (1977) 246 EG 55.

determined it is necessary to apportion the value between land and buildings. The amount apportioned to the land will depend on the facts of the case, but in general the proportion attributable to the land value of houses in high value areas will be greater than that in low value areas in the same way that values per plot for high value houses are greater than values per plot for cheaper housing. There can be no hard and fast rules but tribunal decisions have tended to adopt around 40% for houses in London and around 30% for houses elsewhere as a reflection of this general proposition.

Having determined the land value, the final stage is to determine the modern ground rent. This is found by decapitalising the value by applying the appropriate yield.[15]

Clearly, if both yields are the same then the whole exercise can be short-circuited by simply deferring the land values at the chosen yield. This approach can be taken and the reversion after the 50 years extension ignored where the existing lease is so very long that the reversion adds little or no value.

3.   Reversion to house

Following the expiry of the 50 years deemed extension, the whole property reverts to the freeholder. However, if the existing lease does not expire for a considerable period, by the time account is taken of the 50-year extension the reversion to the standing house may have an insignificant value and can be ignored, so that the modern ground rent is valued to perpetuity (see above). However, if the unexpired term of the existing lease is not over long, the reversion in just over 50 years will almost certainly be of significant value and should therefore be included, as in the case of Haresign v St John the Baptist's College, Oxford (see above).

(c) Market transactions

One of the principal causes for the valuation arguments that have emerged is that the statutory approach is not usually found in the open market, thereby leaving the valuer to draw upon valuation principles to carry out the valuation. The only comparables which will generally be found are of sales to tenants who are also enfranchising and where it may be assumed that the statutory approach will be adopted . However, in practice, a tenant may be prepared to pay over the odds to effect a purchase rather than go to the local Leasehold Valuation Tribunal, then possibly on appeal to the Lands Tribunal with the contingent risk of additional costs, delay and uncertainty of outcome if, for example, he is selling.

The Lands Tribunal (now the Upper Tribunal (Lands Chamber)) has recognised this factor when considering comparables by sometimes discounting the prices

15 (1980) 255 EG 711.

paid, using what is shown as the "Delaforce effect", from the case of *Delaforce* v Evans.[16]

Although not a popular concept with tribunals, the reverse may also be true, with landlords settling for too low a premium because of the cost implications of appeals, since costs are not recoverable by the winning side in LVTs or the Upper Chamber except for a nominal amount of £250 where one side can be accused of "abuse of process".

Example 16–1

The leaseholder of a late 19th-century brick-built four-bedroom semi-detached house with garage, in a fairly good residential area, wishes to purchase the freehold interest. The lease is for 99 years from 1 January 1921, at an annual ground rent of £5.

On 1 January 2012, the leaseholder, who is qualified under the Act, served the necessary statutory notice under the Leasehold Reform Act 1967, claiming the right to have the freehold.

The house occupies a site with an area of 300 m². Comparable residential building land in the area has recently changed hands at £2,750,000 per hectare for development at 38 houses per hectare and with services costing £300,000 per hectare. Similar houses in the area have been sold on the open market with vacant possession at figures ranging from £200,000 to £250,000.

Valuation
In order to value the reversion after the existing lease it is necessary to estimate a modern ground rent. Two approaches are set out below, namely the "standing house approach" and "the cleared site approach". There is no single approach which must always be applied. The best approach is the one best supported by the available evidence.

(a) *Standing house approach*

Standing house value, say	£225,000	
Land apportionment, say 35%	0.35	
Site value	£78,750	
Modern ground rent at 6.5%[17]	.065	£5,119 p.a.

(b) *Capital value of site with all services*, from comparables	£80,000		
Modern ground rent at 6.5%	0.065	£5,200	p.a.

In this example the two methods have produced values within the usual valuation parameters. Method (b) is the preferred method where there is available evidence because it is less contrived, but the site value for a single house would not necessarily be the same as for a single unit derived from a large site. Method (a) is the more

---

16 Delaforce v Evans (1970) 215 EG 315, (1980) 255 EG 711.

common because of the comparative ease of obtaining evidence as to the standing house value. However, when using the standing house approach the house value must assume a house developed to its full potential, modernised and extended as appropriate. Both methods could produce a result where the modern ground rent, known as the section 15 rent, exceeds the fair rent for the actual property. This was discussed in the *Dulwich College* case[17] and found to be acceptable. In this case a section 15 rent of £5,200 would not seem unreasonable.

Term			
Present ground rent		£5 p.a.	
YP 9 years at 7%		6.515	£33
Reversion			
After 31 December 2020, to modern ground rent based on 2011 site value		£5,200 p.a.	
YP 30 years at 6.5%	14.725		
def'd[18]			
9 years @ 5%	0.645	9.492	£49,356
Reversion to vacant possession value		£225,000	
× PV of £1 in 59 yrs at 4.75%		0.065	£14,558
Price payable for enfranchisement			£63,940

(d) Valuations under the Housing Act 1974 and the Leasehold Reform, Housing and Urban Development Act 1993 (LRHUD)

The Housing Act 1974 and the LRHUD Act 1993 extended the rights of enfranchisement to houses generally without limit. The rateable value may be adjusted so as to ignore tenants' improvements and a certificate may be obtained from the Valuation Officer (i.e. the District Valuer) for this purpose.[19] Thus, a house in London with a rateable value of, say, £1,070 may come within the 1967 Act if, for example, the tenant has put in central heating and added a garage, all of which has added more than £70 to the rateable value.

(e) Valuations to determine enfranchisement price – "higher value method"

In respect of houses above the 1967 Act limit as set by the 1974 Act, the valuation approach for enfranchisement differs in two important respects from the "original method" already considered. The first is that the price may reflect the

---

17 Carthew v Estates Governors of Alleyn's College of God's Gift (1974) 231 EG 809
18 Section 118 of and Schedule 8 to the Housing Act 1974 as amended by S.141 of and Schedule 21 to Housing Act 1980 adding S.1(4A) to the Leasehold Reform Act 1967 (over page).

tenant's bid, so allowing for marriage value. The second is that, as with S(9) valuations, the valuation could reflect the tenant's right to remain in possession under Schedule 10 to the local government Act of 1989[19] and not their right to extend the lease by 50 years at a modern ground rent; this does not mean that this right will have a valuation effect and it will depend on the evidence of each individual case as to whether it will. This was considered in the Eyre Estate case[20] and to date no tribunal has given a discount in a 9(1) case. This basis of valuation is now the standard method and the provisions are in section 9(1A) of the 1967 Act. The valuation is therefore significantly different.

Example 16–2

The same facts apply as in Example 16-1, save that the RV after adjustment for tenant's improvements was £1,150 in London and the vacant possession value ignoring tenant's improvements is £850,000 for the freehold and £230,000 for the unexpired lease.

Vacant possession value – freehold				£850,000
Leasehold interest			£230,000	
Freehold interest				
Rent reserved	£5			
YP 9 yrs at 7%	6.515	£33		
	£850,000			
Reversion to PV of £1 in 9 yrs at 4.75%	0.659	£559,600	£559,633	£789,833
Marriage value				£60,167
Divided equally				0.5
				£30,084
Freeholder's interest as above				£559,633
Enfranchisement price				£589,717

Although the tenant has the right to remain in occupation after the expiration of the lease, a discount is now unlikely to apply because the standard view is that a tenant would not wish to remain as an assured tenant.[21] However, it is open to tenants to argue that they would have remained in the premises even at an assured rent if the reversion was to be within a year or two of the date of the notice, and this would be a matter of evidence. The valuation date in all cases is the date when the notice is served on the landlord.

(f) Extension of the existing lease (lower value houses)

Where a leaseholder is qualified under the Act and gives the landlord written notice of a desire to extend the lease, then, except as provided by the Act, the

19 Re 167 Kingshurst Road [2012] UKUT 4 (UT).
20 Eyre Estate Trustees v Shack [1995] 1 EGLR 213.

landlord must grant and the leaseholder must accept a new tenancy for a term expiring 50 years after the term date of the existing tenancy, with a rent review after 25 years.

With the exception of the rent, the terms of the new tenancy will be significantly the same as the terms of the existing tenancy, although some updating will be allowed. Rules for the ascertainment of the new rent are laid down in section 15 of the 1967 Act, but the main point to be noted is that the rent must be a ground rent in the sense that it represents the letting value of the site and excludes anything for the value of the buildings on the site. This modern ground rent, which is payable as from the original term date, can be revised if the landlord so requires after the expiration of 25 years. The rent will be calculated in exactly the same way as for calculating the enfranchisement price, and the valuation date is the original term date of the lease (or 25 years later), with the calculation done at that time.

If tenants have opted for a lease extension, they can subsequently serve a notice on the landlord requiring the freehold to be sold. However, the valuation will always be under section 9(1A) and marriage value will be payable in all such cases.

(g) Landlord's overriding rights

In certain circumstances the leaseholder's right to purchase the freehold or to extend the lease under the 1967 Act may be defeated. If the landlord can satisfy the court that possession of the house is required for redevelopment or for own occupation, or occupation by an adult member of the landlord's family, the court may grant an order for possession on the term date of the lease. The tenant is entitled to receive compensation for the loss of the buildings. The compensation will be the value of the 50-year lease at a modern ground rent, which will normally be calculated by deducting from the vacant possession value of the freehold what would have been the cost of enfranchising on the last day of the lease. Where there is development potential, this is reflected in the enfranchisement price by assuming that at the end of the current lease there will be no renewal but compensation will be payable. The landlord's reversion will therefore be to the vacant possession value of the property with its development potential less the compensation to the tenant deferred for the term.

Example 16–3
The same facts as in Example 16-1, but assuming that the house stood on 4,000 m^2 of ground worth £800,000 for development and the extra garden only added £30,000 to the value of the freehold house if there was no possibility of development.

Rent reserved	£5	
YP 9 yrs at 7%	6.915	£33
Reversion to site value	£800,000	

Less
Compensation to tenant

Value of freehold house	£225,000	
Enfranchisement price if no landlord's rights, say 33% × £225,000	£78,950	£146,250
		£653,750
× PV of £1 in 9 yrs at 5.00%	0.645	£421,413
Enfranchisement price		£421,446

(h) Retention of management powers

In circumstances specified in section 19 of the 1967 Act, landlords could retain certain powers of management on enfranchisement in order to maintain adequate standards of appearance and regular redevelopment in an area.

For a more detailed consideration of the Leasehold Reform Act 1967 see Chapter 3 of *Valuation: Principles into Practice* (6th edition), ed. Richard Hayward (Estates Gazette).

(i) The LRHUD Act 1993 as amended by the Commonhold and Leasehold Reform Act 2002

This Act has extended the right to enfranchisement to all houses regardless of their rateable value, and it also allows leaseholders of flats to collectively purchase the freehold of their block or to individually extend their leases by 90 years. Only qualified lessees can exercise these rights, in the same way as under the earlier enfranchisement legislation. These include investors, except in the case of collective enfranchisement where they own more than two flats. The lessee must have owned the lease for not less than 24 months for a lease extension, but there is no time requirement of ownership for joining in a collective enfranchisement. The qualifying leases must have originally been longer than 21 years at a low rent or longer than 35 years regardless of the rent payable (Housing Act 1996).

The purchase price is calculated in the same way as for the higher rateable value houses, with 50% of the marriage value payable to the freeholder by virtue of the Commonhold and Leasehold Reform Act 2002. The provision for flats is that the existing lease is surrendered and a new lease is obtained at a peppercorn rent for the original term plus 90 years. There are no limits on the number of 90-year extensions that can be obtained. Where there are superior leaseholders, the ground rent is only reduced to a peppercorn rent between the qualifying lessees and their immediate landlord; all other ground rents remain in place. The marriage value is split between the landlords in proportion to the value of their existing interests.

Instead of acquiring lease extensions individually, the lessees can collectively purchase the freehold of their block. The price is nominally more than the sum of all of the leasehold extension prices for all of the participating flats in the block, plus any payments for additional land which the landlord requires to be purchased, plus any further payment required to compensate the landlord for any other losses sustained as a result of the acquisition, plus the investment value of the non-participating flats and hope value. In order to purchase, not less than half of the lessees must sign the notice and qualifying leases must account for not less than two-thirds of all the flats in the block. Where a lessee owns more than two qualifying leases, such a lessee may not be a participator in a freehold purchase and therefore does not count in the two-thirds total. Where the block has a commercial element, this must not exceed 25% of the floor area as otherwise the tenants cannot purchase.

Where there is a head lease of a block of flats and some or all of the flats are not sublet on qualifying leases, the head lessee can obtain lease extensions of all the flats that are not the subject of qualifying sub-leases, even though the head lessee cannot acquire the freehold.

The leasehold reform legislation is very complicated and the reader should refer to *Valuation: Principles into Practice* by Richard Hayward, published by EG Books in 2008, for more details. The subject is constantly evolving and the valuer must keep up to date. For example, immediately before publication of this book, in *Earl Cadogan* v *Grandeded* [2008] UKHL71 (being the final part of the Sportelli litigation), the House of Lords decided that in a collective enfranchisement the landlord is entitled to hope value for non-participating lessees and, in a group of cases known as "nailrile", the Lands Tribunal gave significant directions relating to yields and the valuation of head leasehold interests.

5. Smaller residential properties, HMOs and tenements

(a) General method of valuation

The first question to be asked in advance of any mathematical approach to a valuation is whether the property is saleable, with or without improvement or repair, with vacant possession to owner-occupiers. If the answer is in the affirmative, then the value will be more a function of the potential for vacant possession than a capitalisation of the net income. Thus a country cottage let to a widow of 85 living alone will be significantly more valuable than if let to a couple in their 40s at the same rent. However, in the case of HMOs (houses in multiple occupation) and tenements, a sale of an individual flat for owner occupation may be unlawful or unlikely and the rental income is much more important for a determination of value. Properties registered as HMOs cannot now be changed into a single house or converted into self-contained flats without planning consent as such a use now constitutes a planning restriction, since multiple occupation is a use that local authorities are keen to maintain. The market value in virtually every HMO property will be less than the vacant possession of a comparable vacant house because the latter could always be used as an HMO and therefore provides a competitive property. HMOs always give high gross yields because of (1) their difficulty of management; (2) the incidence of local authority involvement with regulations for fire prevention and health and safety issues; (3) the wear and tear on the property; (4) the likelihood of bad debts; and (5) the nature of the tenants. This list is not exhaustive.

The gross rent will be determined by either the rent officer or by market forces and will in all cases reflect the age, character, locality and state of repair; it will usually be paid either weekly or monthly.

In free market conditions slight differences in style and accommodation – for instance, a few feet of front garden with a dwarf brick wall and railings, or the fact that the front living room opens out off a small hall-passage, instead of direct on to the street – may cause a considerable difference in rental between houses within a short distance of each other.

Where vacant possession is the determining factor in the valuation no allowance need be made for bad debts because this could lead to a windfall early possession. In other cases the differential between a fair rent and a market rent will also ensure payment of the former or give rise to a windfall increase in income, and so alleviate the need for bad debt provision. In other cases a bad debt allowance of, say, 10% might be appropriate because of the low quality of covenant provided by many tenants in this part of the market.

If, owing to overcrowding or tenant ignorance, a house is producing what may be regarded as an excessive rent, the excess should be disregarded. This is because it will certainly be lost should the local authority exercise its powers under the Housing Act 1985 to pay housing benefit at an assessed level and not necessarily at the same level as the rent payable, bearing in mind that some or all of the rent for this low-quality accommodation is often paid by either local or national government. It is also now the intention of the Government to limit the amount of housing benefit to a fixed amount, regardless of the size or location of the property, and this could have a profound effect on values where the "buy to let" market is significant.

Repairs, insurance and management will be the main outgoings and, in some cases, the landlord may also pay rates and water and drainage charges.

Yields vary widely according to the circumstances of the particular case and the state of the local market. In the case of freeholds they may range from 5% or less, where vacant possession is likely in the comparatively short term and where the security is good, to figures of 15% or more where property is old, in poor repair and occupied by an unsatisfactory type of tenant. Where the rate of interest is low, this is because of the large increase in capital value which occurs if vacant possession is obtained and if the property can be sold into the "owner-occupier" market. In this connection it may be noted that the market for what may be described as "tenements", i.e. large old buildings let out to a number of tenants, is usually poor and uncertain compared with that for small, fairly modern houses let to single families. This is particularly so at the present time, owing to the high cost of repairs in the case of the "tenement" type and to the large increase in capital value if vacant possession is obtained of a house let to a single family.

There are few investors who are interested in holding poor quality residential properties as long-term investments, unless they are capable of being improved when vacant possession is obtained. In most cases where a tenanted property is being valued, it is suggested that a normal investment valuation should be performed, i.e. the net income should be multiplied by the appropriate YP, and that the resultant value should be checked as a percentage of vacant possession value. In the case of houses saleable with vacant possession and let at fair rents, the value will be found to fall within a range of 50–80%, depending on how the market assesses the possibility of vacant possession being obtained. Where the property is let on an assured tenancy (i.e. at market rent but with the tenant having security of tenure), the percentage of vacant possession value will be higher because of the higher yield available; the value could be as high as 80–90%, dependent on yield. In other cases the property may be regarded as a poor investment with unlimited outgoings, and a high yield will be required. Where there is certainty of obtaining possession, as in the case of an assured shorthold tenancy (AST), the vacant possession value should be deferred for the period of waiting, and thus the value would not normally exceed 90–95% of vacant possession value unless the property is being sold into the "buy to let" market. In this case the existing AST letting could be regarded as an advantage due to the saving of letting costs, and full vacant possession value might be paid.

(b) Outgoings

The principal outgoings common to all residential lettings are repairs, insurance and management. In some cases the landlord may also pay council tax and water and drainage charges. Under the Rent Act 1977, any increase in rates due either to changes in assessment or poundage can be passed on to the tenant.

Apart from any contractual liability to repair, in the case of leases granted after 24 November 1961 for terms of less than seven years for dwellinghouses to which the statutory protection of business or agricultural tenancies does not apply, the Landlord and Tenant Act 1985, sections 11–16, implies a covenant by the lessor to keep in repair the structure, exterior and installations for water, gas and electricity supplies, for space and water heating and for sanitation (contracting out is prohibited except with the approval of the County Court).

It is generally preferable to make the allowance for repairs by deducting a lump sum per property, which should be checked by reference to past records and to the age, extent and construction of the premises (see Chapter 6). For example, the allowance in the case of a house stuccoed externally and requiring periodic painting will naturally be higher than in the case of a similar house with brick facings in good condition. It has been customary in the past to estimate the amount to be allowed for repairs as a percentage of the rent, but this method is unreliable and illogical.

Where in the past the work of repair has been left undone, it may be necessary to include in the valuation a capital deduction for immediate expenditure to allow for the cost of putting the property into a reasonable (but not necessarily a first-class) state of repair. This is called an end allowance.

The annual cost of insurance can usually be determined with sufficient accuracy either by using the actual premium paid or by assessing it on the basis described in Chapter 20. Fire and terrorism insurance premiums are becoming significant sums so that, where the landlord is responsible, care must be taken to ensure that the appropriate deduction is made.

Management may cost from 10% upwards plus VAT, according to the circumstances and the services rendered.

In the case of ASTs, the landlord will frequently instruct letting agents to find tenants. AST lettings are usually between six months and two years and agents frequently charge a percentage of the total rent payable over the whole term, often 10% plus VAT. In addition, there are frequently void periods between lettings, and the landlord will often have to redecorate and refurbish to obtain the best rents from what is, by definition, a floating market. Experience has shown that average annual outgoings on AST lettings are between 30% and 50% of the gross annual rent.

Reference so far has been made to the estimation of net income by calculating each outgoing separately. It is useful as a check to determine what percentage they represent of the gross rents.

(c) State of repair

Careful consideration must be given to the actual state of repair as affecting the annual cost of repairs, the life of the building and the possibility of heavy expenditure in the future. Particular attention should be paid to the structural condition, the presence or absence of dampness in walls or ceilings, the proper provision of sanitary accommodation and of cooking and washing facilities.

Regard should be had to the possibility of the service of a dangerous structure notice in respect of such defects as a bulging wall.

Where premises are in such a condition as to be a nuisance or injurious to health, the local authority may require necessary works to be done. They may also serve a notice to repair a house that is in substantial disrepair or unfit for human habitation. The standard for deciding whether a house is "unfit for human habitation" is that laid down by section 604 of the Housing Act 1985.

Where a house cannot be made fit for human habitation at a reasonable cost, the local authority may serve a demolition order. In respect of any part of a building unfit for habitation, including any underground room, or any unfit house from which adjoining property obtains support, or which can be used for a non-residential purpose, or is of special historic or architectural interest, they may serve a closing order instead of a demolition order.

In lieu of making a demolition order, a local authority may purchase an unfit house if they consider that it can be rendered capable of providing accommodation of a standard that is adequate for the time being.

Special provision is made in the Housing Act 2004 (Part 2) as regards houses in multiple occupation (HMOs), requiring all such properties to be registered and licensed. If a house is registered as an HMO, then it is probable that planning consent will be required to return it to a single dwelling and therefore the valuation may have to assume that the property will be a permanent investment and that it cannot be valued on a vacant possession basis as a house even if full vacant possession can be obtained.

Where, on inspection, a property is considered not to comply with statutory provisions currently in force, allowance should be made in the valuation for the cost of compliance.

The allowance for annual repairs included in the outgoings is usually based on the assumption that the property is in a reasonable state of repair, at least sufficient to justify the rent at which it is let. If it is not, a deduction should be made from the valuation for the cost of putting it into reasonable repair, less any grant payable.

(d) Clearance areas

There are many houses, let either as a whole or in parts, which are of considerable age, indifferent construction, or so situated as likely to be included in a clearance area, in consequence of which the local authority may acquire the houses and buildings for the purpose of demolishing them and re-planning the area.

Particular regard must be had to the possible application of this procedure, and to the statutory provisions applying to individual houses in a clearance area, when valuing old or defective property.

It must be remembered that a house, although not unfit in itself, may be included in consequence of its being in or adjoining a congested or overcrowded area.

(e) Duration of income

In addition to the factors already considered, regard must also be had to the length of time during which it is expected that the net income will continue.

In the case of houses of some age, it is sometimes suggested that an estimate should be made of the length of life of the property, and the net income valued for that period with a reversion to site value. However, as explained in Chapter 9, it is more usual for the factor of uncertainty of continuance of income to be reflected in the rate per cent adopted.

In such cases a very high net yield may well be expected, whereas, in the case of more modern properties, the yield may be as low as 5%, or even lower if capital appreciation can be expected.

Where, however, there are strong reasons for assuming that the life of the property will be limited, allowance should be made for that factor in the valuation.

If demolition of existing buildings is likely in the future under the provisions of the Housing Act 1985 (Part IX), the income may properly be valued as receivable for a limited term, with reversion to site value.

Example 16–4

A terrace house in a suburban area built about 70 years ago comprises ground, first and second floors, with two large and one small room on each floor. There is a WC on each floor and each of the small rooms has been adapted for use as a kitchen with a sink and ventilated food storage provided. The house is situated in a neighbourhood where the development is open in character. The general structural condition is good, with the exception of the flank wall of the back addition, which is badly bulged, and the front wall where there are extensive signs of rising damp.

There are three tenants: the ground floor producing £30.00 per week, the first floor £42.00, and the second floor £30.00. These rents are the maximum rents permitted under the Rent Act 1977 and exclude council tax.

Sewerage and water charges total £400 p.a.

What is the present market value?

Valuation			
Gross income:		per week	per annum
Ground floor	£30.00		
First floor	£42.00		
Second floor	£30.00	£102.00	£5,304
Less Outgoings:			
Sewerage and water	£400		
Repairs and insurance (incl. VAT)	£1,000		
Management at 10% + VAT	£625		£2,025
Net income			£3,279
YP in perp. at, say, 10%			10
			£32,790
Estimated cost of repair to flank wall and			£15,000
damp in front wall, net of grant and			
inclusive of VAT, say			
Value, say			£17,790
say			£18,000

Note: This is a very poor investment. There is no early prospect of vacant possession and the only potential is for a regulated tenant to be replaced by an assured shorthold tenant, thus increasing the income, which is reflected in the comparatively low yield of 10%. However, the value must still be "tested" against its VP value as there may be some logic in keeping the property vacant on a piecemeal basis until full VP is obtained. If there is a large potential gain to be made, then the value would be much higher than £18,000. If the VP value was £140,000, then the market value would still be in excess of £35,000 equal to say 25% of VP.

In the case of a part-possession house there could be a good demand from owner-occupiers if they can secure finance. In the above example the first and second floors could become vacant, leaving only a single regulated tenant in the ground floor.

The small room on the top floor could be converted into a bathroom and thus these two floors would provide four rooms plus a kitchen, a bathroom and two WCs – perfectly good owner-occupier accommodation. However, the accommodation is not self-contained. It could be possible to make it so, in which case the value would be the YP value of this part less the cost of works plus the residual value of the ground floor. However, the purchase could require a profit to reflect the effort of creating the upper flat, and one view might be that the ground floor is included at no value to reflect that value. If self-containment is not possible, regard would be had to the age and health of the ground floor tenants and an enterprising owner-occupier could accept living in a non-self-contained flat for a limited period so as to obtain a house which, when vacant possession is eventually obtained, would otherwise be outside their financial ability to buy. The value might be as high as two-thirds of the vacant possession value, allowing for the state of disrepair.

6. Larger residential properties

(a) Generally

When valuing houses with vacant possession, as already pointed out in section 1 of this chapter, the capital value must be fixed by direct comparison with prices actually realised on sale.

In the case of larger houses which are let, the method recommended earlier in this chapter in relation to smaller houses, of performing a normal investment valuation and comparing the resultant figure with the vacant possession value, is again appropriate.

(b) Factors affecting value

The factors in question may be briefly summarised as follows:

1. *Size and number of rooms.* Prospective buyers are primarily in search of a certain amount of accommodation. They are likely, in the first instance, to restrict their enquiries to properties having the number of rooms of the size they require.
2. *Position.* The price they are prepared to pay for this accommodation will be influenced by all the factors associated with position; proximity to shops, travelling facilities, places of worship, open spaces, golf courses and schools; the character of surrounding property, building development in the neighbourhood, the presence and cost of public services and the level of council tax.
3. *Planning etc.* In viewing the property itself, the prospective buyers will also consider such points as the arrangement, aspect and lighting of the rooms, the adequacy of the domestic and sanitary offices, the methods of heating, the

presence or absence of a garage, the size and condition of the garden and the prospects for improving and enlarging.

The value of the accommodation provided by a house will be increased or diminished according to the way in which, in the details enumerated, it compares with other properties of a similar size.

4. *Age and state of repair.* Changes in taste and fashion and the greater amenities provided in modern houses tend to reduce the value of older houses, but this is not always the case as "period houses" sometimes have an additional value.

5. *Energy performance.* As stated earlier, this will become a more important factor in the valuation of houses and flats as new laws are promulgated relating to the need to conform with energy performance levels. As from 2018, all properties with EPC (Energy Performance Certificate) levels F and G will not be lettable and this is likely to be extended to properties for sale. No property in the UK can legally be sold without an up-to-date EPC and with the modern cost of heating, purchasers will increasingly be concerned with the running costs of their properties.

The state of repair must be considered both as regards the cost of putting the premises into a satisfactory state of repair now and the cost of maintenance in the future.

A number of the points already mentioned in connection with smaller properties must be considered.

The principal points include: the condition of the main structure and roof; the penetration of damp, either by reason of a defective damp-proof course or insufficient insulation against wet of the external walls; the presence of conditions favourable to dry rot; the presence or absence of such rot or of wet rot or attack by woodworm or beetle; the arrangement and condition of the drainage system; the condition of external paintwork; internal decorations.

Houses with a large expanse of external paintwork or complicated roofs with turrets and domes will cost more in annual upkeep than houses of plainer design. The annual cost of repairs will also be influenced by the age of the premises.

When houses are let, the incidence of the liability for repair as between landlord and tenant is an important factor in their valuation. In the case of the larger houses let on a long lease, the tenant usually undertakes to do all repairs; but in the case of short-term lettings of small or moderate sized houses, the bulk of the burden has usually fallen on the landlord. The implied repairing obligations placed on the landlord by the Landlord and Tenant Act 1985, sections 11–16 (already referred to under small houses), must also be borne in mind.

(c) Methods of valuation

At the present day, formal estimates of market value for residential property are seldom made except for mortgage, divorce or probate purposes. If a house is vacant, it will almost certainly be put on the market with vacant possession and its capital value arrived at by direct comparison with recent sales, without reference to rental value. Comparisons are sometimes made on the basis of so much per unit of gross internal floor space (GIA), but care must be taken in using this approach. Purchasers in the UK did not traditionally base their purchase price on a comparison with other houses or flats on a rate per square metre basis. But this is becoming a more common practice as more and more estate agents' details contain floor plans with a calculation of Gross Development Area (GDA), and floor plans are now mandatory in all leases. It can be very dangerous to value on this basis as gross floor area alone may not be the sole criterion of value. Non-usable areas of accommodation, such as a long narrow hall, may not add to the overall value but the area of such accommodation is included in the GIA, and thus an analysis will produce a low rate per square metre which will not be applicable to a house or flat without this "space waster". Similarly, the effect on value of a luxurious standard of finish will cause difficulties. Where houses and flats are valued by direct comparables of like with like, small differences in room size will have little effect on value. Where direct comparables are not available, then the best available comparables should be used and suitably adjusted for location, size, etc., but this cannot be done on the basis of any pure mathematical formula.

Utilising a rate per square metre basis is more appropriate when comparing flats in a single block or on an estate of similar blocks. This is frequently the approach used in leasehold reform valuations, where tenant's improvements are ignored, so that a single standard of finish is applied to all of the flats. In a collective enfranchisement of a large block of flats, it is almost never going to be possible to value each flat individually on the statutory basis if there are numerous flat types. A pragmatic approach such as a rate per square metre is the best method available, but even in this example it may be necessary to adjust for such matters as aspect and floor location.

However, flats in central London, which are popular with overseas buyers, are frequently valued on a rate per square metre basis because this is the approach used in Europe and the Far East, but usually only as a guide. The marketing of the flat will produce the correct value as this is the only way that deviations from the "norm" due to the matters set out in (b) above can be established.

New house builders often apply a rate per square metre to their proposed houses, taken from their experience of houses elsewhere on other estates they have developed. This approach works because of the factual similarities of their product in terms of room sizes, standard of finish, etc., and this is a short-hand way of using the comparables. However, the end result must then be scrutinised to ensure that the resultant value compares with other local properties, since there can well be a locational addition or deduction.

Whilst the Upper Chamber has in the past rejected valuations using this mathematical approach, preferring the entirety approach of comparing the subject property to other specific properties, this has not been the Chamber's universal approach and most London leasehold valuation tribunals are content with this method of valuation in the absence of more direct comparables.

Example 16–5
A small semi-detached, brick-built modern house in good repair, on a plot with a 30-ft frontage, contains two reception rooms and a WC on the ground floor, two large and one small bedroom, bathroom and WC on the first floor. What is its market value?

Valuation
Direct comparison of capital value: Similar houses in the neighbourhood sell for prices ranging from £255,000 to £275,000; by direct comparison of position, size of rooms, condition of repair and the amenities provided, the value is estimated to be £260,000.

In the case of tenant-occupied houses where tenants are protected by the Rent Acts or hold the house under a lease or agreement with a reasonable term to run at a low rent, the tenants themselves may be anxious to purchase the property. In these circumstances the price paid is likely to be the result of bargaining between the parties. The tenant is obviously not going to pay full vacant possession value; the landlord is not likely to be content with investment value only; the result will usually be a purchase price between these two extremes. In effect this is similar to the marriage valuation referred to in Chapter 9. The tenant's interest may have a nil market value (it may not be saleable by law) and the price paid will be the landlord's investment value and a percentage of the difference between this value and the vacant possession value.

(d) Sales records

The systematic recording of the results of sales is essential to the valuer; the form such records can take varies widely. There are now a number of free websites which provide sales information (e.g. *www.nethouseprices.com* and *www.mouseprice.com*), and some also retain the actual sales particulars online. The Land Registry is also available to check on particular property sales as well as details of tenure. However, there has been an increasing tendency to own higher value properties in "single purpose" companies (SPVs) so that the shares in the company are sold and not the property. This device was used to avoid DVLT and to preserve anonymity. However, the Budget of 2012 has sought to address this "tax avoidance" scheme by increasing DVLT to 15% for sales to a company and further measures (as yet undisclosed) are being considered to prevent the sale of these SPVs by means of a punitive tax charge. The purpose of mentioning this here is to highlight the fact that there may be many transactions that are not recorded by the Lands Registry and therefore local knowledge of the market is still important.

7. Blocks of flats held as investments

(a) Generally

The valuation of a block of flats does not differ in principle from the valuation of properties, already considered.

The problem is only complicated by the greater difficulty in forming a correct estimate of gross income and of outgoings, although where flats are let on ASTs or are vacant, it is the vacant possession value that will dominate the valuation.

(b) Gross income

In most cases where flats are let on assured or assured shorthold tenancies they are let at inclusive rents, with the tenants not being liable for any outgoings, although extra charges for special services may be encountered.

In estimating gross income, it should be borne in mind that fair rents can be reviewed every two years, but this is now subject to capping, or, if tenants are not protected, existing agreements may permit rents to be reviewed more frequently.

Among the many factors affecting the present rental value is the presence of lifts, central heating, constant hot water, the adequacy of natural lighting, the degree of sound insulation between flats, convenience of proximity to main traffic routes – offset by possible nuisance from noise.

(c) Outgoings

In estimating the allowance to be made for outgoings on an existing block of flats, it is necessary to study the tenancy agreements carefully in order to determine the extent of the landlord's liabilities. In recent years there has been a marked tendency to make tenants responsible for all possible repairs, but the provisions of the Landlord and Tenant Act 1985 (S.11–16) must be kept in mind.

Taxes and charges

The present cost of council tax and water and sewerage charges can easily be ascertained. Generally, flats are let at rents exclusive of council tax. VAT payable on items such as repairs is not recoverable and these items should be included gross of VAT.

Repairs

It is difficult to give any general guide to the allowance to be made for repairs, since the cost will depend upon many factors.

An exterior of stucco work, needing to be painted approximately every five years, will cost much more to maintain than one of plain brickwork, which is only likely to involve expenditure on repointing every 25–30 years. Regard must also be had to the entrance hall, main staircase and corridors; those having marble or other permanent wall coverings will have a low maintenance cost; other types of decoration requiring considerable expenditure to keep in a satisfactory condition will have a much higher maintenance cost.

Services

Until the Rent Act 1957, rents were frequently inclusive of services. Now these are frequently covered by a separate service charge payable in addition to the registered rent in the case of regulated tenancies. Such service charges are not usually variable between registrations but are fixed according to their costs, and may now be subject to capping. It is necessary for the valuer to check that the charge is sufficient to cover costs and depreciation. Where services are included in rents paid, the cost must be deducted from the gross rents together with other outgoings to find the net income. Where sections 11–16 of the Housing Act 1985 apply, the variable service charge element can only cover non-repair items.

The Housing Acts, the Landlord and Tenant Acts 1987 and 1996 and the Commonhold and Leasehold Reform Act 2002 give tenants paying service charges a large measure of protection. They must be consulted in advance of expenditure above given limits and given copies of specifications; very strict rules apply and if these are not followed a landlord may find himself unable to recover some or all of his expenditure. Lessees who are liable to pay for repairs are entitled to receive a written summary certified by a qualified accountant, justifying the amount charged. Service charges must be such as a court considers reasonable both as to standard of services and amounts charged. In a recent Upper Tribunal case (Garside v RFYC & Maunder Taylor) the tribunal decided that the timing of works also had to be reasonable in all the circumstances. Before ordering any "qualifying works", landlords or their agent must obtain not less than two estimates, one of which must be from a genuinely independent source, and notify the tenants (including any association representing the tenants). The Housing Act 1996 provides for a Local Valuation Tribunal to adjudicate on a tenant's liability for a service charge before a landlord can commence forfeiture proceedings for non-payment.

Similar protection is extended to "secure tenants" (i.e. tenants in the public sector) by the Housing and Building Control Act 1984, section 18 and Schedule 4, amplified by the Housing and Planning Act 1986, sections 4–5. The latter Act (section 4) inserts an elaborate code into the Housing Act 1985, Part V, concerning information which tenants are entitled to receive from landlords when exercising the "right to buy". This code, contained in sections 125A, B and C inserted into the 1985 Act, deals with "improvement contributions" as well as service charges.

(d) Net income and yield

A reasonable deduction from gross income having been made in respect of the estimated outgoings, a figure will be arrived at representing prospective net income. However carefully this figure may have been estimated, it is likely to vary from time to time, particularly in relation to expenditure on repairs. Valuers will have regard to this fact in selecting the rate per cent of YP on which their valuation will be based.

The rate per cent yield on the purchase prices of blocks of flats varies considerably. Consideration has to be given to the ratios of regulated tenants, assured tenants and ASTs. Regard must also be had to the potential for sales with vacant possession. The preferred method of valuation is to divide the flats into categories, valuing each category as a percentage of vacant possession value and with an end allowance for putting the block into such condition as will be required by a purchaser with vacant possession. Allowance must also be made for the costs of sale (e.g. agent's fees, Energy Performance Certificates (EPCs), etc.) and a profit for the wholesale buyer. Yield is not usually important as such an investment is not usually held for income.

Example 16-6

You are instructed to value a large freehold property comprising 76 identical flats in eight blocks, some of three storeys and some of two, erected about 20 years ago. The vendor's agents inform you that the flats are all let, 40 at recently reviewed fair rents, 10 are on assured tenancies, 20 are ASTs and the balance are vacant requiring modernisation. The assured rents average £7,000 p.a. and the ASTs £7,500 p.a. The standard form of tenancy agreement provides that the tenants are liable for internal decorative repairs only.

The following particulars are also supplied. Total rent roll, £443,600 (including 18 garages producing £3,600). Outgoings last year: repairs, £38,000; lighting, £4,000; gardening, £5,500; insurances, £19,000. (All costs are inclusive of VAT where payable.)

As a result of your inspection the following additional facts are established.

The tenants pay the council tax on both flats and garages, with the rents for the latter averaging £10 per week. The agreements are for three years.

The gardens are extensive. The general condition of repair is satisfactory.

The flats are brick-built with tiled roofs, concrete floor and a modern hot water system; there is no lift.

There is good demand for flats with vacant possession at an estimated price of £175,000 per flat excluding a garage.

The landlord provides no services other than lighting of common parts and upkeep of the gardens.

The outgoings seem to be reasonable.

Total rent roll			£449,360
Less Garages			£9,360
			£440,000
Add for 4 vacant flats			30,000
Less Outgoings: Lighting		£4,000	£470,000
Garden		£5,500	
Insurances		£19,000	
Repairs (incl. of VAT), say		£38,000	
Management, say		£51,750	£118,250
Net income			£351,750
YP in perp. at 8%			12.5
			£4,396,875
18 garages, gross		£9,360	
Less			
Outgoings	£900		
Repairs			
Management, say	£1,100	£2,000	
Net income		£7,360	
YP in perp. at 12.5%		8	£58,880
			£4,455,755

But

1. Assume that flats let on regulated tenancies worth 60% × YP value,

    $$= 40 \times 60\% \times £175,000 = £4,200,000$$

2. Assume that assured tenancies are worth 80% × YP value

    $$= 10 \times 80\% \times £175,000 = £1,400,000$$

3. Assume that ASTs and YP flats are worth 85% x YP value, allowing 15% for cost of sale, profit, etc.

    $$= 24 \times 85\% \times £175,000 = £3,570,000$$
    $$\text{Total} \quad £9,170,000$$

4. Allow refurbishment costs for four flats £20,000

    £9,150,000 (average £120,395 per flat)

5. 18 garages as above   58,880
    £9,208,880

Therefore there would be no prospect of the blocks being sold at an 8% yield. The actual yield will be in the region of 3.85%.

8. Places of worship

Schedule 6 to the Leasehold Reform Act 1967 amends the Places of Worship (Enfranchisement) Act 1920 which gives rights to persons holding leasehold interests in places of worship or ministers' houses to acquire their freeholds.

> Where premises are held under a long lease to which this Act applies and are held upon trust to be used for the purposes of a place of worship or, in connection with a place of worship, for the purpose of a minister's house, whether in connection with other purposes or not, and the premises are being used in accordance with the terms of the trust, the trustees, notwithstanding any agreement to the contrary (not being an agreement against the enlargement of the leasehold interest into a freehold contained in the lease granted or made before the passing of this Act), shall have the right as regards to their freehold interest to enlarge that interest into fee simple, and for that purpose to acquire the freehold and all intermediate reversions; provided that:
>
> • if the premises exceed two acres in extent, the trustees shall not be entitled to exercise the right in respect of more than two acres thereof;
> • if the person entitled to the freehold, or an intermediate reversion, requires that underlying minerals are to be excepted, the trustee shall not be entitled to acquire interest in the minerals if proper provision is made for the support of the premises as they have been enjoyed during the lease, and in accordance with the terms of the lease and of the trust;
> • this Act shall not apply where premises are used or are proposed to be used for the purposes of a place of worship in contravention of any covenant contained within the lease and which the premises are held, or in any lease superior thereto; and
> • this Act shall not apply where the premises form part of the land which has been acquired by or invested in any municipal, local or rating authority or in the owners thereof for the purposes of a railway, dock, canal or navigation under any Act of Parliament or Provisional Order having the force of an Act of Parliament and the freehold reversion in the premises is held or retained by such owners for those purposes.

The leases to which this Act applies are leases (including underleases and agreements for leases or underleases), whether granted or made before or after the passing of this Act, for lives or a life or for a term of years where the term as originally created was for a term of not less than 21 years, whether determinable on a life or lives or not.

The procedure for acquisition of the reversionary interest is set out in section 2 of the Act and the basis of compensation is the same as if the trustees acquiring were an authority authorised to acquire the premises by virtue of a compulsory purchase order, made under what is now the Acquisition of Land Act 1981, subject to a number of amendments.

Chapter 17

Commercial properties (1)

Landlord and Tenant Acts and
rent reviews

1. Introduction

There has been government intervention in the letting of commercial property for
many years. The extant legislation provides that: leases may not end when they
should end under the terms of the lease; the rents landlords might wish to charge
cannot be charged; tenants may be compensated if they are forced to leave at the
end of their leases, and may be compensated for improvements when they leave.

A valuation of commercial property cannot be prepared unless the possible
impact of the legislation has been considered. The legislation is found mainly
in the Landlord and Tenant Acts of 1927 and 1954 as amended by the Law of
Property Act 1969 and the Landlord and Tenant (Licensed Premises) Act 1990
and, more recently, by secondary legislation in the form of the Regulatory Reform
(Business Tenancies) (England and Wales) Order 2003, which came into effect on
1 June 2004.

2. Landlord and Tenant Act 1927

(a) Landlord and Tenant Act 1927 (Part I)

This Act gives tenants of premises let for trade, business or professional purposes
the right to compensation for improvements made during the tenancy. These provisions are subject to amendments made by Part II of the 1954 Act which are
comparatively unimportant in practice and in their effect on values. The tribunal
for the settlement of questions of compensation for improvements under the Act is
the County Court, but claims are referred by the Court to a referee, who is selected
from a special panel. The following notes relate solely to the possible effect on
value of the tenant's right to claim.

A tenant's claim for compensation for improvements[1] is limited to (a) the net
addition to the value of the holding as a whole which is the direct result of the

---

1 Landlord and Tenant Act 1927, sections 1–3.

improvement, or (b) the reasonable cost of carrying out the improvement at the termination of the tenancy, allowing for the cost of putting the existing improvement into a reasonable state of repair, whichever is the smaller. In determining the compensation, regard must be had to the purpose to which the premises will be put at the end of the tenancy and to the effect that any proposed alterations or demolition or change of user may have on the value of the improvement to the holding. In principle, the basis for determining the compensation payable to the tenant is the benefit that the landlord will derive from the improvement.

In order that an improvement may carry a right to compensation under the Act, tenants must first have served notice on the landlord of their intention to make the improvement, together with full particulars of the work and a plan. If the landlord does not object, or, in the event of objection, if the tribunal certifies that the improvement is a "proper" one, then the tenant may carry out the works. Tenants who carry out work in advance of obtaining the certificate lose their right to compensation. Alternatively, landlords may offer to carry out the improvement themselves in consideration of a reasonable increase in rent.

Compensation is only payable if the tenant quits the holding at the end of the term during which the improvements were made. This is the main reason why very few claims under the Act are served on landlords and why very few notices are actually served.

However, when valuing any commercial property, the valuer should establish whether any compensable improvements have been made by the tenant. The way potential claims for compensation are taken into account when preparing a valuation will depend on the circumstances. An example is given later in this chapter.

(b) Landlord and Tenant Act, 1927 (Part II)

Section 18 of the Act in Part II makes provision to limit the amount payable by tenants to their landlord as damages for dilapidations on the conclusion of the lease. Most leases provide for a tenant to leave the premises in repair at the conclusion of the lease. Failure to comply results in the tenant paying the landlord compensation for the accrued dilapidations. This topic is considered in part 6 of Chapter 13.

3. Landlord and Tenant Act 1954 (Part II) (as amended by the Regulatory Reform Order 2003)

Subject to certain exceptions and qualifications, Part II of the Act (as amended by subsequent legislation) ensures security of tenure where any part of a property is occupied by a tenant for the purposes of a business carried on by the tenant – provided it is not carried on in breach of a general prohibition in the lease. The term "business" in the 1954 Act means any trade, profession or employment and includes any activity carried on by a body of persons corporate or unincorporate.

The Act therefore covers both ordinary shops, factories and commercial and professional offices, and premises occupied by voluntary societies, doctors' and dentists' surgeries, clubs, institutions, etc.[2] Security of tenure under the Act was extended to licensed premises by the Landlord and Tenant (Licensed Premises) Act 1990. The scope of premises to which the 1954 Act applies is wider than that of the 1927 Act, which applies to premises let for trade, business or professional purposes.

The Act extends not only to lettings for fixed terms, e.g. 5-, 10–21- or 99-year leases, but also to periodic tenancies, e.g. quarterly, monthly or weekly tenancies. But the following types of tenancy are excluded:

* tenancies of agricultural holdings;
* mining leases;
* tenancies within the Rent Act 1977, or which would be but for the tenancy being a tenancy at a low rent;
* "service" tenancies, i.e. tenancies granted by an employer only so long as the employee holds a certain office, appointment or employment; and
* tenancies for a fixed term of six months or less, with no right to extend or renew, unless the tenant and the tenant's predecessor (if any) have been in occupation for more than 12 months.

The general principle upon which this part of the Act is based (section 24(1)) is that a tenancy to which it applies continues until it is terminated in one of the ways prescribed by the Act. Thus, if it is a periodic tenancy, the landlord cannot terminate it by the usual notice to quit. If it is for a fixed term, it will continue automatically after that term has expired, on the same terms as before, unless and until steps are taken under the Act to put an end to it.

2 A tennis club registered under the Industrial and Provident Societies Act 1893 was held by the High Court (Queen's Bench) to be a business letting within the 1954 Act in *Addiscombe Garden Estates Ltd* v *Crabbe* [1958] 1 QB 513; but sub-letting part of the premises as unfurnished flats with a view to making a profit out of the rentals was held by the Court of Appeal not to be a business letting within the Act in *Bagettes Ltd* v *GP Estates Co Ltd* [1956] Ch 290. In the former case, the use of the premises was "an activity carried on by a body of persons", but in the latter case it was ordinary residential use. In some circumstances a residential letting could be a business, where the landlord preserves a major degree of control over sub-tenants and provides services (see *Lee-Verhurst Investments* v *Harwood Trust* [1973] 1 QB 204). Yet occupation of premises by a medical school for the purpose of student residences was held by the Court of Appeal to be a protected business use in *Groveside Properties* v *Westminster Medical School* [1983] 2 EGLR 68. A government department can have a protected business tenancy even if the premises are occupied on its behalf by another body, rent free: see *Linden* v *DHSS* [1986] 1 EGLR 108. A lessee of an enclosed market was held not to have a protected tenancy where all of the "stallholders" themselves had protection even though the lessee had control of the walkways and common parts; *Graysim Holdings Ltd* v *P&O Property Holdings Ltd* [1996] 1 EGLR 109. This case also cast doubt on the *Lee-Verhurst* decision.

One way in which a tenancy can be determined is by the parties completing an agreement on the terms of a new tenancy to take effect from a specified date. A tenancy may also be terminated by normal notice to quit given by the tenant, or by surrender or forfeiture. In the case of a tenancy for a fixed term, the tenant may terminate it by three months' notice, either on the date of its normal expiration or on any subsequent quarter day (a section 27 notice) or by vacating before the term date.[3]

But Part II disputes which cannot be resolved by agreement must be referred to the Court, i.e. the County Court for the area in which the premises are situated (or to the High Court in exceptionally complex and difficult cases) under section 29 of the 1954 Act, as set out at the end of this chapter.

Apart from the above cases, the methods available to landlord or tenant to terminate a tenancy to which Part II of the Act applies are as follows.

Under section 25 of the Act the landlord may give notice, in the form prescribed by the Act, to terminate the tenancy at a specified date not earlier than that at which it would expire by effluxion of time or could have been terminated by notice to quit. Not less than six months' or more than 12 months' notice must be given. A landlord must give six months' notice to terminate a quarterly, monthly or weekly tenancy. The notice must require the tenant to specify by counter-notice whether or not the tenant is willing to give up possession and must state whether, and, if so, on what grounds the landlord would oppose an application for a new tenancy.

After notice has been served the parties may negotiate on the terms of a new tenancy or may continue negotiations begun before the notice was served. If they cannot agree the terms, or if the tenant wishes to remain in the premises but the landlord is unwilling to grant a new tenancy, the tenant can apply to the County Court for a new tenancy; the Court is bound to grant it unless the landlord can establish a case for possession on the grounds specified in the Act.

On the other hand, a tenant holding for a fixed term of more than one year can, under the Act, initiate proceedings by serving a notice on the landlord (a section 26 notice) in the prescribed form requesting a new tenancy to take effect not more than 12 months, nor less than six months, after the service of the notice, but not earlier than the date on which the current tenancy would come to an end by effluxion of time, or could be terminated by notice to quit. This notice must state the terms that the tenant has in mind. If the landlord opposes the proposed new tenancy, the tenant can apply to the Court for the grant of a new tenancy, the terms of which can be agreed between the parties or, in default of agreement, will be determined by the Court. The valuation date is the date of the Court Order and the lease will commence three months and four weeks after the Court hearing (the four weeks is to allow for any appeal).

3 See *Esselte AB* v *Pearl Assurance plc* [1997] 1 EGLR 73.

The Court must have regard to the following points in fixing the terms of a new tenancy[4]:

- The tenancy itself may be either a periodic tenancy or for a fixed term of years not exceeding 15 (section 33).
- The rent is to be fixed in relation to current market value and the following are to be disregarded in determining the rent:
  - (a) the fact that the tenant is a sitting tenant;
  - (b) any goodwill attached to the premises by reason of the carrying on of the tenant's business on the premises;
  - (c) any improvement carried out by a person who was the tenant at the time of improvement, but only if the improvement was made otherwise than in pursuance of an obligation to the immediate landlord and, if the improvement was not carried out during the current tenancy, that it was completed not more than 21 years before the application for the new tenancy; and
  - (d) in the case of licensed premises, any additional value attributable to the licence where the benefit of the licence belongs to the tenant.
- The terms may include provision for varying the rent even where there was no review clause in the old lease (section 34(3)).
- Other terms of tenancy must be determined having regard to the terms of the current tenancy and to all relevant circumstances (section 35).

In practice, most terms will be agreed between the parties so that the Court is required only to settle those not agreed. For example, they may agree that the new lease shall be for 25 years rather than the maximum period of 15 years which the Court can fix. Commonly, all terms are agreed other than the rent payable. The courts, in "having regard to the terms of the current tenancy", tend to retain the same covenants as in the existing lease unless sound reasons can be advanced for a change (*O'May* v *City of London Real Property Co Ltd*).[5] This does not mean that the terms of the lease are fossilised; new terms can be introduced, such as the requirement to pay interest on late payment of rent or VAT, or to take into account the provisions of the Landlord and Tenant (Covenants) Act 1995.

Subject to certain safeguards, the Court will revoke an order for the new tenancy if the tenant applies within 14 days of the making of the order, so that a tenant is not bound to accept the terms awarded by the Court.

During the period while an existing tenancy is continuing only by virtue of the Act, and provided the landlord has given notice to determine the tenancy

4 Landlord and Tenant Act 1954, sections 34 and 35, as amended by the Law of Property Act 1969, sections 1 and 2.
5 [1982] 1 EGLR 76.

or the tenant has requested a new tenancy, either party may apply to the Court for the determination of a an interim rent.[6] The interim rent is determined in accordance with the same rules that apply to a rent for a new tenancy (section 24C). But where the terms of the tenancy are changed, the old basis excluding the reference to a periodic tenancy will apply or, for example where no new lease is granted, the old basis in full will apply. The full old basis led to a general approach whereby the rental value established for the new fixed term lease is amended by a given percentage to convert it to a year-to-year basis, to allow for the valuation date (i.e. the commencement day of the interim period), and then reduced by a further percentage so as to have regard to the existing rent, known as cushioning (examples are *Janes (Gowns) Ltd* v *Harlow Development Corporation*[7] and *Ratners (Jewellers) Ltd* v *Lemnoll Ltd*).[8] However, the percentage additions or subtractions are not fixed and the courts may vary the approach depending on the facts of the case. For example, in *Charles Follett Ltd* v *Cabtell Investments Ltd*,[9] the new lease rent was fixed at £106,000. This was reduced to £80,000 as being the rent on a year-to-year basis. The further reduction so as to have regard to the existing rent was fixed at 50%, producing an interim rent of £40,000. This was because the rent was not a reasonable rent but "the rent which it would be reasonable for the tenant to pay" and this, coupled with the need to "have regard to the rent payable under the (old) tenancy" (in this case £13,500 per annum), justified such a large percentage deduction. This level of reduction is wholly unusual and the traditional deduction was approximately 12.5%. These arguments apply where the new rent is above the old rent but, if it is below, then the landlord's argument would be one based on a need to cushion the fall.

A landlord can successfully oppose an application for a new tenancy on the following grounds:

(a) that the tenant has not complied with covenants to repair;
(b) that the tenant has persistently delayed paying rent;
(c) that there are substantial breaches of other covenants by the tenant;
(d) that the landlord can secure or provide suitable alternative accommodation for the tenant, having regard to the nature and class of business and to the situation and extent of, and facilities afforded by, the tenant's present premises; and
(e) that, in the case where the tenancy was created by a sub-lease, the rental value of the whole is greater than the rental value of the parts (the effect on capital value is irrelevant);

6 Regulatory Reform (Business Tenancies) (England and Wales) Order 2003 (SI 2003/3096) inserting sections 24A–24D into the 1954 Act.
7 [1980] 1 EGLR 52.
8 [1980] 2 EGLR 65.
9 [1987] 2 EGLR 88.

(f) that the landlord intends to substantially demolish or reconstruct the premises;[10] and

(g) that the landlord requires the premises for own occupation – but a landlord is debarred from using this ground if the landlord acquired the premises less than five years before the end of the tenancy. This five-year limitation does not apply when purchasers wish to reconstruct premises for their own occupation,[11] but section 31A would most definitely come into play.

Sometimes the landlord is not ready to take possession for redevelopment because their plans are not yet complete. In such cases the landlord can ask the Court to impose a redevelopment clause in the new lease which enables the landlord to determine the lease when ready to develop. The Court of Appeal has considered this in a number of cases and has held that, provided the landlord can show a prima facie case, such a clause should be included.[12]

In cases where the landlord wishes to occupy but the period of ownership is too short, i.e. less than five years, the Court may grant only a short term.[13]

Under sections 37 and 37A, a tenant who is refused a new tenancy is entitled to compensation from the landlord if either the landlord or the Court refused it on grounds (e), (f) or (g) above, and there is not an agreement which effectively excludes compensation. Such compensation is in addition to any which the tenant may be entitled to under the Landlord and Tenant Act 1927, in respect of improvements.

If the business has been carried on in the premises for less than 14 years, the compensation is equal to a multiplier fixed by statutory instrument. If it has been carried on for 14 years or more, the compensation is twice the multiplier. The multiplier is applied to the rateable value of the premises; the current multiplier is 1.[14]

In general it is not possible to contract out of Part II of the Act, but section 38A(1) authorises a landlord and tenant, who so wish, to enter into an agreement to exclude the provisions of the 1954 Act regarding security of tenure (sections 24 to 28). Schedules 1 and 2 of the Regulatory Reform Order 2003 set out the necessary procedure, starting with a notice served by the landlord, plus the prescribed forms which have to be used. Section 38(2) of the 1954 Act allows the

---

10 There are many decided cases relating to this ground. Note that the Law of Property Act 1969, section 7, inserts an additional section (31A) into the 1954 Act which entitles a tenant to a new tenancy despite reconstruction works by the landlord if terms reasonably facilitating the performance of those works can be included, or if "an economically separable part of the holding" can be substituted for the entirety of the holding.

11 See *Atkinson* v *Bettison* [1955] 1 WLR 1127 and *Fisher* v *Taylor's Furnishing Stores Ltd* [1956] 2 QB 78 (both Court of Appeal decisions).

12 See *Adams* v *Green* [1978] 2 EGLR 46 and *National Car Parks Ltd* v *The Paternoster Consortium Ltd* [1990] 1 EGLR 99.

13 *Upsons Ltd* v *E Robins Ltd* [1956] 1 QB 131.

14 Landlord and Tenant Act 1954 (Appropriate Multiplier) Order 1990 (SI 1990/363).

parties to exclude the compensation provisions where the tenant or the tenant's predecessors in business have occupied the premises for less than five years. Some leases contain a covenant which excludes payment of any compensation, but such a covenant is void once occupation reaches five years.

Similar provisions apply under section 38A(2) to agreements between the landlord and the tenant stating that the latter consents to surrender the tenancy at a future date (i.e. not a surrender taking immediate effect). Schedules 3 and 4 of the Regulatory Reform Order 2003 set out the necessary procedure, starting with a notice served by the landlord, plus the prescribed forms that have to be used.

The effect of all these contracting-out provisions on the value of business premises which do fall within Part II of this Act has, generally speaking, been slightly to lower rents obtainable from sitting tenants when leases are renewed. There is no doubt that the rights under the 1954 Act considerably strengthen the hand of a tenant in negotiations for a new lease or in negotiations for the surrender of an existing lease in return for a new lease. Nevertheless, tenants generally prefer to have a lease for a definite term of years rather than to rely solely on their rights to a new lease under the Act. Where the goodwill of a business relates to particular premises (e.g. a retail business), the tenants who have only a few years of their lease unexpired and who wish to sell the goodwill of their businesses, or to invest in the property or the business, nearly always have to surrender their existing short leases to secure a sufficiently long term to realise the full potential of their business. This is particularly so where costs of removal are expensive or when goodwill can be lost by a move. In other cases, tenants prefer the flexibility of manoeuvre which a short lease coupled with security of tenure gives them. Purchasers frequently require the security of a term of years rather than the rights under this Act.

With few exceptions, it would seem that reversions after an existing lease should be valued on a capitalised rental value based on the terms, referred to above, to which a court must have regard when settling disputes between the parties for new tenancies. If a landlord's interest is valued on the assumption that a new tenancy will not be granted, allowance should be made for compensation which may have to be paid to the tenant.

There is little doubt that the security of tenure afforded by Part II of the Act increases the value of a tenant's interest. If the tenant's interest is valued upon the basis that the tenant will have to give up possession at the end of the lease for one of the reasons set out above which attract the payment of compensation, an appropriate addition should be made. Conversely, the effect is to decrease the value of a landlord's interest. Only in a comparatively few cases can reversions be valued on the basis of vacant possession.

In general, the 1954 Act has had a much greater effect in practice than the 1927 Act.

The following example illustrates the effects of both the 1927 and the 1954 Acts.

Example 17–1

A shop in a poor secondary location is let on a lease granted 15 years ago with three years unexpired at a rent of £4,000 p.a. Improvements were carried out seven years ago at a cost of £30,000 by the tenant, who received the landlord's consent under the provisions of the 1927 Act. Without these improvements the MR on normal terms would be £5,000 p.a., but in the improved state the MR is £7,500 p.a. The RV is £6,000. Prepare a valuation of the freehold interest.

Valuation

As the tenant has a right to a new lease disregarding the value of the improvements, the rent of £7,500 p.a. should be ignored for the probable period of the new lease, a maximum of 15 years if fixed by the Court. The landlord might not wish to grant a lease for longer than this, so as to obtain the rent from the improvements at the earliest opportunity (21 years from improvement being made). Strictly speaking, the 21 years will end 11 years into the new lease $(7 + 3 = 10)$, but the next most probable negotiation date would be at the end of the new lease. There will be rent reviews every five years, but the normal assumption is that rent review clauses contain the standard disregards and so, at rent review, the review will be to the then MR disregarding improvements The valuer must base the valuation on the most probable future position.

Term – rent payable under lease		£4,000 p.a.	
PV£1p.a. or YP for 3 years at 9%		2.53	£10,120
Reversion – to new lease (say 15 years) at MR, ignoring improvements		£5,000 p.a.	
PV£1p.a. or YP 15 years at 9%	8.06		
PV£1 in 3 years at 9%	0.77	6.20	£31,000
Reversion – to MR reflecting improvements		£7,500 p.a.	
PV£1p.a. or YP in perp. at 9% deferred			
18 years		2.36	£17,700
			£58,820
Value, say			£58,500

If it is known that the landlord wishes to occupy the premises when the present lease terminates, the value to the landlord but not MV would be:

Term – rent reserved under lease		£4,000 p.a.	
PV£1p.a. or YP for 3 yrs at 9%		2.53	£10,120
Reversion – to MR		£7,500	
PV£1p.a. or YP in perp. at 9% deferred			
3 years		8.58	£64,350
			£74,350
Deduct:			
Compensation under the Landlord and Tenant Act 1954 – say 2 × RV £6,000		£12,000	

Compensation under the Landlord and
  Tenant Act 1927

(a) Increase in rental value of landlord's reversion	£2,500	p.a.
PV£1p.a. or YP in perp. at 9%	11.11	
say	£ 27,775	
(b) Cost of carrying out improvements at end of lease, say	£45,000	
Less depreciation, say 30%	£13,500	
	£31,500	

Take smaller of (a) or (b)		£27,775	
		£39,775	
PV£1 in 3 years at 9%		0.77	£30,627
			£43,723
Value, say			£44,000

This valuation is on the assumption that the landlord is not debarred by the five-year restriction and the tenant has been in occupation for over 14 years and so is entitled to 2 × RV in compensation. The cost of the compensation is normally deferred, as here, at the property ARY.

The value of £44,000 is the value to the landlord who wishes to occupy. This represents the worth to the landlord as only the landlord can obtain possession for occupation. The market value remains £58,500 since this is the value for investor purchasers of the freehold interest.

If the landlord intended to demolish the premises, the valuation would be:

Term – rent reserved under lease	£4,000	p.a.	
PV£1p.a. or YP for 3 years at 9%	2.53		£10,120
Reversion to site value, say	£200,000		
Deduct:			
Compensation under the Landlord and	£12,000		
Tenant Act 1954			
Net value of reversion	£188,000		
PV£1 in 3 years at 9%	0.77		£144,760
			£154,880
Value, say			£155,000

In this case, no compensation would be payable under the 1927 Act as the demolition would eliminate the effect that the improvements would have on the value of the holding.

4. Health and safety at work, etc. Act 1974

This Act "to make further provision for saving the health, safety and welfare of persons at work, for protecting others against risks to health and safety in connection with the activities of persons at work" provided[15] for the progressive replacement by a system of health and safety regulations and approved codes of practice of such legislation as the Factories Act 1961 and the Offices, Shops and Railway Premises Act 1963, which remain in force.

Health and safety regulations may be made for any of the general purposes of Part I of the Act[16] and Schedule 3 sets out some of these matters in detail, including such things as structural condition and stability of the premises, means of access and egress, cleanliness, temperature and ventilation, fire precautions and welfare facilities.

These requirements must be borne in mind when leases are being granted and freehold and leasehold interests are being valued. In appropriate circumstances, deductions may have to be made to allow for the cost of complying with the regulations.

5. Rent reviews

Rent reviews begun to be incorporated in leases from about 1944 to reflect the increases in market rents occurring at that time. Previously landlords would grant leases for as long as possible at a fixed rent. By the 1960s landlords were increasingly aware that if the rent was fixed it became less valuable in real terms with inflation, and that with the steady increase in market rents value was accruing to the tenant who was not, in essence, a property investor.

Two solutions are possible. One is to shorten the length of the lease so that the landlord can adjust the rent more frequently on the grant of a new lease. This is unattractive to investors and to many occupiers because, for the investor, it increases the possibility of voids since tenants have regular opportunities to vacate, whilst for the occupier it reduces any long-term certainty of occupation. Not only that, but the frequent grant of new leases is costly. The other solution is to retain the long term of years, but provide for rent increases during that term. Initially, such increases were often effected by fixing the rent for the first few years at the agreed rental value, and then fixing a higher rent for the subsequent periods. For example, a shop might be let for 21 years at £1,200 p.a. for the first seven years, rising to £1,500 p.a. for the following seven years and £1,800 p.a. for the final seven years.

This approach went some way to overcoming the problem, but it soon became apparent that it was far from ideal because it would be purely fortuitous if the uplift represented MR at that time. If it was less, then the landlord was still losing ground; if it was more, then the tenant would be unhappy at having to pay above MR.

This led to the introduction of a machinery for the rent, at intervals, to be re-determined at the then prevailing MR. The machinery was contained in rent review

15 Section 1.
16 Section 15.

clauses. In the absence of precedents, rent review clauses took many different forms reflecting the differing views of lawyers, some of whom initially raised doubts as to whether such clauses were valid.

In time, however, the practice of providing for rent reviews grew and became accepted. In occupation leases the reviews tended to be at seven-year intervals, but this came down to five years, and in the early 1970s pressure grew for three-yearly reviews because of the high levels of inflation. The market settled in the main for five-year reviews, with the consequential effect that leases tend to be granted for multiples of five years. Ground or building leases have seen a similar change with, initially, the review period being 33 years or 25 years, but now five-year reviews are normally to be found.

The introduction and acceptance of rent reviews took several years, but by 1969 they were the rule rather than the exception. This was recognised in the Law of Property Act 1969 with the introduction of section 34(3) into the 1954 Act, which states that where the Court determines the rent of a new tenancy it can also include provision for rent reviews. The period 1970 to 1973 saw a sharp rise in property values, both capital and rental, until an economic crisis brought on by the miners' strike and the "three-day week" brought it to an end. This led to the introduction of a rent freeze on business properties in 1972, which lasted until 1974. In that period, rent review clauses became ineffective, so that new rentals were agreed but could not be claimed by the landlords. This meant that, following the lifting of the freeze, rents rose in 1974 and into 1975 when capital values were declining rapidly. It is not possible to prove, but experience does suggest that, in that period, tenants began to be far more sensitive to the fact of increasing rents and since then rent reviews have attracted considerable attention. This has been magnified by the fact that the first effects of the practice of five-year reviews, begun in the early 1970s, coincided with the falling in of many leases granted in the 1950s at fixed rents, as well as joining with the period of business depression. Tenants, and particularly companies such as multiple retailers, who are tenants of many properties, when faced with sharp increases in their rent bills in such circumstances naturally seek to restrict increases to the minimum.

The debate over rent reviews was intensified in the years after 1988 when rents generally stopped rising, and in many instances fell, in some cases very sharply to half or less of the 1988 levels. If the purpose of rent review clauses is accepted as ensuring that the rent payable follows changes in market rents, then there should be no greater problem when rents are falling than when they are rising. However, the usual practice, right from the introduction of rent review clauses, was that the rent payable from the time of the rent review should not be less than the rent payable in the period before the review. These "upward only" review clauses cause no friction when values are rising but, after 1988, they acted significantly against tenants. A tenant with a rent review in, say, 1994, when rents may have halved from the rent payable under the lease, was forced by the upward only provisions to continue to pay twice the then prevailing rent. This type of situation resulted in a large number of over-rented properties and leases with sharply negative values (see Chapter 9 for valuations of such interests).

In due course debate switched to the courts in relation to rent review provisions in renewals under section 34(3) of the Landlord and Tenant Act 1954. The sentiments that "what is sauce for the goose is sauce for the gander" are often expressed and upwards/downwards reviews have been determined in a number of cases, starting with *Janes (Gowns) Ltd* v *Harlow Development Corporation* in 1979. There is no fixed rule and the eventual result depends upon the evidence in the local market at the date of the court hearing. The final result is often a trade-off between the landlord wanting the interest to have a long lease and the tenant not wanting to be locked into what might be an uneconomic rent, i.e. an upward/downward review.

Inevitably, with closer attention being paid to the actual wording of the rent review clause in conjunction with the other terms of the lease, valuers developed valuation theories and ideas which have led to a body of valuers specialising in rent reviews. Disputes between the parties have become more common, leading to a marked increase in the use of independent determination and also to a considerable number of cases heard by the courts. There have thus emerged certain principles applicable to rent reviews.

A typical rent review clause will cover the following areas:

(a) the machinery for putting into effect the review;
(b) the factors to be taken into account and assumptions to be made in arriving at rental value; and
(c) the machinery for determining disputes in the absence of agreement.

Most of these aspects can be considered in turn.

(a) Machinery for review

In general, the review is triggered by the service of a notice by the landlord on the tenant, following which the parties' valuers have a period of time in which to reach agreement. Failing agreement, the matter is referred to independent determination, which can be by arbitration or by reference to an independent expert. (Mediation is preferred to court proceedings and the RICS Dispute Resolution Service now runs courses for those wishing to become RICS Accredited Mediators.) Although a time scale is laid down for all these matters, the time scale is not usually so restrictive as to compel absolute compliance therewith,[17] unless it is clearly stated or indicated that time is to be of the essence,[18] or

17 *United Scientific Holdings Ltd* v *Burnley Borough Council* [1977] 2 EGLR 61.
18 *Drebbond Ltd* v *Horsham District Council* [1978] 1 EGLR 96 and *Mammoth Greeting Cards Ltd* v *Agra Ltd* [1990] 2 EGLR 124.

some other terms of the lease (such as a tenant's option to break the lease following review) make it essential for time to be of the essence.[19] Sometimes there is no mechanism for the review but merely a provision for automatic change.

(b) Rent payable

The rent review clause sets down the factors that are to be taken into account in arriving at the rent. These take many forms and each clause must be read carefully to see which apply to any particular case. In general they have tended to vary between two extremes. At one extreme, the various factors affecting rent are set out in great detail, including a definition of the rent payable – "rack rent", "open market rent", "open market rent as between willing lessor and willing lessee", etc.; assumptions as to use; whether to disregard tenant's improvements; whether to disregard tenant's goodwill; whether to assume the lease has its full term to run or merely the actual unexpired term; assumptions as to general lease covenants; assumption that the rent-free period which would be given on a new letting has expired. At the other extreme, the clause is limited to saying that "market rent" or some similar phrase shall be adopted.

The assumptions may have a considerable effect on rent. In interpreting rent review clauses, the landlord and tenant are treated as hypothetical parties[20] so that the status of the actual parties is ignored unless the wording so requires.[21] For example:

(i) Rent description

Generally, where the phrase used is one commonly adopted and widely understood no problems arise. However, where less familiar terms are used they may have a considerable impact on the rent finally determined. For example, "a reasonable rent for the premises" led to the court fixing a rent reflecting the improvements carried out by the tenant, even though they would be ignored on renewal of the lease.[22]

---

19 See, for example, *Al Saloom* v *Shirley James Travel Service Ltd* [1981] 2 EGLR 96 and *Stephenson & Son* v *Orca Properties Ltd* [1989] 2 EGLR 129. For a full discussion of time of the essence, see 284 EG 28.

20 *FR Evans (Leeds) Ltd* v *English Electric Co Ltd* [1978] 1 EGLR 93. In this case the Court held that, as each party was willing, the actual market conditions which showed an absence of any demand were to be ignored so as to reflect the existence of a tenant willing to take the premises. A more realistic view of market conditions can be taken following *Dennis & Robinson Ltd* v *Kiossos Establishment* [1987] 1 EGLR 133.

21 *Thomas Bates & Son Ltd* v *Wyndham's (Lingerie) Ltd* [1981] 1 EGLR 91.

22 *Ponsford* v *HMS Aerosols Ltd* [1978] 2 EGLR 81.

(ii)   User

Normally the use assumption follows the user covenant of the lease under the general lease covenant assumptions, but sometimes an assumption is imposed which is different. In either case, where a lease severely limits the use to which a property may be put, this will tend to reduce the rent below the level acceptable where the tenant has flexibility of approach. Thus, the rental value of a shop whose use is strictly limited to the retail business of cutlers is clearly worth less than one where the tenant may adopt other activities with the landlord's consent, which is not to be unreasonably withheld.[23] This contrasts the difference between restricted user clauses and open user clauses. The class of person who may use a building may be restricted. Thus, offices held under a lease that limits occupation to civil engineers are likely to be worth less than the same offices if they can be occupied by anyone in the market, since the competition for such leases is so restricted. Such was the situation in the *Plinth Property* case[24] when it was agreed that the "open" use rental value was £130,455 p.a. whereas in the market limited to civil engineers it became £89,200 p.a. The possibility that landlords might well relax the provisions was ignored.

Where the use is even more restricted to that of a named tenant only, then it is assumed that there is a "blank lease" with the name to be inserted, and the generally prevailing rental values for that type of property will be adopted.[25] If there is an assumed use for rent review purposes which displaces the actual use covenant in the lease, it is assumed that planning permission exists for that use,[26] although the need to carry out works to achieve the assumed use would need to be reflected in the rent.[27]

(iii)   Improvements

It is widely accepted that tenants should receive the benefit of any improvements that are carried out by them not as a condition of the lease, a view reflected in successive Landlord and Tenant Acts, including section 34(1c) and (2) of the 1954 Act which requires their effect on rents to be disregarded (unless carried out more than 21 years ago). The wording of the 1954 Act is itself commonly adopted in rent review clauses.

A problem arises as to how the valuer should disregard the effect of improvements. Approaches commonly adopted include assessing the rental value as if the improvements had never been carried out, or assuming that the improvements are to be carried out at the time of review at current building costs, amortising the cost

---

23 *Charles Clements (London) Ltd* v *Rank City Wall Ltd* [1978] 1 EGLR 47.
24 *Plinth Property Investments Ltd* v *Mott, Hay & Anderson* [1979] 1 EGLR 17.
25 *Law Land Company Ltd* v *Consumers' Association Ltd* [1980] 2 EGLR 109.
26 *Bovis Group Pension Fund Ltd* v *GC Flooring & Furnishing Ltd* [1984] 1 EGLR 123.
27 *Trusthouse Forte Albany Hotels Ltd* v *Daejan Investments Ltd* [1980] 2 EGLR 123.

over the hypothetical term of years, allowing for the tenant's right to a new lease or compensation, and deducting the amortised cost from the rental value which reflects the improvements. A further approach is to adopt a "gearing" procedure whereby if, at the time of the improvements being carried out, the rental value were increased by 50%, then the rent at review, ignoring the effect of the improvements, is taken to be two-thirds of the rental value reflecting the improvements. The courts have rejected these approaches and have suggested that the parties put themselves in the same factual matrix at the date of review as they were when the lease was granted,[28] but, notwithstanding such criticisms, the methods are still commonly applied as providing a practical valuation solution even though not, perhaps, a true legal interpretation.

(iv) Goodwill

Although valuers are commonly required to ignore goodwill attributable to the carrying on of the tenant's business, this is not normally a matter which raises specific valuation issues. It would generally apply only to special properties, such as hotels and restaurants, where the history of trading by the tenant generates rental valuation evidence by analysis of the trading record of the tenant.

(v) Length of lease

A rent review clause commonly provides that, on review, the actual unexpired term is to be adopted in determining the rent. Thus, on a 20-year lease with five-yearly reviews, at the first review it is to be assumed that a new lease of 15 years is being offered (with five-yearly reviews), at the second 10 years, and at the third five years. This is unlikely to have any significant effect, apart from the final review when such a short term might be argued to be unattractive to tenants. Where the tenant has carried out major improvements which are to be ignored, then the effect might be significant at later reviews as the time left to recoup the expenditure by adjustment of rent could require a considerable reduction. However, the possibility of the tenant obtaining a new lease under the 1954 Act should be reflected and this will tend to negate the tenant's argument for a lower rent.[29]

On the other hand, some leases do provide that it shall be assumed that on each review a lease for the original term is being offered. However, it seems that this is to be interpreted as assuming that the lease offered is for the actual term of years but commencing when the actual lease started,[30] unless the facts

28 *GREA Real Property Investments Ltd* v *Williams* [1979] 1 EGLR 121 and *Estates Projects Ltd* v *Greenwich London Borough* [1979] 2 EGLR 85.
29 *Pivot Properties Ltd* v *Secretary of State for the Environment* [1980] 2 EGLR 126.
30 See, for example, *Ritz Hotel (London) Ltd* v *Ritz Casino Ltd* [1989] 2 EGLR 135; *British Gas plc* v *Dollar Land Holdings plc* [1992] 1 EGLR 135.

of the case require otherwise.[31] This of course coincides with the reality of the case.

(vi)   General lease covenants

The rent review clauses will generally provide that the other terms of the lease are to be assumed to be included in the hypothetical new lease. Thus covenants as to repair, insurance and alienation are to be the same as in the lease subject to review. The more restrictive or onerous they are to one party, the more will the rent be adjusted upwards or downwards to reflect this quality, as is the case on the grant of a lease.

(vii)   Expiry of rent-free period

In the years after 1988 and at various times since, landlords have had to offer inducements to tenants to take leases. These often took the form of a long rent-free period, way beyond the traditional few months at the start of a lease when tenants are fitting out premises. The rent payable at the end of the rent-free period became known as the "headline rent". So it might be said that "offices have been let at £10 per sq ft (£108 per m²) with a rent-free period".

Two developments emerged from this introduction of long rent-free periods. One was the problem of analysing the terms of the letting to seek to obtain a rent on review which was equivalent to the headline rent obtained on the letting of comparable properties. The approach is considered below.

The second was the interpretation of rent review clauses where landlords sought to obtain the headline rent on a rent review, but which rent was recognised as being above the true rental value if rent is payable without a rent-free period, as is the case with rents arising from a rent review.

As an example, the rent review clause of a lease provided that the rent on review should be determined "... upon the assumption that ... no reduction or allowance is to be made on account of any rent-free period or other rent concession which in a new letting might be granted to an incoming tenant". This clause, and other clauses with similar wording, was considered by the Court of Appeal when it heard four cases together. The Court adopted a "presumption of reality" in interpreting the clauses that seek to avoid business commonsense. As a result, in three cases[32] it was held that the rent payable on review was not the headline rent, but a rent derived from analysis of the comparable transactions (see below). In one case,

31 *Prudential Assurance Co Ltd* v *Salisburys Handbags Ltd* [1992] 1 EGLR 153.
32 *Prudential Nominees Ltd* v *Greenham Trading Ltd* [1995] 1 EGLR 97; *Co-operative Wholesale Society Ltd* v *National Westminster Bank plc* [1995] 1 EGLR 97; *Scottish Amicable Life Assurance Society* v *Middleton Potts & Co* [1995] 1 EGLR 97.

however,[33] it was held that a headline rent was payable on the particular wording of the rent review clause.

The general conclusion is that a headline rent will not be payable unless no other interpretation of the rent review clause is possible. If the short fitting-out period is to be disregarded, then suitable wording to this effect will be successful.

(viii)   Whether rent review in hypothetical lease

One problem area concerns a phrase commonly adopted, whereby the hypothetical lease is assumed to contain the same covenants as in the existing lease "other than as to rent" or some such exclusion. It seems probable that the purpose of this exclusion is to indicate that the rent itself will be different. However, the courts have considered the meaning of this exclusion clause in the various forms it takes, and in particular whether it meant that the actual clause requiring the rent to be reviewed should be assumed to exist or not.

Initially, the courts adopted a strict literal interpretation and decided that ignoring the provisions as to rent meant that the rent review clause itself should be ignored.[34] Later cases adopted a more commercial view so that, unless the words in the clause clearly required the provisions as to rent review to be disregarded, a rent review clause should be assumed.[35] In a yet later decision on appeal to the Court of Appeal, the more literal approach was favoured but this was for rather special reasons.[36] Finally, two cases in 1991[37] appear to have settled the matter, requiring the commercial view to be adopted, unless the wording of the clause makes it beyond question that no rent reviews shall be adopted. Indeed, a judge in the *Arnold* case stated that the Arthur Young decision in 1984 was clearly wrong.[38]

These examples illustrate that the valuer needs to weigh up the effects of the assumptions in determining the rent, as the rental valuation is of the hypothetical world which the assumptions create rather than the real world if the property were being put on the market with full flexibility of action. Even where the clause requires the rent to be the market rent as between willing lessor and willing lessee, if the parties are not willing then a hypothetical willing lessor and lessee must be assumed to exist.[39]

---

33 *Broadgate Square plc* v *Lehman Brothers Ltd (No 2)* [1995] 2 EGLR 5.

34 *National Westminster Bank plc* v *Arthur Young McClelland Moores & Co* [1985] 2 EGLR 13.

35 *British Gas Corporation* v *Universities Superannuation Scheme Ltd* [1986] 1 EGLR 120.

36 *Equity & Law Assurance Society plc* v *Bodfield Ltd* [1987] 1 EGLR 124.

37 *Arnold* v *National Westminster Bank plc* [1993] 1 EGLR 23 and *Prudential Assurance Co Ltd* v *99 Bishopsgate Ltd* [1992] 1 EGLR 119.

38 See n.34 *supra*. However, even this may not turn out as badly as would have been thought. In *Prudential Assurance Co Ltd* v *Salisburys Handbags Ltd* (n.31 *supra*) the absence of an assumed term of years allowed the court to hypothesise a term equal to what was normally offered in the market.

39 See n.19 *supra*.

6. Valuation consequences

The considerable attention given to aspects of rent reviews has led to some stan-dardisation of the contents of rent review clauses. Nonetheless, they still vary considerably and valuers needs to pay close attention to their actual wording in each case as the rent review clause represents the instructions to valuers as to how they should prepare their valuation. As a general rule, the more onerous or restrictive the covenants are on the freedom of action of the tenant, the lower the rent, and vice versa. The consequences on the rent payable can be considerable. For example:

(a) Restricted user

Although there are two cases where the effect on rent of a restricted user has been determined, each case depends on the fact of that particular case. For example, a lease of a shop where the user is restricted to greengrocer might have minimal effect where the shop is one in a small neighbourhood centre. On the other hand, such a restriction where the shop is close to a supermarket selling a full range of food products including fruit and vegetables could justify a huge reduction from the generally prevailing level of shop rents.

Normally there will be no direct evidence to support any rent adjustment, so that it becomes a matter of valuation judgment.

There are cases where a restrictive user can justify an increase in rent. This will be so where the retail units in the area or centre are under the control of a single landlord who maintains a pattern of use through an estate use plan so that the tenant is guaranteed a monopoly, or only a limited amount of competition.

(b) Period between reviews

Another problem that may arise in the welter of different lease covenant assump-tions is the actual period between reviews. Clearly, if the lease in question has five-year rent reviews and the rental evidence is of similar leases with five-year rent reviews, then a direct comparison can be made. But what if the review period is shorter, say three years, or is actually longer, say seven or 21 years, or is assumed to be longer by the interpretation of the phrase "other than as to rent" referred to above, when no rent review is to be assumed? As a general rule it can be said that the longer the period between the reviews, the higher should be the rent, and vice versa. The mathematical analysis using discounted cash flow techniques is considered in Chapter 10.

Landlords tend to resort to a mathematical solution which puts them in the same position on a discounted rent basis whatever the review period. Where the period is relatively long, this can produce a rent fixed well above the rents paid with the common five-year review pattern. In the Arthur Young case referred to above,[40] the Court recognised that the uplift was 20.5%. In practice, there is strong tenant resistance to significant increases as rents reach levels which tenants argue they cannot afford, particularly in the early years after the review. They may accept an uplift of up to 10%, but argue that tenants would not afford more than this. A "rule of thumb" approach has emerged whereby 1% for each additional year beyond five years is added to the rent on a five-year basis. These arguments can only be tested by marketing properties on different rent review bases, but this is rarely attempted.

(c) Analysis of incentives

When the market supply outweighs demand, as happened particularly after 1973 and 1988, landlords offer incentives to applicants to persuade them to take their space. These incentives take various forms but typically include long rent-free periods and capital payments to the new tenant – "reverse premiums". In such market conditions these lettings are the only evidence usually available to the valuer since rent reviews, being upward only, are not activated. The problem for the valuer is how to analyse the transactions.

Example 17–2

A landlord has recently let 10,000 sq ft (930 m^2) of offices on a lease for 10 years with a review to full rental value after five years. No rent is payable for the first three years, the rents payable in years four and five being £100,000 p.a. The offices are similar to offices for which a rent review is now due.

The valuer needs to determine what rent the tenant would have paid if there had been no rent-free period, which is the assumption of the rent review of the similar offices. It is assumed that the headline rent of £10 per sq ft (£108 per m^2) is not payable under the rent review clause (if it is, a substantially different answer would be achieved).

Various approaches are available to the valuer. One is to find the annual equivalent of £100,000 p.a. payable in years four and five, which would be:

$$\frac{£100,000 \text{ p.a. } \times \text{ YP 2 years } \times \text{ PV 3 years}}{\text{YP 5 years}}$$

(At a yield of 8% this produces a rent of £35,463.)

A DCF comparison, which establishes a common NPV between an assumed day one rent and the actual letting, would produce the same result.

However, such analyses are derived from an assumption that the present value of either approach should be the same. This is a landlord's point of view. A tenant

---

40 See n.34 *supra*.

might not agree with such an analysis since the initial rent-free period is a form of business loan which might be of particular value, especially for a newly established business.

In practice, an arithmetical average approach has become commonly adopted. Thus, taking the period up to the rent review, the analysis would be:

$$\text{Rent payable (Years 4 and 5)} = 2 \times £100,000 = £200,000$$

$$\text{Total years to review} = 5$$

$$\therefore \text{Annual average} = £40,000 \text{ p.a.}$$

Similar approaches can be adopted for the analysis of a reverse premium. Suppose, for example, the landlord in the above example also offered an initial sum of £200,000. Here, even more than with the rent-free period, the attraction to a tenant of such a capital sum clearly produces a different approach between the parties and thus a different analysis.

There are yet further aspects to consider. Although the landlord is receiving no rent for three years, the fact that the property is occupied means that he avoids outgoings such as rates, insurance and repairs so that, by the letting, the landlord converts the negative cash flow of an empty property into nil costs. The tenant takes on this burden, so that the comparison between the two lease approaches should probably be on the basis of the cost of occupation rather than rent only, particularly where a discounted cost/value analysis is adopted.

No single accepted form of analysis has yet emerged, although the problem has been well detailed in articles[41] and is covered by the RICS UK Valuation Standards in UKGN6 *Analysis of commercial lease transactions*. As in other aspects of valuation, the valuer needs to follow Red Book guidance but must formulate his or her opinion on the basis of their market awareness.

It may well be that the simple arithmetical average, though divorced from valuation principles of discounting, produces as reliable an answer as other methods given that the many factors involved make an approximate day one rent the best that can be achieved. It is up to the valuer to apply judgment to the result of the analysis to form a view as to its reasonableness.

(d) Comparable evidence

In a normal market the valuer will have evidence of actual market lettings. These form the most reliable guide as to the level of values. Clearly, the valuer needs to know the details of the whole agreement.

---

41 For example, R. Goodchild [1992] 08 EG 85; J. Lyon [1992] 23 EG 82; J. Rich [1992] 43 EG 104; T. Asson [1994] 36 EG 135; D. Epstein [1993] 14 EG 90, 15 EG 120, 16 EG 83, 17 EG 76; and Inaugural Professorial Lecture by Neil Crosby at Oxford Brookes University on 6 October 1993.

In the absence of market lettings, the valuer may have details of rent review agreements negotiated between the parties which can be analysed to produce comparable evidence.

The only other sources of evidence are the rents awarded by experts or arbitrators in the case of disputed rent reviews, and rents determined by the courts for new leases granted under the Landlord and Tenant Act 1954. Such transactions provide the least reliable evidence, and arbitrators' determinations cannot be relied on as evidence when presenting a case in support of a rental value to such parties.[42]

(e) Machinery for settling disputes

Where the parties cannot agree, then the clause usually provides for the matter to be referred to an independent third party, commonly a valuer agreed between the parties or one appointed by the President of the Royal Institution of Chartered Surveyors. The valuer may be required to act either as an arbitrator or as an independent expert; normally the rent review provisions in the lease will be specific as to which, but some leave the choice to one or the other party. An arbitrator must determine the new rent as between the parties, but an expert can determine a rent using their own knowledge and experience; it is that expert's opinion.

7. Market value

A valuer instructed to estimate market value will need to pay careful attention to any rent review clauses where there is a rent review in the lease and where the valuation may require a reversion or top slice income to be reflected in the calculations. Clearly, in these circumstances, the rent at reversion may be different to market rent as defined by the RICS.

8. Court procedure for the grant of a new tenancy or the termination of a current tenancy

Section 29 of the 1954 Act (as amended[43] with effect from 1 June 2004) provides as follows:

1. Subject to the provisions of this Act, on application under section 24(1) of this Act, the court shall make an order for the grant of a new tenancy, and accordingly for the termination of the current tenancy immediately before the commencement of the new tenancy.

---

42 *Land Securities plc* v *Westminster City Council* [1992] 2 EGLR 15.
43 Regulatory Reform (Business Tenancies) (England and Wales) Order 2003 (SI 2003/3096).

2. Subject to the following provisions of this Act, landlords may apply to the court for an order for the termination of a tenancy to which this Part of this Act applies without the grant of a new tenancy:

    (a) if they have given notice under section 25 of this Act that they are opposed to the grant of a new tenancy to the tenant, or
    (b) if the tenant has made a request for a new tenancy in accordance with section 26 of this Act and the landlord has given notice under subsection (6) of that section.

3. The landlord may not make an application under subsection (2) above if either the tenant or the landlord has made an application under section 24(1) of this Act.

4. Subject to the provisions of this Act, where the landlord makes an application under subsection (2) above:

    (a) if the landlord establishes to the satisfaction of the court any of the grounds on which the landlord is entitled to make the application in accordance with section 30 of this Act, the court shall make an order for the termination of the current tenancy ... without the grant of a new tenancy, and
    (b) if not, it shall make an order for the grant of a new tenancy and before the commencement of the new tenancy.

5. [The tenant does not want a new tenancy].

6. [The tenant consents to the withdrawal of an application made under subsection 2].

Chapter 18

Commercial properties (2)

Types of property

1. Introduction

The main categories of commercial property are:

(a) Retail.
(b) Industrial.
(c) Warehouses.
(d) Offices.

There is considerable overlap between categories such as retail and warehouse and in each there are several sub-categories. For example, retail property can be subdivided into:

- Single shops, market stalls, barrows, kiosks, which can be found in:

  - high streets,
  - secondary locations,
  - corner units,
  - small parades.

- Shopping centres:

  - in town centres with multi-storey car parks, or
  - on edge-of-town sites close to motorways with surface or multi-storey car parks.

- Superstores, which can be:

  - predominantly food stores, or
  - single store companies selling food, homeware, electricals and almost everything that was previously available in town centres and shopping centres but is now available at a one-stop shop.

- Department stores.
- Retail parks with warehouse-style buildings on one or two floors selling direct to the public a range of merchandise including electrical, pet requirements, DIY, electronics, carpets, etc.
- Factory outlets and factory outlet villages.

Thus "retail" covers several types of property where goods are sold to the public, with some being similar to warehouses but open to the public and, therefore, unlike a non-retail warehouse where goods are centrally assembled prior to distribution to trade purchasers.

A similar overlap exists between offices and industrial. Offices themselves range from small shop-type units in shopping streets, used by building societies or estate agents, to floors above shops occupied by solicitors, accountants or employment agencies providing a local service, to free-standing buildings given over to office activities, which themselves range from the compact to the largest multi-storey office blocks found in city centres. More recently, business parks have often been developed close to motorway links or airports. A business park will consist of several buildings where office activity runs alongside research and manufacturing activity but where the whole building is of a more-or-less uniform standard of fitting and finish, commonly termed "high-tech" or B1 buildings. This manufacturing activity, generally in electronics or similar activities, can properly be described as industrial, but differs substantially from other, more traditional industrial activities, in small to large factories, and even more from the large "heavy" industrial plants given over to engineering and other large-scale manufacturing activities. A recent new use for such premises is the private hospital facility that undertakes contract operations for the NHS.

The traditional dividing line between these various activities of retail, industrial, warehouse and office is, in the interests of business efficiency, becoming less rigid – a fact acknowledged by the introduction of the Town and Country Planning (Use Classes) Order 1987 (SI 1987/764). Nevertheless, most properties tend to relate principally to one of these activities.

The feature common to these four classes of property is that, generally speaking, they are occupied for the purpose of carrying on an industry, trade or profession in the expectation of profit; and it is the profit that can reasonably be expected to be made in the premises or corporately which, in the long run, will determine the rent a tenant can pay. The principal factors affecting each class are considered in detail later and fall under three broad headings:

(1) the quality of the premises;
(2) the quantity of the accommodation; and
(3) the location of the premises.

There are some types of property which, although capable of inclusion in one or other of the above categories, are seldom let to a tenant for trading purposes and in respect of which there may be little or no evidence of rental transactions to support an investment valuation. Typical examples are premises given over to a special purpose, such as chemical, gas or electricity plants or works, and premises occupied for charitable purposes, other than charity shops.

Such properties are likely to come onto the market only when the use for which they were built has ceased. In this case their value for an alternative use to which they can be put must be considered, subject to planning permission for that change of use or for redevelopment. Where any have to be valued for the particular purpose for which they are used, resort may be had to the depreciated replacement cost approach described in Chapter 19.

2. Retail premises

(a) Location

The range of properties in this category has already been considered but, for all properties, the chief factor affecting values is that of position. The prospective buyer or tenant is likely, in nearly every case, to attempt some estimate of the trade in those premises in that position. The accuracy of such an estimate will depend on the retailer's experience. Many of the major retailers, that is those with multiple branches, can, with a fair degree of accuracy, assess the turnover they are likely to achieve in a given location. From their estimate of turnover they deduct the cost of goods to be sold and all their overheads; allowing for their profit margin and interest on capital. They will then arrive at a residue which they can afford to pay in rent, rates and repairs.

Shop premises can generally be used for a variety of trades without the necessity of obtaining planning permission. In an open market, premises which are capable of occupation for a number of different purposes will appeal to a number of different tenants; there will be competition and the prospective occupier whose estimate of the margin available for rent, etc., is largest, is likely to secure the premises by tenancy or purchase.

A valuation of shop premises is rarely based on an analysis of the probable profits of a particular trade. But the general factors likely to influence prospective occupiers in their estimate of turnover and margin available for rent will have to be taken into account by the valuer.

Shops in a prime position, such as the main thoroughfare of an important town with large numbers of passers-by, will command a higher rent than those in secondary positions. The purchasing power of the shopper must also be considered; for instance, customers' retail spend is likely to be greater in Bond Street, London, than in the high street of a provincial town.

In some instances, large variations in value may be found within a comparatively short distance. A location at the corner of a main thoroughfare and a side street may be more valuable than one a short distance down a side street. The three main qualities that determine value are said to be "location, location and location", and this is particularly so in the case of shops.

Some of the most important points regarding retail location are: the class of the area; the type of street and the type of shopper; the position of the unit in the street; the proximity to any multiple stores or other "magnet" such as a department store; proximity to any breaks such as a town hall, bank or cinema; the relationship to the "prime" pitch – the location with the highest footfall; and proximity to car parks or public transport. Edge-of-town or out-of-town centres create their own locational advantage, provided there are adequate transport links. Overall, consideration must be given to the catchment area and its spending power.

(b) Type of premises

The next factor is the quality of the premises and their suitability for retailing; the need for goods to be displayed may be an important factor for some trades and, where it is, the frontage may be critical to rental and capital values. Other factors that may need to be considered include: the quality of the shop front – although most traders like to fit their own, this can be important in secondary locations; the size and shape and condition of the interior; the adequacy of the lighting; the presence or absence of rear loading facilities; the character of the upper floors, whether separately occupied or only usable with the shop; the general condition of repair; and the ancillary staff facilities.

In the case of newly erected premises, shops are often offered to let on the basis that the tenants fit out the shop and install their own shop fronts at their own expense; what is offered is a "shell", often with the walls and ceiling to be plastered and the floor in rough concrete. In the case of new developments, capital incentives and rent-free periods may need to be offered to tenants to induce them to take units. The more important the trader is for the success of the centre, the more it is likely to be offered. In established districts premises will be offered to let complete with shop front. If the latter is modern and likely to be suitable for a number of different trades, the value of the premises may be enhanced. If, however, the shop front is old-fashioned, or only suitable for a limited number of trades, it may need to be replaced by or for the incoming tenant and this may affect rent; however, most multiple retailers have their own house style and so the nature and condition of the shop front, electrics and other fittings will be immaterial as they will undertake a full fit-out. Sometimes a forecourt may add value as, provided ownership is retained, it can be used for the display of goods or by cafes, restaurants, wine bars and the like for additional seating.

In some cases it may be important to consider the possibility of structural alterations in order to extend at the rear, convert upper floors to retailing or, in a retail warehouse unit, insert a mezzanine floor. Any potential of this nature, given sufficient local demand, will increase rent and capital values. On the other hand, larger premises may benefit from subdivision to provide a few "standard" units or, for very large premises, the creation of a shopping precinct.

In valuing older premises, it will be important to consider the possibility of redevelopment or refurbishment, subject to planning permission, to extract added value.

In the case of upper floors, it is necessary to consider the best purpose to which they can be put, subject to planning permission. In high rent locations this space can be used for retailing; elsewhere the highest rent may come from use as offices or from residential use. The extent to which basement areas are of value will depend upon location. In high rent areas basements will be used for retailing or storage but in low rent areas they may be more of a liability than a benefit.

The comments so far have concentrated on the traditional shop unit found in shopping streets, but similar considerations apply to out-of-town superstores, to retail parks and to shopping centres, although the emphasis will be different.

For example, the quality of location for out-of-town premises will be tested more against accessibility for car-owning customers than footfall, although public transport may be important. Thus the location should be well placed for the road network, with sites near to motorway junctions being preferred. Extensive car parking will be required, ideally at ground level. In the case of in-town shopping centres, a good road access is less easy to achieve but extensive, generally multi-storey, parking is essential.

Similarly, the type of premises is important. Retail warehouses require large open lofty spaces, although a high standard of internal finish is not generally necessary. On the other hand, food superstores require a high standard of design both internally and externally. Indeed, the cost of fitting out such stores can match or exceed the cost of the building itself. Similar requirements as to design and finish apply to shopping centres, not only to the shop units themselves but increasingly to the common parts. Most of the early shopping centres with open malls have now been covered in and upgraded with better floor surfaces and improvements to lighting, general design and to public areas and facilities. The effect on rent of such works can be significant, with tenants paying considerably more in recognition of the improvement to trading potential. Retailing is a rapidly changing business and the valuer involved in this sector needs to be alert to the changing patterns and trends in the industry and to the performance of all retailers.

(c) Type of tenant

This is of considerable importance to the investment value of retail property.

A letting to a large multiple concern is generally considered to give better security of income, since the covenant to pay rent by such an undertaking can, normally, be relied on for the duration of the lease even if the profits from the particular unit are less than was expected or the lease is assigned to another trader (subject to the provisions of the Landlord and Tenant (Covenants) Act 1995). This contrasts with premises let to individuals or small companies with limited resources, where the chance of business failure is higher. The cessation of business by retailers such as Woolworths, MFI and others from 2009 onwards has reaffirmed the need to assess tenants' covenants with extreme care. Credit checks are now part of due diligence when advising on new tenants and when undertaking valuations.

The tendency for prime properties to be occupied by multiple concerns and secondary properties by tenants of weaker covenant contributes to the sharp difference in yield between prime property at low yields and secondary or tertiary property at high yields.

(d) Terms of the lease

At one extreme the tenant may be responsible for all repairs and insurance and at the other the landlord. In between these extremes a variety of different responsibilities is found. Until recently, responsibility has generally been placed on the tenant, but with shorter leases there may be more pressure for landlords to take on responsibility for external repairs. Where the landlord has any responsibility for outgoings, an allowance must be made in the valuation.

The length of the lease is important. An investor will prefer a long lease with provision for regular rent reviews, but this may not be possible, and landlords may have to accept either short leases or leases with break clauses.

The terms of letting of the upper parts (if separate from the shop) and the presence of any protected tenancies under the Rent Acts must be noted. In the case of shopping centres (and similarly in the case of Business Parks, Industrial estates and multi tenanted office buildings) the unweighted or weighted average unexpired lease term is important in terms of income security.

(e) Market rent

When comparing the rent of shops in very similar positions regard must be had to size. Lock-up and small or medium size shops may provide sufficient floor space to enable a large variety of businesses to be carried on profitably and may be in great demand, whereas the rent of larger shops, for which there is less demand, may not increase in proportion to the additional floor space. For example, a lock-up shop with a frontage of 7 m and a floor space of 140 m^2 may command a rent of £14,000 p.a., equivalent to £100 per m^2; whereas nearby, in a similar position, a larger shop with a frontage of 10 m and a depth of 20 m may only command a rent of £16,000 p.a., equivalent to a rent of £80 per m^2.

In recent years, as patterns of retailing change, demand has grown for large units so that it could be that, in prime positions, the rent for larger units will be proportionately higher than that for small units. The valuer will need to assess the relative demand for different sizes of units within the shopping area before preparing a valuation.

Although various "units of comparison" can be used for comparing rents paid for other units with the unit to be valued, such as frontage or standard unit comparisons, the method most commonly used is that of "zoning", as described in detail in Chapter 20 Where zoning is used, the areas are NIA in accordance with the RICS *Code of Measuring Practice*.

In valuing extensive premises comprising basement, ground, and a number of upper floors, there may be some difficulty in deciding upon the comparable rent of the different floors. Zoning allows for this by adopting a proportion of the Zone A rent; thus, a first-floor retail area may be valued at one-sixth of the Zone A rent, whereas first-floor storage might be one-tenth. The actual rent per m^2 would need to be considered to assess whether it is uneconomic for such a use. For example, if the Zone A rent is £1,500 per m^2, then one-tenth, or £150 per m^2, might be regarded as excessive for storage space on the first floor if warehouse or storage space in the area has a rental value of only £40 per m^2. The solution to problems of this type is illustrated by the following example.

Example 18–1

Estimate the MR of shop premises comprising basement, ground, first and second floors, in a large town. The premises are old but in a fair state of repair; the basement is dry, but has no natural light; the upper part is only capable of occupation with the shop. The shop has a frontage of 7 m and a depth of 14 m. The first floor is retail and the second floor comprises staff rooms.

The net floor space is: basement, $110 \ m^2$; ground floor, $98 \ m^2$; first floor, $80 \ m^2$; second floor, $30 \ m^2$.

The following is the comparable evidence of recent lettings of modern units:

(1) Premises similar in size let at £36,000 p.a. Areas: basement, $100 \ m^2$; ground floor, $100 \ m^2$; first floor, $70 \ m^2$ retail space; second floor, $60 \ m^2$ storage. The ground floor has a frontage of 5 m and a depth of 20 m.
(2) A lock-up shop, 4 m frontage and 5 m depth, let at £9,000 p.a.
(3) A shop and basement, frontage 15 m and depth 28 m, with a basement of $100 \ m^2$ let at £86,500 p.a. The upper part is let separately, the first floor of $140 \ m^2$ at £7,000 p.a. and the second floor, $130 \ m^2$ at £5,200 p.a.
(4) The basement under lock-up shop No. 2 and three more shops is dry and well lit with direct street access; floor space is $120 \ m^2$ and it is let for storage purposes at £3,600 p.a.

In practice these lettings might be on different terms agreed at different times. Adjustments would be required to the rents to reduce them to net figures, adjusted for any rent movements.

Analysis

Analysis of the first letting is best deferred until other lettings of various parts of the premises have been dealt with:

Letting	Floor	Area in m^2	Rent, £	Rent, £ per m^2
2	Ground	20	9,000	450
4	Basement	120	3,600	30
3	First	140	7,000	50
3	Second	130	5,200	40

The evidence from these lettings can be used to analyse the letting of shop No. 3 as follows:

Rent for shop and basement		£86,500
Less		
Basement		
100 m^2 at, say, £25 per m^2		£2,500
Rent for shop		£84,000
Adopting 6 m zones:		
Zone A	15 m × 6 m = 90 m^2 – in terms of Zone A =	90 m^2
Zone B	15 m × 6 m = 90 m^2 – in terms of Zone A/2 =	45 m^2
Remainder	15 m × 16 m = 240 m^2 – in terms of Zone A/4 =	60 m^2
Total area in terms of Zone A ("ITZA")		195 m^2

95 m^2 ITZA at £84,000 gives a Zone A value of £431 per m^2.

This evidence can then be used to check the letting of shop No. 1, as follows:

Basement:	100 m^2 at £25	£2,500
Ground floor:	55 m^2 ITZA at £431	£23,705
First floor:	70 m^2 at £71.83	£5,028
Second floor:	60 m^2 at, say, £25	£1,500
		£32,733

Actual rent £36,000, equating to £480 per m^2.

Note: The ground floor is Zone A (5 × 6) + Zone B (5 × 6)/2 + remainder (5 × 8)/4 = 66.5 m^2 in terms of Zone A. The first floor, being retail, is valued at A/6. The Zone A rent is higher than in Shop No. 3, which is larger than usual.

Market rent estimate
Bearing in mind age, position and other relevant factors, the estimate might be as follows:

Basement:	110 m^2 at £20	£2,200
Ground floor:	66.5 m^2 ITZA at £450	£29,925
First floor:	80 m^2 at £75	£6,000
Second floor:	30 m^2 at £20	£600
		£38,725
MR, say £39,000 p.a. net if all comparables are net.		

The preceding comments on zoning apply to shops in town centres and shopping parades. However, the zoning method has limitations and is not generally applied to larger units such as supermarkets, chain stores and department stores. In these cases ground-floor retail areas will be valued at an overall rate per m^2 and upper-floor retail areas again at an overall rate per m^2, possibly related to the ground-floor rate, or on the basis of a percentage of turnover.[1]

In the case of retail warehouses or out-of-town food stores, the MR will generally be determined by comparison with similar units in the area or even in comparable locations in other urban centres. The rental analysis will produce rents per m^2 applied to the total retail floor space (GIA), perhaps with a different rent for storage space. Unlike town-centre shopping centres, the factors in determining rent will depend more on accessibility, customer parking and visibility.

A further form of rental analysis may be found where the rent payable is related to the level of trade or turnover, commonly known as a "turnover rent". Generally, the rent payable has two elements, a base rent payable whatever the turnover, and a further rent that is payable, being a proportion of the turnover in excess of a specified amount.[2] This is the method most frequently adopted in factory outlet centres (FOCs) such as those at Bicester and Ashford.

Where shopping centres are spread over more than one floor the rent may be different on different floors. All will depend on the design of the centre as in some cases there is no dominant floor, as for example in Brent Cross, London, where, because of the topography of the site, each of the two main floors is directly accessible from the surface-level car parks and each is therefore the "ground floor". In some developments the multi-storey car parks are accessed directly from each floor so that again there is no dominant floor. However, in others the basement and upper floors have no direct access point from the outside and they rely on shoppers using the lifts and escalators; in these cases the rents for these floors may be 50% or even less of the dominant floor rent. Even in shopping centres there can be different rents according to location, as there may be side malls with lower footfalls than the main avenues of the centre. In all these illustrations rent is considered to be a function of turnover, so that the greater the footfall the higher the rent.

(f) Market value

Similar factors need to be considered when analysing and comparing retail yields. Shops let to good tenants can generally be regarded as a sound security since rent ranks before many other liabilities, including debenture interest and preference dividends in the case of limited liability companies.

Present-day yields may vary widely from around 5–9%. The lower rates reflect the above-average and above-inflation rate of growth anticipated for shops in

---

1 See Chapter 7 of *Valuation: Principles into Practice* (5th edition) ed. W.H. Rees and R.E. Hayward.
2 For a detailed examination of turnover rents see Paper Five, "Turnover Rents for Retail Property", by R.N. Goodchild in the *Property Valuation Method Research Report* published by the RICS and South Bank Polytechnic in July 1986.

prime positions which has been seen over recent decades, with occasional periods of low growth or even falling values as experienced, for example, in some centres in the early 1990s and, in some areas, post 2008.

Where the upper part of a freehold shop premises is let separately, e.g. for residential purposes, it is probably better to value it separately at a different yield from that used for the shop. However, in the case of leaseholds, this would involve an apportionment of the ground or head rent, and a more practical method may be to deduct the capitalised head rent, at a fair average yield for the whole premises, as an end allowance.

Evidence of actual transactions is the best basis for assessing market yields.

3. Industrial premises

(a) Introduction

The rent for this type of property is based on a rent per m^2 of gross internal area (GIA) (see Chapter 5, section 4). The range of industrial properties is extensive and varies from shop or residential property or railway arch, converted for use for storage or manufacturing purposes, to well constructed, well lit, up-to-date premises with many amenities.

Location is one of the main factors affecting value, including access to motorways or link roads and railways, proximity to markets and, in particular, proximity to the supply of suitable labour. These factors must be considered when comparing similar properties.

(b) Construction

In terms of construction, the following factors and others will need to be considered:[3]

* Good natural lighting is essential in many trades; premises in which the use of artificial light can be kept to a minimum might command a higher rent than those where a good deal of work has to be done under artificial light.
* Adequate working heights are essential and in new single-storey factories 6–7 m to eaves is regarded as a minimum.
* Floor strength is an important factor as many trades require a high load factor for machinery.
* Fire risk, fire safety, fire safeguards such as sprinkler systems, fire escapes. Fire insurance is high in this sector, and a design that reduces fire risk reduces insurance premiums.

---

3 Full check lists can be found in the RICS guidance note *Contamination, the environment and sustainability*, 3rd edition, 2010.

- Compliance with the requirements of the Factory Acts in respect of sanitation, ventilation, light, etc. and all other Health and Safety legislation.
- Available services, including the type of heating plant, electricity supply and equipment, a gas supply and mains water.
- The layout of the premises should be such as to give clear working space, to facilitate dealing with the delivery and dispatch of goods, and the handling of goods within the premises without undue labour. Ideally the space should be uninterrupted by columns or other roof supports so as to provide large clear areas.
- There must be an adequate standard of artificial lighting.
- The mode of construction may also be important in relation to the annual cost of repair. The rent will be diminished, for example, where the roof is old and, by reason of age or construction, is likely to leak. Most modern industrial processes require single-storey accommodation so as to provide continuous production lines. Therefore, upper floor space is usually of little value except to house ancillary uses such as offices. These upper floors represent usually no more than 10% of the total gross internal floor area and are measured gross. It is a disadvantage if the whole of each floor of each building is not at the same level, or not at the same level as the exterior, i.e. if there are steps or breaks in the floor because of the extensive use of fork-lift trucks.
- Multi-storey factories are rarely found except in town centres and, to be effective, these must be provided with adequate goods and passenger lifts. They are usually divided into small units and are often referred to as "flatted factories". In these cases the letting terms should provide for a service charge.
- In large factories the provision of adequate canteen, welfare and car parking facilities is essential.
- Loading and unloading facilities should allow easy access for fork-lift trucks and large vehicles. In some urban areas access to premises is often poor and, at worst, off-street loading and unloading may be impossible. This has a significant adverse impact on values and many such buildings must be considered to be obsolete; indeed, many have been converted into flats and sold under the generic term "lofts".
- Finally, but not least, environmental or green issues must be considered, such as:
  - contaminants; flood risk; rainwater runoff and storage; waste disposal provisions.

(c) Market rent and market value

Many industrial properties are owner occupied and there may be more evidence of capital values than rental values. In these cases valuation by direct comparison will be possible. Market values arrived at on this basis should be checked against the value as an investment. An investment valuation may be needed in the case of a sale and leaseback transaction. If the building is over-specialised it may have little value to an alternative use and may have to be considered as a development site. If it is being valued for accounting purposes, then a Depreciated Replacement Cost method could be used (Chapter 19).

In town centres older multi-storey buildings are often let off in floors, with the tenants paying inclusive rents and the landlord being responsible for the maintenance of common staircases, lifts, and sometimes the provision of heating or the supply of power. The method of valuation here is to estimate the net income by deducting the outgoings from the gross rents. Where necessary, the actual rents will be checked against comparables. Typically, an allowance will be made for that percentage of the building which is vacant – a void allowance (see Chapter 6). The cost of services may, however, be recovered by a service charge. Service charges are also found on industrial estates where, although individual factories will be let on full repairing and insuring terms, the landlord provides other services, the cost of which is recovered through a service charge. In some cases service charges may be capped and there could be a shortfall.

Currently, the yields for modern single-storey factories are around 7–10% but are significantly higher for poor-quality premises. The range of rents shows a similar wide disparity, from about £150 per m^2 for the best down to £20 per m^2 or less for the poorest.

Any comparison must take account of all the above factors. It may be that one factory in a similar position, area and construction has a modern heating system, sprinklers and good road access, whereas the other lacks these amenities. Such differences are usually taken into account by modifying the rent/price per m^2. But where the provision of additional facilities is possible at a low cost it may be more realistic to value the property as if the improvement had been carried out and then deduct the cost as an end allowance.

It may be that with extensive manufacturing processes there will be both new and old buildings, some of which may have little market value. They may have been erected for the purpose of a particular production process and, although the GIA area may be extensive, they may have little market value. MV is not a function of size but a function of supply and demand.

The following example is typical of fairly modern factory premises.

Example 18–2

Value, for the purpose of sale, a large factory on the outskirts of an important town.

The premises are owner occupied and acquired by them in 1988; the premises are mainly single storey, with offices, canteen and kitchens in a two-storey block fronting the access road. The building is a steel frame structure, externally clad and lined internally. Height to eaves is 7 m. There is good top light. The floor is of concrete, finished with granolithic paving. There is an efficient heating system. Gas, electricity and main water supplies are connected. The site has a frontage of 100 m to a main road and has a private drive-in with two loading docks. There is adequate lavatory and cloakroom accommodation. The site area is 1.20 ha.

Temporary storage buildings with breeze-slab walls and steel-truss roofs carried on steel stanchions have been erected since the property was bought. The available areas are: main factory, 3,000 m^2; boiler house, 70 m^2; loading docks, 120 m^2; outside temporary stores, 200 m^2; ground floor offices, 200 m^2; first floor offices, 120 m^2.

There are a number of similar factories in the vicinity varying from 1,000 m^2 to 2,500 m^2, let within the last five years at rents varying from £40 to £100 per m^2. The rents have risen steadily.

Market value
Valuations of industrial premises are usually based on the GIA. From the inspection, it will be possible to estimate the market rent by comparison. A reasonable basis for a valuation might be as follows:

		m^2		£
Ground floor:	Main factory floor	3,000	at £80	£240,000
	Loading docks	120	at £80	£9,600
	Outside stores	200	at £20	£4,000
	Offices	200	at £110	£22,000
First floor:	Offices	120	at £100	£12,000
MR, net				£287,600
PV£1p.a. or YP in perp., say 8.5%				11.76
				£3,382,176
Less costs of acquisition at 5.75%				£183,901
Value, say				£3,198,275

Alternatively, the GIA can be valued at an overall rate of £79 per m^2 plus the outdoor stores taken at £20 per square foot or £82.50 per m^2. This can be done provided that the ancillary space is of average size for an industrial building of this size. The more common approach is to take an overall rate for the good useable space, notwithstanding that the offices are finished to a higher standard than the main industrial area.

Notes:
(1) The increased value placed on floor space used as offices reflects the higher standard of finish and amenity compared with the main factory. The office content, just under 10% of the total floor space, accords with the norm.
(2) The total area covered by buildings, allowing for outside walls etc., is probably about 4,000 m^2, so that on a site of 1.20 ha and working on a site coverage of 40%, there is room for extension. Depending on the particular circumstances, this might justify some specific addition to the value but, in this case, has been accommodated in the rounding up.

4. "High-tech" industrials

Reference was made earlier to the types of building that are provided to house high-technology-based industry. Although many of the locational requirements that apply to light industry also apply to this type of industry, there are some differences to be noted. The first is general locational environment, where higher standards are required. The second is proximity to residential areas and other facilities appropriate to the level of staff employed, which again must be of a higher standard. The third, which is non-locational, is a much higher standard of internal working environment. "High-tech" buildings tend to be of two or three storeys and the internal finish is of office rather than industrial standard. The essential feature, however, is flexibility, so that any part of the building can be used for any of the various activities which form part of the total process. Finally, a higher quantity and standard of car parking is required.

Rents for these types of premises range, at the time of writing, from £80 to £250 per m^2 overall and yields tend to be slightly lower than for the best industrial, from about 6%, but there may be an owner/occupier premium.

A matter which has led to some difficulty has been the terms of leases. Landlords seek the traditional pattern of 20 or 25 years with five-year reviews, but the type of occupier concerned often desires a shorter term commitment; this is particularly true of foreign-owned companies.

There is no consensus of opinion as to whether the rent should be applied to the GIA or NIA. The more the premises appear to be offices, the more likely it will be that NIA will be used.

5. Business and science parks

Reference has already been made to the industrial estate, which has been a feature of the industrial scene since the early part of the 20th century. The business park is a more recent development which exploits the greater flexibility between business uses permitted by the 1987 Use Classes Order. Access to a motorway tends to be the chief locational requirement but, in so far as "high-tech" and office uses will be accommodated, their specific locational requirements must also be met. The science park caters primarily for research and development activities, and the essential locational requirement is proximity to a university or other academic institution with which the occupiers cooperate.

6. Warehouses

A warehouse[4] is, by definition, a place where people house their wares. At one time warehouses were found mainly around docks. Goods were brought to the

4 The comments here refer to wholesale warehouses, not to the previously considered retail warehouse.

docks to be loaded on to ships, or they lay there, having been unloaded from ships, waiting to be distributed around the country. Indeed, in many of the older docks there are former spice warehouses, sugar warehouses, and others designed for specific commodities which have been converted to "loft" spaces.

After a decline, warehouses have now taken on a greater significance in the development of the economy so that today they represent an important element in the economic structure. For example, the growth of major retailing groups has led to demands by them for large warehouses to which the various goods they offer for sale can be delivered by the manufacturers and there assembled with other items for distribution to their stores.

Similarly, shippers, having moved towards containerisation, require warehouses where they can have goods delivered and where they can join with others to fill the containers for various destinations. Indeed, the many developments in the manufacturing, retailing and distribution industries have led, in recent years, to considerable demand for warehousing facilities which were not contemplated previously and which led to the development of warehouses, and whole warehouse estates, to meet these demands. In consequence, the warehouse, or shed as it is often termed, has become a major attraction for investors, in contrast to former times when it was generally of secondary importance and interest.

The investor in warehouses is looking for various qualities when judging any particular warehouse; qualities which will also attract the occupier. The principal qualities are:

(i) Location

Ideally, it should be well located within the general transport network close to the motorway system.

(ii) Site layout

A warehouse is a building to which products are brought and subsequently removed. This involves lorries bringing in goods and removing them. Thus there will be a flow of vehicles, and the best planned warehouses have facilities to accommodate lorries waiting to unload, good unloading facilities, and easy means of access. Contrast this with old warehouses where lorries have to park in narrow streets when loading or unloading or have to manoeuvre into a narrow access to a loading yard.

(iii) Design

A modern warehouse will have easy access for vehicles, and loading bays, or docks, so that lorries can readily be brought up to the premises to load or unload. It will be single storey to avoid raising goods by lift or crane or gantry. It will have clear floor space to allow fork-lift trucks to transfer goods around the property unhindered by columns or walls. The headroom will allow bulky goods to be moved around without hindrance and will allow smaller items to be stocked to the height limit of the fork-lift trucks or other mechanical devices.

If these qualities are brought together, it is seen that a modern warehouse is normally a single-storey building, of clear space, with minimum height to eaves of around 7 m, and up to 12 m or more within the loading/unloading facilities to accommodate the largest lorry, high floor loading capacity, and located close to a motorway and urban centres. Such a building is totally different from the "traditional" warehouse, usually multi-storey, close to docks or a railhead, in an area with poor street access and no off-street loading/unloading facilities. It is important to keep this sharply contrasting picture in mind when warehouses are mentioned.

The valuation of a warehouse reflects these various qualities. In determining the rental value, the actual quality judged against the desirable qualities must be considered. The common approach is to apply a rent based on the GIA. Thus, in a locality, warehouses that exhibit the most desirable qualities might command a rent of up to £150 per m^2, whereas old multi-storey warehouses command rents of £20 per m^2 on the ground floor with reducing rents on the upper floors – if any tenants can be found for them.

This appears to be the general approach, so that cubic content tends not to be considered in assessing rental value even though it would appear that the capacity for storage should be a critical factor. Warehouses with low headroom tend to attract lower rents than warehouses with high overall clearance, but nonetheless the cubic capacity is rarely determined and noted. An exception to this is specialist warehouses. For example, cold stores, which are essentially warehouses of a special type, are generally described in relation to their cubic content. In the case of occupiers using containers or pallets, the required height is a multiple of the height of the containers or pallets – the difference between a height of 4.5 containers to 5 containers is 11% in cubic capacity, but effectively 25% in storage capacity. Apart from such special cases, a warehouse is typically described and considered in relation to its floor area.

The yields required by investors in warehouses tend to be at or around the yields required for industrial premises of a similar age and quality. There are, however, two qualities that might lead to the acceptance of a lower yield. First, the amount of wear and damage that a warehouse suffers by use will tend to be less than that of an industrial unit. Secondly, a warehouse is an adaptable building. It is usually a simple building, referred to familiarly as "a large shed", which enables it to be switched to other uses speedily and for a low cost. Hence, if the demand for warehousing falls away, a modern warehouse can readily be adapted to industrial use, or to retail use. Indeed, there has been a strong demand for the adaptation and use of warehouses as retail stores (e.g. cash and carry) for discount stores, superstores or "retail warehouses", which were considered earlier in this chapter. In the case of older multi-storey warehouses, many of these can be converted to residential use in the form of flats or lofts, giving them an extended life and a higher, residential value.

Thus, the warehouse is again an important building in the economy, both to users who require well planned storage space for their operations, to investors who feel that they offer a sound investment because of this demand by users, frequently large companies with a sound covenant, and to developers who are happy to meet the demand since the actual development of warehouses generally presents fewer problems than most other forms of development.

The most recent development has been that for the storage of electronic data. This is often as a second or third back up for financial institutions needing to ensure that they can continue with salary and pension payments or other sensitive financial business in the event of an unplanned failure in their main data storage facilities. Their valuation is very specialised and is the subject of an RICS guidance note titled *Valuation of data centres*.

The preponderance of owner occupation in factories and warehouses is greater than for offices and shops; thus the unit of comparison may be value per sq m rather than a rent per sq m.

7. Office properties

(a) Introduction

Offices range from converted houses or floors above shops to modern buildings in large towns or on landscaped edge-of-town sites. The largest will contain considerable floor space and will provide in the one building many amenities, which may include restaurants, club rooms, shops, a post office, gymnasiums, etc. and are equipped with central heating, air conditioning and lifts. In the latter case the services provided by the landlord may be considerable, and may include not only lighting and cleaning of the common parts of the building but also the cleaning of offices occupied by tenants. The cost of such services is usually covered by a service charge payable in addition to the rent. Recently there has been the development of out-of-town offices in spacious, well landscaped sites, sometimes referred to as campus offices.

The method of valuation usually employed will be to arrive at a fair estimate of the gross income, deduct the outgoings borne by the landlord, and capitalise the net income. Apart from retail premises, offices tend to have a significantly higher rent than other uses, with rents for high-quality offices ranging from around £100 per m² up to £1,000 per m² or more in the best locations.

In practice, some offices are distinguished from others because they are open to and are visited by members of the public or clients. Typically these are banks, post offices, solicitors and estate agents and are located in or close to retail centres, sometimes in retail units converted for their use. The valuation approach to such offices follows that for retail property as set out above.

(b) Market rent

The annual value is usually considered in relation to area, and in advertisements of floor space to let in office buildings the rent is often quoted at so much per m² or per sq ft, based on the NIA. Notwithstanding that a code of measuring practice has been produced by the RICS, disputes often arise as to the exact NIA or net useable area. In the case of new buildings, the premises are often re-measured after they have been fitted out and areas are agreed between landlord and tenant for rent review purposes.

Actual lettings will require careful consideration and analysis and may often appear to be inconsistent. This may be accounted for by the fact that tenancies were entered into at various dates and that the landlord will have sought to make the best bargain possible with each tenant. In recent years, incentives have commonly been offered to new tenants. Such incentives include long rent-free periods and reverse premiums, particularly for office lettings. The need to have the full details of a comparable letting is obvious, and the problems of analysis to provide a rent on a unit basis were examined in Chapter 5.

A further factor is connected with the question of varying areas of floor space. It may be, for example, that the whole of the fifth floor of a building is let to one tenant, whereas the sixth floor is let to seven or eight different tenants, so that the total net income derived from the higher floor may be greater than that from the one below.

The valuer's estimate of market rent will depend upon the circumstances of each case and upon the valuer's judgment of the best way in which the building can be let. The valuer will be guided primarily by the level of market rent obtaining in the area, based on analysis of recent lettings of similar premises let on similar terms in comparable buildings. In general, offices are let on full repairing and insuring leases. In the case of buildings in multiple occupation, the tenant will be limited to internal repairs but a service charge will be levied in addition to cover external repairs and maintenance of common parts and services.

It is important that the same basis of area is used in analysis as is employed in the subsequent valuation. This will usually be on an NIA basis excluding lobbies, toilets and the like, as detailed in the RICS *Code of Measuring Practice.*

(c) Terms of tenancy

The terms of occupation will vary considerably. Lettings are today almost always exclusive of rates. The cost of external repairs, maintenance of staircases, lifts and other parts in common use, in the case of buildings let in suites, will be covered by a separate service charge. The tenants may be liable for all repairs to the interior of the offices. Where whole buildings are let to single tenants, the lease is normally a full repairing and insuring one.

Where accommodation is let in suites it is quite likely that individual variations will be found in the tenancies within the same building, and careful examination of the terms of tenancy for each letting is necessary.

(d) Outgoings

Where property is let on terms other than full repairing and insuring, the landlord's outgoings must be assessed and, even where there is a service charge, care must be taken in checking the lease as these are sometimes capped and may not cover the full cost, in which case there could be an income shortfall. Outgoings are considered in depth in Chapter 6.

Where there is a cost to the landlord, it is usual to make an allowance for management and this is likely to vary between 3% and 5% on the gross rents.[5]

But, as most management fees are negotiated on a time-and-cost basis, the management fee estimated at, say, 3% of market rent should be checked with what might be a reasonable fee based on the probable management workload.

(e) Net income and market value

The estimates of net income made on the lines indicated above will not necessarily accord with the actual net income of any recent year. It should, however, represent what can be taken to be a fair average expectation over a number of years.

The basis upon which net income is to be capitalised will depend on the type of property, its situation and neighbourhood, the competition of other properties in the vicinity, and the valuer's estimate of the general trend of values for the type of investment under consideration. The yields on a high-class modern building let to first-class tenants will, at the time of writing, range from 5.5% in the best locations to around 8%. Similar buildings in multi-occupation will produce slightly higher yields and older, poorer buildings and conversions even higher than this.

Care should be taken, in analysing sale prices to arrive at yields, to ensure that the methods used are similar to those it is intended to apply to the valuation. For example, reliable evidence of yields can only be deduced from the result of a sale where an accurate estimate of net income is known or can be made. In addition, special factors may affect the sale price and the yield – for example, there may be an allowance for a void period, or the existing lease may expire shortly with a probable void on a re-letting brought into account, or the property may be over-rented. These problems are addressed in Chapter 9.

In some cases an investment that has to be valued may comprise different types of property, which in themselves might be valued at different yields. For example, a property might comprise shops on the ground floor, a basement restaurant, and

---

5 See *Service Charges – Law and Practice* (2012) by P. Freedman, E. Shapiro and B. Slater (Jordans) for a detailed consideration of services and service charges.

offices in the upper parts. The appropriate yields might be 7.5% for the shops, 7% for the restaurant and 8% for the offices. Where the property is held on lease at a ground rent it is difficult to apportion this so as to arrive at a fair net income in respect of each of the different occupations. In practice the valuer, having capitalised the occupation rents individually at appropriate yields, may deduct the capitalised ground rent – capitalised at an average of those yields – as an end allowance. However, if the building represents a single investment, then an overall yield might be used that reflects the relative risks.

Example 18–3

Value for sale the freehold investment in a block of offices on ground and five floors over.

The landlord supplies services comprising lifts, central heating to the offices (but not hot water to lavatory basins, as the tenants are responsible for this) and lighting and cleaning of those parts of the building not let to tenants. The costs of these services, of repairs for which the tenants are not directly liable and of all insurances are recoverable from the tenants by a service charge. The provisions in the leases regarding the service charge have the normal escalator clause which covers rises in costs. The tenants arrange for their own office cleaning and pay for their own lighting. All leases make tenants responsible for internal repairs and incorporate five-year upward-only rent review clauses.

Ground floor
Let to an insurance company for offices at £70,000 p.a. exclusive for 20 years from this year.

First floor
Let at £40,000 p.a. exclusive for 10 years – now having two years to run.

Second floor
Let at £38,000 p.a. exclusive for 10 years – now having two years to run.

Third floor
Let at £43,000 p.a. exclusive for 15 years – now having five years to run.

Fourth floor
Let at £34,000 p.a. exclusive for 10 years – now having seven years to run.

Top floor
Let at £8,000 p.a. exclusive for five years – now having two years to run.

The premises are in good condition and provide an acceptable modern standard of accommodation. The building dates to the 1930s, is of brick and stone construction with a slate roof and was refurbished to a high standard about ten years ago. The NIA of the various floors is:

Floor	Area	Current rent	Market rent	Period to increase
	m²	£	£	years
Ground	350	70,000*	70,000	0
First	310	40,000	46,500	2
Second	300	38,000	45,000	2
Third	285	43,000*	43,000	0
Fourth	285	34,000	43,000	2
Top	100	8,000	15,000	2
		233,000	262,500	

Valuation

The first stage in the valuation is to draw up a schedule collating the information to hand. This is best done in column form. Those floors marked * are the most recently let and form the basis for the other figures. The valuer would rely not only on the rents obtained for this building, but also on one's local knowledge of similar recent lettings of comparable accommodation. The evidence suggests that the upper floors are worth £150 per m² (£43,000 for about 285 m²) and the ground floor has a higher value of £200 per m².

The actual and estimated gross income is now apparent as:

For the next two years	£233,000 p.a.
Then onwards	£262,500 p.a.

Outgoings

All outgoings are included in the service charge, except management.

Gross rents for next 2 years		£233,000	
Less management, say 5%	£11,650		
Net income		£221,350	
PV£1p.a. or YP 2 years at 8%		1.7833	£394,733
Gross rents after 2 years		£262,500	
Less management, say 5%	£13,125		
Net income		£249,375	
PV£1p.a. or YP in perp. at 8%	12.50		
PV £1 in 2 years at 8%	0.8573	10.7162	£2,672,365
£3,067,097			
Less costs of acquisition at say 5.80%			£166,768
			£2,898,957
Market value for sale, say			£3,000,000

An allowance has been made for management. This is unusual where the rents are on the equivalent of FRI terms and where the managing agent's fee will be payable by the tenants under their service charges.

An end allowance for the buyer's costs of acquisition is made, otherwise the true yield will not equate to the required yield. The costs are stamp duty (currently 4% over £500,000), solicitor's and valuer's fees, plus VAT. With VAT at 20% the total is approximately 5.80%. To calculate the adjustment, the sum of £3,067,097 is divided by 1.0580 to get £2,898,957. This is correct as adding back 5.80%, namely £168,139, produces £3,067,097.

Chapter 19

UK valuation standards

Valuation for financial statements; replacement cost (contractor's) method; loan security valuations and fire insurance replacement cost

1. Introduction to the RICS valuation standards

Valuers have been required to provide valuations for financial statements for many years. These purposes include values to be incorporated in the financial statements that are required by law to be produced by an entity; values of a company's property assets to be incorporated in a prospectus when the company is going public or in respect of takeovers or mergers; and values of property unit trusts and the like. These and other valuation purposes, including valuations for loan security, are covered by the RICS *Valuation – Professional Standards (VS)* and in the UK by the *UK Valuation Standards (UKVS)*. These are mandatory for members of the RICS and the Institute of Revenues Rating and Valuation. This chapter is based on the 2012 Red Book; a 2013 edition is planned and may affect parts of this chapter.

The Red Book consists of an introduction and glossary, followed by Valuation Statements (VS) 1 to 6 covering: Compliance and ethical requirements (VS1), mandatory for all valuers; Agreement of terms of engagement (VS2); Basis of value (VS3); Applications (VS4); Investigations (VS5); and Valuation reports (VS6). These are followed by the UK specific standards covering: Valuation for financial statements (UKVS1); Valuations for financial statements – specific applications (UKVS2); Valuation of residential property (UKVS3); and Regulated purpose valuations (UKVS4). These standards include a number of guidance notes and UK guidance notes (GN and UKGN). The Red Book is concerned with standards only; for information relating to other aspects of valuation, readers are referred to the RICS information papers.

The stated purpose of the standards "is to provide an effective framework, within the Rules of Conduct, so that users of valuation services can have confidence that a valuation provided by an RICS member is objective and delivered in a manner consistent with internationally recognized standards including those set by the International Valuation Standards Council (IVSC)". Compliance with the standards is mandatory for RICS members, and sanctions apply in the case of any material breach.

The purpose of The Red Book is to set out what is described as "best practice" to be followed by valuers in the preparation and production of a valuation. It does not provide guidance on how to value in an individual case.

A number of the purposes covered by the Red Book are "Regulated purpose valuations". These are:

* valuations for financial statements under UKVS 1.1;
* valuation reports for inclusion in prospectuses and circulars to be issued by UK companies under UKVS 2.1;
* valuations in connection with takeovers and mergers under UKVS 2.2;
* valuations for collective investment schemes under UKVS 2.3; and
* valuations for unregistered property unit trusts under UKVS 2.4.

Source: RICS UKVS 4 Regulated purpose valuations.

In all these cases more "stringent" requirements must be complied with (VS 1.9; UKVS 4.2; UKVS 4.3).

Valuers undertaking Red Book valuations must be members of the RICS Valuer Registration Scheme and firms will generally be RICS-regulated firms. The notes that follow provide an introduction to UK valuations for financial statements and valuations for loan security purposes for the benefit of student readers; those actively involved in these areas of valuation will need to familiarise themselves with the whole of VS 4, aspects of VS 6 Appendix 5 on Valuations for commercial secured lending, and the whole of the UK valuation standards.

2. Valuations for financial statements

The 2012 Red Book, under VS 4.1 Applications, covers valuations for inclusion in financial statements. This specifies that "Valuations for inclusion in financial statements shall be provided to comply with the applicable financial reporting standards adopted by the entity." Where International Financial Reporting Standards (IFRS) are adopted, then the basis of value will be *fair value*. The Red Book states that:

Valuations based on *fair value* shall adopt one of two definitions:

1. The definition adopted by the IVSC: The estimated price for the transfer of an asset or liability between identified knowledgeable and willing parties that reflects the respective interests of those parties (IVS 2011).
2. The price that would be received to sell an asset, or paid to transfer a liability, in an orderly transaction between market participants at the measurement date (IFRS 13); this is the definition adopted by the International Accounting Standards Board (IASB).

These two definitions are not the same, and when valuers are asked to value for financial purposes they should pay strict regard to VS 1 in respect of professional competence. The Red Book incorporates references to IFRS 13 www.iasb.org and to IVS 300. Full knowledge of these standards and their application are essential where a firm is agreeing terms for the valuation of property for the purpose of financial statements.

Under UKVS 1 the basis of valuation for *"financial statements"*[1] prepared in accordance with UK Generally Accepted Accounting Principles (GAAP) shall be on the basis of either:

(a) property other than *specialised property* – existing use value (EUV), as defined in UKVS 1.3 for property that is owner-occupied for the purposes of the entity's business; or *market value (MV)*, as defined in VS 3.2 for property that is either surplus to an entity's requirements or held as an investment;

(b) for *specialised property – depreciated replacement cost*.

If the property is a *specialised property* then, under UKVS 1.1(b), the depreciated replacement cost method can be used. This is because the property's specialised nature means that there will be virtually no directly comparable sales evidence and no comparable rental evidence, thus precluding the use of the comparative method and the income or investment method.

In summary, valuations for financial statements under IFRS should be on the basis of FV; this may in some circumstances be equivalent to MV. In those cases where statements are prepared under UK GAAP, surplus property and investment property are valued to MV, owner occupied to EUV, and DRC will be used for specialised property (see below).

Existing Use Value is defined in UKVS 1.3 as:

> The estimated amount for which a property should exchange on the *date of valuation* between a willing buyer and a willing seller in an arm's length transaction after proper marketing wherein the parties had knowledgeably, prudently and without compulsion, assuming that the buyer is granted vacant possession of all parts of the property required by the business, and disregarding potential alternative uses and any other characteristics of the property that would cause its *Market Value* to differ from that needed to replace the remaining service potential at least cost.

Put simply, it is MV assuming vacant possession and excluding other alternative uses.

---

1 *Financial statements* are defined in the Red Book glossary as " Written statements of the financial position of a person or a corporate entity, and formal financial records of prescribed content and form. These are published to provide information to a wide variety of unspecified *third-party* users. *Financial statements* carry a measure of public accountability that is developed within a regulatory framework of accounting atandards and the law".

RICS members are kept informed of all updates to the Red Book and will need to check the sections relating to financial statements in the 2012 edition and later in the 2013 edition.

3. Depreciated replacement cost (DRC)

(a) Cost-based valuation methods generally

A cost approach is one of the three internationally recognised valuation approaches or methods. The cost of construction is also the principal or main factor in residual valuations and development appraisals (Chapter 11), and the contractor's basis in rating (Chapter 20). In the UK its use is mainly confined to non-market situations, which is where there is no actual market for the particular type of property, where the valuation is being performed in a hypothetical market or where the market cannot provide reliable direct guidance. In the USA, cost-based methods have been used for many years in parallel with market-based methods, but disillusionment with this practice is now widespread and American appraisers are moving towards limiting the use of cost methods to non-market situations. It will be found in many developing countries where the property market is opaque and undeveloped.

DRC is not a basis of valuation but the method for arriving at the depreciated replacement cost of a specialised property.

(b) The DRC method of valuation for financial reporting

This topic is now covered by GN 6 (see also RICS Red Book glossary) where DRC is defined as:

> The current cost of replacing an asset with its modern equivalent asset less deductions for physical deterioration and all relevant forms of obsolescence and optimisation.

GN 6 12.3 requires all DRC valuations in the private sector to be qualified as "subject to adequate profitability" and, in the case of properties in the public sector (12.4) or not-for-profit organisations, as "subject to the prospect and viability of the continued occupation and use".

(c) Types of property involved

The cost approach is used in the UK for properties for which there is no market and therefore no evidence of either rents or capital values. Properties for which there is no market are referred to as "specialised properties" and are defined in the VS glossary as:

> Property that is rarely, if ever, sold in the market, except by way of sale of the business or entity of which it is part, due to uniqueness arising

from its specialised nature and design, its configuration, size, location or otherwise.

In the private sector the most common examples are heavy industrial plants such as oil refineries, steelworks and chemical works, but in the public sector many buildings fall into this category, including schools, hospitals, fire stations, police stations and museums.

(d) The DRC method

GN 6 sets out the approach(es) to the three inputs to the method, namely:

1. the site value of a specialised property;
2. the cost of the buildings and site improvements of a specialised property; and
3. assessing depreciation.

The principle behind the method is that of market substitution. The valuer is seeking to establish what a willing buyer negotiating with a willing seller would pay for an existing specialised building where there is no market evidence. The approach is therefore to find the answer to three questions:

1. How much would a willing buyer pay for a similar site suitable for a modern equivalent asset?
2. How much would it cost to build that modern equivalent building?
3. How much should the cost at 2 be written down to reflect the fact that the actual building is not brand new i.e. to assess depreciation?

The current site value need not be that of the specific site of the same size in the same location, but needs to relate to a site which today would be of a size suitable for a modern equivalent building. So if it is an old hospital in a congested town centre then that would not today be suitable for a hospital. In addition, the specialised use may require very specific planning consent, referred to by the RICS as *sui generis*, so there will be no evidence of a "comparable" land sale. So value may have to be based on a similar-use category. This could be industrial for a specialised industrial plant; for the public sector it might have to be based on the amount that would have to be paid under compulsory powers; if it is a health centre in a residential area the entity may have to compete against other land uses, such as residential.

The cost of the buildings is based on the current cost of erecting the buildings and site works with the normal pluses including fees and finance charges. Where buildings contain elements that are of no value to the occupier, such as surplus floor space or excessive ceiling heights, these may be excluded; in many cases the cost of erecting a modern equivalent building will be appropriate (see GN 6.7 and 6.8).

A major issue arises in the calculation of depreciation. In the case of newly erected buildings there may be no need to make any adjustment. In other cases the valuer is seeking to assess how much less a hypothetical buyer would wish to pay for the actual building, with all its warts, than for a modern equivalent building suitable for the same use and purpose of the existing building.

GN 6.9 considers three types of depreciation, namely "physical deterioration, functional obsolescence and economic obsolescence". Briefly, physical deterioration refers to wear and tear; functional obsolescence is concerned with the design or specification; and economic obsolescence relates to factors outside the building which can be affecting demand for goods and services produced by the asset. For example, in the downturn in demand for new cars there may be built-in over-capacity in some car plants. These factors account, with the possible exception of historic buildings (GN 8.7, which includes listed buildings) where value does not decrease over time, for the DRC of the existing building being less than the current cost of its replacement.

These three types of obsolescence may be difficult to isolate as they often overlap and coalesce, so care must be taken not to double count. It is suggested that in some cases there will be total obsolescence, that is, if the building is irreparable or, if repaired, the cost would exceed the cost of a modern equivalent building; new technology may make an existing design functionally obsolete – no one would want that building for that purpose; or there may no longer be a demand for the product or service. In these cases there may be no value, or only salvage value, and in most cases the land would be surplus to requirements. If this is agreed with the client then, for financial statements, it would be valued as surplus property at MV. In other cases, one of the following methods is most likely to be used in order to allow for depreciation.

(e) Straight line depreciation (SLD, GN 6.9.23)

This method is probably the most commonly used and assumes that the same amount is allowed for depreciation for each year of the asset's life.

SLD is a technique for reflecting the wearing out of an asset from new until it has a negligible value. This method was adopted by the Inland Revenue Valuation Office Agency for the depreciation of hospital assets in its first valuation of the NHS Estate in 1988, but it is understood that a form of S-curve has been adopted for later revaluations. At its simplest, this method assumes that an asset depreciates at a constant rate from new to the end of its life. Thus, an asset with a life of 50 years is assumed to wear out at a constant annual rate of 2% from its cost or value at the beginning of its life to nil at the end. To use the method, estimates of the total life of an asset must be made, but the valuer should bear in mind that such estimates are very unlikely to prove realistic and the weight given to them should take cognisance of this. Revaluations at regular intervals will help to overcome inaccurate life estimates.

The SLD method requires the valuer to establish the age of the property at the date of valuation and to estimate the remaining useful life. Added together, these figures give the total life of the building. The next step towards calculating the accumulated depreciation at the date of valuation is to determine the average annual percentage rate of depreciation, which is simply 100/total life. Thus, with a building that is 10 years old and with an estimated 15 years of remaining life (total life 25 years), the average annual percentage rate of depreciation is (100/25), i.e. 4%. Applying this annual rate to a 10-year old building therefore gives an accumulated depreciation of 10 years at 4% = 40%. This accumulated depreciation is deducted from the cost of the modern equivalent to give the DRC at the valuation date (10th year) as set out below.

Replacement cost	£2,500,000
Annual depreciation rate (100/25) = 4%	
deduct accumulated depreciation at	
10th year = 10 years at 4% = 40% =	£1,000,000
DRC at 10th year	£1,500,000

The main criticism is that depreciation will rarely follow a straight line.

(f) Reducing balance (GN 6.9.25)

The GN states:

> The reducing balance method of depreciation assumes a constant percentage rate of depreciation from the reducing base. The reduction of the balance at the end of each period by a fixed proportion of itself creates a sagging depreciating curve over the life of the asset. This method effectively "compounds" the total depreciation. This may match reasonable expectations of declining value over time better than the straight-line method.

Very little more is said, which could suggest that this is the least preferred approach.

(g) S-curve approach (GN 6.9.26)

This approach is felt to be the most realistic, as the S-curve represents the way an asset is likely to depreciate. It can reasonably be assumed that most assets depreciate slowly in the early years of the asset's life, depreciate faster in the middle years, and then depreciation slows down in the latter years.

It requires considerable data to establish the S-curve, but where data exists then this may be the best method of depreciation.

Only one method has been shown here showing straight line depreciation as this is still a favoured method provided the asset is revalued on an annual basis so that remaining life can be reassessed at regular intervals. For a fuller discussion on depreciation, readers are referred to the publications listed at the end of the chapter.

(h) Conclusion

DRC is a method of assessing replacement cost of an asset for financial statements; in this conext it establishes MV. DRC is an approach within the scope of the Red Book and must only be undertaken by an appropriately qualified valuer. The asset must be specialised and therefore not normally bought or sold in the market. This precludes the use of MV as a basis of valuation, due to the absence of market comparables. DRC is a method of valuation not a basis of value. The valuer is seeking to arrive at a figure which best represents the price a buyer would pay by comparison with what a modern equivalent asset would cost, i.e. land plus construction. This figure is based on an assessment of the site value for the specialised property to which is added the depreciated cost of the modern equivalent building. The latter is based on current costs including all normal add-ons depreciated to reflect physical deterioration, functional obsolescence and external obsolescence. Depreciation is assessed using straight line, reducing balance or S-curve approaches.

The RICS in GN 6 identifies the specific knowledge a valuer needs for a DRC, which indicates that in addition to experience in valuations using the DRC approach the valuer must have detailed knowledge of the asset, its function and environment and must be very knowledgeable about the asset, how it works, is built, and wears out.

4. Mortgages and loans

(a) Borrowing against property assets

Property is rarely bought out of earnings or savings. In most cases money will be borrowed to complete the purchase. The exception to this might be the case of the institution that is able to meet all claims on its funds out of income and invests surpluses in the acquisition of new assets.

By numbers, the largest market for such loan transactions is that for mortgages to finance the purchase of residential properties. The majority of lenders are either building societies or banks, and the loans take the form of mortgages. The majority of buyers are buying for owner occupation or to let.

However, the largest loans in value are generally for commercial property (see Appendix 5 to the Red Book), but where the amounts are very large other forms of lending, rather than a simple mortgage, may be adopted. (A loan made solely on the strength of covenant of the borrower is not a property loan.)

When any lender provides a loan secured against a property asset, a valuation of the asset will be required. Before considering the approach to valuations for loan purposes, it is helpful to be clear as to the nature of a mortgage and the rights and duties of the parties.

(b) Nature of a mortgage

A mortgage of freehold or leasehold property is a transaction whereby one party, the mortgagor, grants an interest in the property to another party, the mortgagee, as security for a loan.

The transaction is effected by means of a mortgage deed in which the mortgagor usually agrees to pay interest on the loan at a given rate per cent, and may also enter into express covenants as to the repair and insurance of the property. In some cases, the mortgage deed provides for periodical repayments of capital as well as interest (commonly in the case of a home purchase mortgage).

Mortgagors retain the right to recover their property freed from the charge created by the mortgage deed on repayment of the amount due to the mortgagee. This is known as the mortgagor's "equity of redemption".

Since 1925, a legal mortgage of a freehold can only be made either (i) by the grant of a lease to the mortgagee for a long term of years, usually 3,000, with a provision for cesser on redemption, or (ii) by a charge expressed to be by way of legal mortgage (Law of Property Act 1925, sections 85 and 87).

A legal mortgage of a leasehold can only be made either (i) by a sub-lease to the mortgagee of the whole term, less the last day or days, with a provision for cesser on redemption, or (ii) by a charge by way of legal mortgage (Law of Property Act 1925, section 86).

What is known as an "equitable mortgage" of land may be effected without a mortgage deed if it takes the form of (i) a signed written agreement acknowledging the loan and either stating that the specified property is security for its repayment or promising to execute a legal mortgage if required, or (ii) any mortgage of an equitable interest in land. A verbal agreement accompanied by deposit of the title deeds of the property can no longer suffice, since the Law of Property (Miscellaneous Provisions) Act 1989 came into force – see *United Bank of Kuwait plc* v *Sahib*[2] and section 2 of the 1989 Act, which requires the agreement to be in writing to be valid.

As a general rule, a mortgage is a sound form of investment offering reasonable security and a fair rate of interest and, although the mortgage deed probably stipulates for repayment at the end of six months, it is usual for the loan to continue for a much longer period.

So long as the mortgagor pays the interest regularly and observes the covenants of the mortgage deed, the mortgagee will usually be content to leave the mortgagor in possession and control of the property. But if the interest is falling into arrears or

2 [1996] 122 All ER 215.

the mortgagor is unable to meet a demand for repayment of the loan, mortgagees must take steps to protect their security.

Property is also offered as a security for other types of lending, particularly development property where the sums borrowed are used to pay for the development and are repaid on successful completion and disposal of the development. It may also be used by companies to raise funding for expansion of the company's business generally, the loan being a secured loan or a debenture on the company's assets. Very large sums of money are raised in this way.

(c) The mortgagee's security

The mortgagee's security for the money lent depends primarily upon the property and upon the sum it might be expected to realise if brought to sale at any time. In the case of commercial property which is commonly let, such as shops or offices, the security for payment of interest on the loan at the agreed rate depends upon the net income the property is capable of producing. This is known as the rent: interest ratio; a ratio of 1.5:1 might be considered acceptable, i.e the rent now, and for the period of the loan, should be one and a half times the interest payable. In the case of residential loans, the mortgagee typically takes account of the mortgagor's income when assessing the ability to pay. The underlying net rent may be disproportionately low where the asset is valued on a vacant possession basis.

A common advance by way of mortgage is around two-thirds of the estimated market value of the property, thus leaving the mortgagee a one-third margin of safety. Trustees do not usually advance more than two-thirds of a valuation of the property made by a skilled valuer for loans within the Trustee Act 1925. Other investors, such as building societies and banks, often make larger advances, particularly if some form of collateral security is offered or where provision is made for repayment of capital by instalments over a certain period. On the other hand, when there is a risk of future depreciation in value, an advance of three-fifths, or even one-half, may be more satisfactory than the usual two-thirds. In all cases the mortgagee, or the mortgagee's advisor, should consider not only the value of the property in relation to the proposed loan, but also whether the net income from the property is sufficient to provide interest at the agreed rate. The strength of covenant of the borrower will also be a factor. It is the role of the mortgagee and not the valuer to determine the proportion of the value to be offered as a loan. An exception to that is in the case of loans made by trustees, which are covered by the Trustee Act 1925. In such cases the "skilled valuer", who is employed by the trustees, does have the duty to decide on a prudent level for the loan sought.

If the mortgagor defaults in respect of payment of interest, observance of the covenants of the mortgage deed or repayment of the loan when legally demanded, the mortgagee has the following remedies against the property:

(i) Under certain conditions (see Law of Property Act 1925, section 101) they may sell the mortgaged property and apply the proceeds to repayment of the loan and any arrears of interest together with the expenses of sale. Any surplus must be paid to the mortgagor.

(ii) They may apply to the Court for a foreclosure order which will have the effect of extinguishing the mortgagor's equity of redemption.

(iii) They may at any time take personal possession of the income from the property and, after paying all necessary outgoings, may apply the balance to paying interest on the mortgage debt, including any arrears. This surplus, if any, must be paid to the mortgagor or applied to reducing the mortgage debt.

(iv) Under the same conditions as in (i) they may appoint a receiver to collect the income from the property and apply it to the purposes indicated in (iii), including payment of the receiver's commission.

It is evident that these remedies will only be fully effective where, in cases (i) and (ii), the market value of the property exceeds the amount due to the mortgagee, or where, in cases (iii) and (iv), the net income from the property, after paying all outgoings and annual charges having priority to the mortgage, is sufficient to discharge the annual interest on the loan with a margin to cover possible arrears of interest.

Although the property itself is the mortgagee's principal security, it is usual for the mortgage deed to include a personal covenant by the mortgagor to repay the loan. This may be reinforced by the personal guarantee of some third party, so that, in addition to the remedies mentioned above, there is that of action on the personal covenant.

The character and position of the borrower and any guarantor are therefore matters of considerable importance to the mortgagee, both as an additional security for the repayment of the loan and also as a guarantee for the regular payment of interest as it accrues due.

(d) The valuation

The basis of value is normally market value (VS 4.2)[3]. The RICS has been very concerned in recent years about valuations for secured lending, especially in the new homes residential sector.[4] Issues have also emerged in the valuation of commercial property for secured lending, some of which involve mortgage fraud. Those undertaking this essential valuation service need to be wary of any undue pressure exerted by clients and third parties. Most practices will have strict operational procedures in place which should provide for adequate checks and counter-checks

---

3 "Valuations of real property for secured lending shall have regard to IVS 310, Valuations of Real Property Interests for Secured Lending". IVS states that the *basis of value* will normally be *market value*.

4 Valuation of individual new-build homes 2nd edition RICS guidance note RICS.

to ensure compliance with the Red Book; these checks should also ensure that any improper actions by valuation staff will be spotted.

UKVS 3.1 Residential property mortgage valuations and the RICS residential mortgage valuation specification UK Appendix 10 set out the approach that a valuer in the residential sector must follow. There are, however, certain aspects of a valuation for the purposes of a mortgage that will distinguish it from valuations for other purposes. The property is offered as a security for a loan, so the valuation must reflect what can be realised if the lender will need to sell it to recoup the loan. The appendix is a comprehensive statement of practice agreed with the Council of Mortgage Lenders and the Building Societies Association. The specification covers: inspection; basis of value; factors that have a material impact on value; assumptions and special assumptions; reinstatement cost; the form of the valuation report; and treatment of incentives. Appendix 11 is relevant to reinspections, buy to let and other related purposes. These are the subject of consideration in their own right and must be read and adhered to by those engaged in residential valuations.

The valuer will disregard the value of items that can be sold or removed by the borrower. In the case of new residential builds, the valuer must request a developer's *Disclosure of incentives* form. There is additional guidance in *The valuation of individual new-build homes*. The point here is that MV must represent what the property could be sold for at the date of valuation, not what the buyer is paying, and therefore the value of incentives and other factors may need to be discounted because they are not factors that can be re-offered on future sales and a lower current market value may be implied. In the case of business premises, goodwill must be ignored.

The valuer must consider very carefully any factors such as probable action by the local authority which may unfavourably affect the value of the property in the future, particularly if they will require expenditure or will restrain the use of the property. In one negligence case, the valuer ignored the absence of a fire certificate when valuing a hotel, which could have led the local authority to close it down. The valuer should fully consider any possible effects of the development plan for the area. Valuations should usually be on the basis of current permitted planning use, unless planning permission has been granted for development. Development value should only be taken into account to the extent that buyers in the open market would take it into account when considering their bid for a property. Thus, for example, if a house is built on a double building plot with a tennis court occupying the second half, the market would recognise the development potential of the tennis court. The extent of any addition will depend on the certainty of obtaining planning permission. If an owner wishes to raise capital to carry out development, then they will enter into a funding agreement which is different from a straightforward mortgage.

The valuer may also have to consider whether the existing market is unduly influenced by national or local conditions of a temporary nature which may have caused something like an artificial "boom" or "slump" in prices due to a lack of confidence. Any likely capital expenditure on the property, such as accrued dilapidations or the estimated cost of future development or reconstruction, must be allowed for as a deduction.

Any future element of value that is reasonably certain in its nature, such as reversion to market rental value on the expiration of an existing lease at a low rent, may properly be taken into account; but anything that is purely speculative in its nature should be disregarded. The valuer must be cautious when capitalising "full" rents and must consider whether they will be maintained or are likely to fall; this is particularly important where the property is not of a first-class type. The valuer must beware of including in the valuation elements of potential value which may never come about.

Information is sometimes tendered as to the price paid for the property by the mortgagor, and the valuer should request details of any recent transactions. If there was a recent actual or agreed purchase, then it may provide a useful indication of the market value. However, the valuer will need to be sure that an excessive price was not paid or that the property was not purchased at a bargain price. This is particularly so where the valuer's opinion of value is significantly different from the price paid. The valuer must not ignore such transactions, but should treat them with care.

Notwithstanding the need for caution that the valuer must recognise when preparing a valuation for mortgage purposes, the valuation is not necessarily any different to a valuation, say, for advice on a sale or purchase. If there are no special factors as described, then a valuation for mortgage purposes should be no different from a valuation for advice on a sale or purchase.

The 2012 Red Book recognizes "… that for some purposes a prospective valuation may be required in addition to a current value(such as market value). Any such valuation should comply with the applicable jurisdictional and/or national association standards." In the UK a lender may require a valuer to give an opinion on the projected market value (UKVS 3.3) for a residential property. This is defined as:

> The estimated amount for which a property is expected to exchange at a date, after the *date of valuation* and specified by the valuer, between a willing buyer and a willing seller, in an arm's length transaction, after proper marketing wherein the parties had each acted knowledgeably, prudently and without compulsion.

This does not assume anything other than a normal, albeit a future, disposal. Most valuers would not give a date more than a few months beyond the valuation date. But "Where a valuation is prospective, any limitations on use of the valuation and the conditions and assumptions that applied in developing the opinion must be clearly set out." (2012 Red Book Appendix 6 1.2 g).

The method of valuation used for mortgage valuations will generally be the market approach (comparative method) in the residential market and the income approach or market approach (comparative method) for commercial property.

Lenders usually require to have sight of the comparables used in formulating the opinion of value. Where development land is being offered as security for a loan, it may be possible to use the residual method with a sensitivity analysis, but evidence of comparable land sales will usually be preferred. DRC cannot be used because, by definition, it is used only for properties that are not normally bought and sold and, by default, are not suitable as security for a loan.

(e) Second or subsequent mortgages

It is possible for there to be more than one mortgage on a property, the second or subsequent mortgages being mortgages of the mortgagor's equity of redemption. Second or subsequent mortgages are also termed "mezzanine finance". Provided they are all registered, the second and subsequent mortgagees will each have a claim on the property in their regular order after the first mortgagee's claim has been satisfied.

It is evident that there will be little security for such an advance unless care is taken to ensure that the total amount advanced, including the first mortgage, does not exceed what may reasonably be lent on the security of the property.

(f) Secured lending for commercial property

The valuation of commercial property for secured lending is covered by VS 4 and appendix 5. (The UK residential standards operate specifically to the UK market.)

VS incorporates the requirements of IVS 310 on *Valuations for loan security*, which covers: the property interest; incentives; valuation approaches; property types; investment property; owner-occupied property; specialized property; trade-related property; development property; and wasting assets. The basis of value is normally market value but, in accordance with EU directives relating to bank solvency ratios, a second valuation approach may be adopted based on mortgage lending value (MLV); this will be encountered in the EU as a request from European banks. MLV is considered in the Red Book in Appendix 8.

An important consideration in the case of investment property is the "rent:interest ratio", which might typically be 1.5:1, and currently lenders are very concerned with the issue of valuation certainty. A useful reference is the previously noted RICS guide *Reflecting uncertainty in valuations for investment purposes*, which considers the issue of uncertainty in relation to loan security valuations.

This section should be read in conjunction with all relevant valuation standards.

5. Fire and terrorism insurance: reinstatement cost

(a) Generally

The calculations prepared by valuers for the purpose of establishing reinstatement cost for fire insurance purposes are not valuations, but they are frequently prepared by valuers and are included here for completeness. They are not covered by the Red Book other than as part of UK Appendix 10 on residential mortgage valuations, which specifies that only if specifically requested will a figure be provided and, further, that it will be in accordance with the Building Cost Information Service (BCIS) guidance. BCIS is available online and can be used online for assessing the reinstatement cost of typical residential properties.

When insuring against the loss of a property by fire, the sum insured is broadly the estimated cost of reinstatement of the property. Such reinstatement and cost assessments have traditionally been termed fire insurance valuations, and valuers regarded the preparation of such "valuations" as coming within their field of expertise.

These "valuations" frequently cause confusion to non-valuers since they are usually different from the market value of the freehold interest, sometimes lower, other times higher. As a result, the RICS has promoted the use of "cost of reinstatement" or "reinstatement cost assessment" and the abandonment of "valuations". This work is best carried out by building or quantity surveyors who have detailed knowledge of the construction and the cost of buildings. Valuers preparing replacement cost estimates are acting as cost advisors and not as valuers.

(b) Cost of reinstatement

The basis is the estimated cost of reinstating the asset damaged or destroyed to its former condition. Full reinstatement cost is determined on the basis of building costs prevailing at the time of reinstatement (see *Glennifer Finance Corporation Ltd* v *Bamar Wood & Products Ltd*)[5] and to this must be added the other incidental costs such as architect's fees, site clearance, loss of income to a landlord during the rebuilding period, cost of alternative accommodation during rebuilding in the case of a house, or reductions in takings in a shop. So far as the estimation of reinstatement cost is concerned, many house insurance policies incorporate automatic updating of the sum insured on the basis of the RICS Building Cost Information Service (BCIS).

Although the basic proposition in relation to fire insurance cost estimates is comparatively simply put, it is an area fraught with difficulties.

(c) Effect of under insurance

It is essential that a building should be insured for the full current cost of rebuilding to whatever standard may be required at the time of rebuilding. In many cases, a building will not be completely destroyed by fire, but if the insurance policy contains an average clause the insurers may pay only such proportion of the cost of repairs as the full cost of reinstatement bears to the insured value. Thus, if a building costing £400,000 to reinstate is insured for only £200,000 and there is

---

5 (1978) 37 P&CR 208.

damage by fire costing £50,000 to repair, the insurer may agree to bear only £25,000 of the cost. In cases such as churches, where the full cost of reinstatement results in an impossibly high premium and in any event the building, once destroyed, may not be reinstated, it may be possible to arrange a fire-loss policy to cover major repairs up to a particular percentage of the full cost of reinstatement.

(d) Public authority requirements

It may not be realistic to assume that a building completely destroyed can be reinstated in its present form, quite apart from any new-for-old aspect. If, for example, a building does not meet current building regulations, the cost of providing a new building may be in excess of the cost of reinstatement. In the extreme case it may be that planning permission to rebuild would not be given, and such a refusal of permission does not attract compensation. In the former case a sufficient addition should be made to the reinstatement cost to cover additional requirements, but the latter case would probably be dealt with by a separate policy to cover the loss of site value.

(e) Cost and value

Since there is no definite relationship between cost and value, it may be that a building which would cost, say, £1,500,000 to reinstate may have a market value of, say, £1,000,000. In these circumstances, for the reasons given in section 5(c) above, insurance cover cannot be confined to value although it may be possible to insure against the average clause being applied.

(f) Incidental costs

Mention was made in section 5(b) of some of the incidental costs, such as architects' fees and loss of income, which may arise. In most cases it is necessary to identify such costs separately from building costs, and some may have to be the subject of separate insurance policies.

(g) Value added tax on building costs

Complete rebuilding of a building destroyed by fire is treated as new work. The position regarding value added tax (VAT) depends on the circumstances. If the property is residential, for example, there are several possible scenarios. In the case of an owner-occupier, the owner may recover VAT on building materials under the DIY rules. If the property was tenanted, the VAT position depends on whether the landlord will continue to let the replacement property on further periodic tenancies. If so, the VAT on building costs will not be recovered. On the other hand, if the intention is to sell the landlord's freehold interest in the replacement property, then the building process will become a zero-rated supply with no

VAT on building costs. Whatever the VAT position on complete rebuilding, any repair following partial damage will attract VAT. Again, the ability to recover VAT depends on the circumstances.

In preparing a cost estimate for fire insurance purposes, the manner in which VAT should be treated must be agreed with the insurer. As for the insured, it would be prudent to establish that the policy does provide for the reimbursement of any VAT that may be incurred on building work and related fees. For most organisations it is important that risk areas other than just fire are covered by building insurance policies. Additional premiums may be required to cover aginst terrorism, loss of trade and loss of rent. The issue is important when valuers are required to make allowance in a valuation for the cost of insurance. A check on actual premiums and policies is advisable.

6. Further reading

RICS Valuation – Professional Standards (the 2012 Red Book) (RICS London, 2012), including GN 6 *Depreciated replacement cost method of valuation for financial reporting*.

The Cost Approach to Valuation by W. Britton, O. Connellan and M. Crofts (RICS and Kingston Polytechnic, Kingston, 1991).

RICS (2009) Valuation Information Alert, *Guidance to Lenders and Valuers when Conducting Loan Security Valuation Reviews*. This is essential reading for those involved in "loan security reviews in circumstances when values are changing rapidly".

RICS (on line, regularly updated) Building Cost Information Service, www.rics.org

Chapter 20

Valuations for rating

Introduction

This chapter is concerned with the practice and procedures applicable to rating lists in respect of non-domestic rates, and valuation lists in respect of Council Tax in England and Wales. The law is stated as at 1 July 2012.

SECTION A: NON-DOMESTIC PROPERTIES

1. General

Non-domestic rates are a form of tax levied, as a general rule, upon the occupiers of property in respect of the annual value of their occupation in order to assist in defraying the expenses of local government.

A term popularly used for non-domestic rates is 'business rates'. This term is not very helpful and, unfortunately, somewhat of a misnomer as there is no requirement that a business should be run in a property for the occupier to be liable to pay rates. Basically, if a property is occupied and is classed as being neither wholly domestic nor exempt from non-domestic rates, then the property should be assessed for rating.

To ascertain the rate liability of a particular occupier, two things must be known. First, the rating assessment of the occupied property expressed in terms of rateable value, and second, the national non-domestic multiplier or uniform business rate (UBR) applicable to all non-domestic hereditaments. This is fixed by central government and indexed to allow for inflation. As a result, once the rateable value is established, a business is able to budget with some degree of accuracy for its rate liabilities within the duration of a revaluation cycle.

In recent years the multiplier has varied as follows:

	England		Wales
2007/08	44.4p	reduced by 0.3p for small properties	44.8p
2008/09	46.2p	reduced by 0.4p for small properties	46.6P
2009/10	48.5p	reduced by 0.4p for small properties	48.9p
2010/11	41.4p	reduced by 0.7p for small properties	40.9p

2011/12   43.3p   reduced by 0.7p for small properties   42.8p
2012/13   45.8p   reduced by 0.8p for small properties   45.2p

The City of London is able to set a different multiplier from the rest of England. For 2011/12 this was 43.7p (Small Business Rate 43p).

The making of assessments for rating purposes has, since the early 1950s, been the responsibility of valuation officers of the Inland Revenue, now HM Revenue and Customs, and it is with these assessments that this chapter is primarily concerned. Whilst the main principles of procedure and rateable occupation will be considered, it is not proposed to deal here with the history and development of the rating system, with the law of rateable occupation in detail or with the minutiae of procedural details.

2. Procedure

The main statute governing assessment and valuation procedure is the Local Government Finance Act 1988, enlarged by a multitude of Statutory Instruments. However, this legislation maintains many of the principles of valuation and rating practice that have evolved over the four hundred years of rating.

The collection of rates is undertaken by local authorities. They are called 'billing authorities' in this role and are the district, metropolitan or unitary councils outside London, but in London the Common Council of the City of London and the various London Borough Councils are the billing authorities for their respective areas.

The basis of the tax is "rateable value". Rates are a property tax which uses as its value base rental rather than capital value. Rating assessments represent the rental value of each rateable property at a standard valuation date; these valuations are called "rateable values".

The rating assessments of all rateable properties, known as "hereditaments", within a rating area are entered in the rating list for that area. Since 1990 new rating lists have been compiled every five years. The 1990 lists came into force on 1 April 1990, with successive lists at five-yearly intervals from then, and the next lists are due to come into force on 1 April 2015. The common valuation date for these lists is not the date they come into force but a date two years earlier. So for the 2010 lists the date used is 1 April 2008 and for the 2015 lists the date is 1 April 2013. This date is known as the "antecedent valuation date". as it comes 'ante', or before, the date the lists are compiled.

The date for each revaluation is specified in a statutory order. For the 2010 rating lists the order is The Rating Lists (Valuation Date) (England) Order 2008 No. 216. There is no statutory requirement for the date to be set two years before the compilation date, but for revaluations since 1990 this has been the practice.

The adoption of an antecedent valuation date has been taken up by other jurisdictions. For example, in Hong Kong the antecedent valuation date is set six months before the list comes into force. Unlike England and Wales, Hong Kong has achieved an annual rather than five-yearly sequence of revaluations.

Having a valuation date prior to the lists coming into force allows valuation officers to take a backward look when preparing their valuations and judge the state of the market at the antecedent valuation date, rather than having to take a forward look if preparing lists with a valuation date in the future, which would be the case if the valuation date was the date the lists came into force. This is likely to result in more accurate valuations in the compiled rating lists.

Briefly, the procedure for the preparation of new rating lists is as follows. Valuation officers send out forms of return, for completion by owners and occupiers, asking for details of such matters as the rent paid, length and date of grant of lease, repairing covenants, etc. Rented properties may then be inspected to check that the valuation officers' survey records are correct for analysis. Rents are adjusted and analysed to a common unit of comparison, and valuation scales for valuing hereditaments are prepared and valuations undertaken. The modern availability of computer support has very much changed the way lists are prepared and much of the process is automated. Drafts of the new lists the valuation officers propose to compile are copied to the billing authority, at whose offices the local list for its area is available for inspection by the end of September, before coming into force on the following 1 April. The lists are easily viewable on the Valuation Office Agency's website www.voa.gov.uk.

Once a list has come into force, the local valuation officer has a statutory duty to maintain it in a correct form. This the officer does by altering it when any defect, such as an incorrect value in respect of an existing hereditament or the need to include a new hereditament, is brought to his attention. Valuation officers are required to serve notice on the ratepayer and billing authority stating the effect of any such alteration within six weeks of altering the list.

Challenges to a rateable value assessed by the valuation officer can only be made by specified persons and within certain time limits, having regard to set criteria.

The process of challenging or appealing against a rateable value is done by a person proposing a change to the list. A formal challenge is called a "proposal". Only an "interested person" can make a proposal for the alteration of a rating list. An interested person means primarily the occupier of the property in question, any other person having a legal estate or an equitable interest entitling that person to future possession, but not a mortgagee not in possession, or anybody having a qualifying connection with either of these. A proposal can be made at any time against the assessment from when it first appears in a rating list up to the date a new list is compiled. There are two exceptions to this:

- Proposals citing tribunal or court decisions can be made during an extra period of six months. For the 2010 lists this means up until 30 September 2015.
- Valuation officers are able to alter the lists on their own initiative for another year after the general right to make proposals ends, i.e. up until 31 March 2016. Proposals challenging these alterations can be made after a new list is compiled, provided they are made within six months of the valuation officer's alteration.

There are detailed requirements for making a proposal contained in the the Non-Domestic Rating (Alteration of Lists and Appeals) (England) Regulations 2009, SI 2268, and for Wales the Non-domestic Rating (Alteration of Lists and Appeals) Regulations 2005 (SI 758). A written letter that satisfies these requirements will be a valid proposal. Alternatively, there is a printed proposal form issued by the Valuation Office Agency, or proposals can be made online at the Valuation Office's website www.voa.gov.uk. The Valuation Office Agency is the supporting organisation within HM Revenue and Customs for valuation officers.

If the valuation officer cannot accept that the proposal is well founded or a settlement cannot be agreed with the proposer, the dispute is referred to the Valuation Tribunal as an appeal.

Valuation Tribunals are empowered to hear all aspects of an appeal and give a decision. There is a single Valuation Tribunal for England and a single Tribunal for Wales, though they sit in many places and have many members and chairmen.

If either the valuation officer or the proposer is dissatisfied with the decision of the Valuation Tribunal, they may further appeal to the Lands Chamber of the Upper Tribunal (formerly known as the "Lands Tribunal") provided that they were represented at the Valuation Tribunal hearing. On questions of fact and valuation, the decision of the Lands Chamber is final, but on matters of law, a right of appeal lies to the Court of Appeal and thence to the Supreme Court.

3. Exemptions and reliefs

There are a number of rather disparate exemptions and reliefs which have been granted and added to over the years. There was an expectation that a follow-up to recommendations of the Lyons Inquiry into Local Government in 2007 would be a careful examination of the rationale and justification behind the various exemptions and reliefs, but in the event the government decided not to proceed with this. The statutory exemptions appear in the Local Government Finance Act 1988, Schedule 5. Although a detailed analysis of exemptions and reliefs is not appropriate for inclusion in this book, classes of property that are exempt include agricultural land and buildings, fish farms, places of public religious worship, public parks and property used for the welfare of disabled persons. Crown or government property and hereditaments occupied by local authorities are not exempt. Reliefs are available to, *inter alia*, charities and similar organisations and for small businesses and rural shops, pubs and petrol filling stations.

In the case of exemptions, no rateable value will appear in the rating list; in the case of reliefs, specific relief will be given from the calculated rate bill, usually by a percentage allowance determined by statute.

4. Elements of rateable occupation

Although not strictly connected with the methodology of rating valuation, this chapter would not be complete without some reference to the rateable occupier.

Before any rating valuation can be undertaken, it is necessary to establish the extent of the hereditament. Essentially, the ultimate establishment of that hereditament is a combination of the identification of both the unit of rateable occupation and the rateable occupier.

For an occupier to be rateable it is necessary to establish four essential elements. There must be:

(a) actual occupation;
(b) exclusive occupation;
(c) beneficial occupation; and
(d) the occupation must not be too transient.

(a) Actual occupation

In general terms, whether or not there is actual occupation will be self-evident. However, a number of points should be noted.

Title does not limit occupation and a single occupation can extend over differing titles. It is not necessary for there to be legal entitlement to possession for there to be rateable occupation; a squatter has been held liable for rates.[1]

(b) Exclusive occupation

The occupation has to be exclusive. In cases where it would appear that a number of persons have a right of occupation, it will be necessary to ascertain who has the paramount control in relation to the land under consideration.

In *Westminster City Council* v *Southern Railway Co*[2] Victoria Railway Station was under consideration. Here various kiosks, bookstalls etc. were let to individual traders, but the railway company exercised a degree of control over their occupations to the extent that at certain hours of the day they were physically prevented from occupying their premises, due to the fact that the station was closed. The House of Lords held that notwithstanding that they were denied possession at certain times of the day, the kiosks, etc., were in the occupation of the traders and their occupation was exclusive because it was the traders' purpose which was most being achieved, and their occupation was therefore the paramount one.

(c) Beneficial occupation

The occupation must be of value or be of some benefit to the occupier. However, this does not necessarily mean that the occupier will make a profit.[3]

---

1 See R v Melladew [1907] 1 KB 192.
2 [1936] AC 511.
3 See London County Council v Erith Churchwardens and Overseers [1893] AC 562.

Similarly, where a local authority is under a statutory duty to provide some benefit to the community, it will still be regarded as having beneficial occupation even though clearly it operates at a loss.[4] This situation, however, must be contrasted with properties such as public parks, which have been held not to be in the beneficial occupation of the local authority that holds the title to the land or is required to maintain them.[5]

(d) Transience of occupation

Transience of occupation really has two elements to it: transience in the context of time and transience in the context of the degree of use made of the land.

From *London County Council* v *Wilkins (VO)*,[6] concerning the rateability of contractors' site huts, a rule of thumb arose that an occupation had to exist for longer than 12 months to be rateable. However, in *Dick Hampton (Earth Moving) Ltd* v *Lewis (VO)*,[7] concerning quarrying operations only lasting six months, the 12-month rule of thumb was dismissed and emphasis was placed more on the nature of the occupation and use of the land itself.

In summary, therefore, once a person has been identified who can satisfy the four elements set out above in respect of an occupation of land, the next step is to establish the rateable hereditament in respect of which the valuation has to be made.

5. Principles of assessment

It is of prime importance that the extent of the hereditament to be valued is established at the outset. Without first determining the hereditament it is impossible to ensure that the correct unit of property is valued. Section 64, Local Government Finance Act 1988, maintains the definition of "hereditament" found in Section 115, General Rate Act 1967, namely "property which is or may become liable to a rate being a unit of such property which is, or would fall to be, shown as a separate item in the [rating] list". In general, for a property to be a hereditament there must be a single rateable occupier and the property must be a single geographic unit.

Having established there is a hereditament, there are a number of general principles to be observed in preparing valuations for rating, the most important of which are:

(i) The hereditament being valued must be assumed to be vacant and to let. The statutory basis for rating is rental value and having a rent presupposes some

---

4 Governors of the Poor of Bristol v Wait (1836) A&E 1 (the subject property was a workhouse).
5 See Lambeth Overseers v LCC [1897] AC 625.
6 [1957] AC 362.
7 [1975] 3 All ER 946.

form of tenancy and, indeed, the definition of rateable value forms the terms of this tenancy. As is usually the case in valuation work, the assumption is made that the hereditament is available vacant on the market so that the bid of the likely tenant on the statutory terms can be ascertained.

(ii) The hereditament must be valued *rebus sic stantibus*, which means taking the thing as it is. A hereditament has to be valued in its actual existing physical state, "as it stands and as used and occupied when the assessment is made".[8] This rule does not, however, prevent the assumption of changes of a minor nature from being envisaged . The rule also requires a hereditament to be valued as being restricted to its existing mode or category of use; thus a shop cannot be valued as an office or public house. Mode or category of use does not mean a very precise use, more a general category of use, so a shop must be valued as a shop rather than any particular type of shop.

The principle of *rebus sic stantibus* has been the subject of considerable judicial development over the years. Reference may be made to *Willliams* v *Scottish and Newcastle*[9] for a comprehensive resumé of the point.

(iii) Prior to *Garton* v *Hunter (VO)*,[10] if a property was let at what was plainly a rack rent, then that was the only permissible evidence. However, this case decided that "the actual rent is no criterion unless it indeed happens to be the rent the imaginary tenant might reasonably be expected to pay" and, as a result, other evidence of value should now be examined. This includes rents passing on comparable properties which may, indeed, show the rent of the actual hereditament to be out of line with its fellows and not truly representing the likely open market rent.

(iv) Assessments on comparable properties may be considered in the absence of better evidence once the basis of value of a new list has been established. This is called the 'Tone of the List'.

(v) A question of considerable importance is the extent of the rateable hereditament and whether separate assessments should be made in respect of different parts. The rateable unit will normally be taken as the whole of the land and buildings in the occupation of an occupier within a single curtilage. However, the Court of Appeal in *Gilbert (VO)* v *(S) Hickinbottom & Sons Ltd*[11] ruled that a bakery and a building used for repairs essential for its efficient operation could constitute a single hereditament although separated by a highway, as they were functionally essential to each other. Similarly, in *Harris Graphics*

---

8 Lord Parmoor—*Great Western & Metropolitan Rail Companies* v *Hammersmith Assessment Committee* [1916] 1 AC 23.
9 [1978] 1 EGLR 189.
10 [1969] 2 QB 37.
11 [1956] 2 QB 240.

Ltd v *Williams (VO)*[12] a proven functional connection between two geograph-
ically separate buildings on the same industrial estate was held to justify a
single assessment.

It is not necessary for parts to be structurally severed to be capable of separate
assessment if the parts are physically capable of separate occupation.[13] From the
valuation viewpoint, separate assessments usually, but not necessarily, result in a
higher total rateable value.

6. Statutory basis of assessment

All non-domestic hereditaments are assessed to rateable value in accordance with
the statutory provision set down in Schedule 6, Local Government Finance Act
1988 (as amended by the Rating (Valuation) Act 1999). This states that:

> **2.–** (1) The rateable value of a non-domestic hereditament none of which
> consists of domestic property and none of which is exempt from local non-
> domestic rating, shall be taken to be an amount equal to the rent at which it is
> estimated the hereditament might reasonably be expected to let from year to
> year on these three assumptions–
>
> (a)   the first assumption is that the tenancy begins on the day by reference to
> which the determination is to be made;
> (b)   the second assumption is that immediately before the tenancy begins the
> hereditament is in a state of reasonable repair, but excluding from this
> assumption any repairs which a reasonable landlord would consider
> uneconomic;
> (c)   the third assumption is that the tenant undertakes to pay all usual tenant's
> rates and taxes and to bear the cost of the repairs and insurance and the
> other expenses (if any) necessary to maintain the hereditament in a state
> to command the rent mentioned above.
>
> (8A) For the purposes of this paragraph the state of repair of a hereditament
> at any time relevant for the purposes of a list shall be assumed to be the
> state of repair in which, under sub-paragraph (1) above, it is assumed to be
> immediately before the assumed tenancy begins.

The "rent" is not necessarily the rent actually paid, but the rent which the
likely tenant for the hereditament, a hypothetical tenant, might reasonably be
expected to pay. All possible occupiers, including the actual occupier wishing

---

12 [1989] RA 211.
13 See *Moffat (VO)* v *Venus Packaging Ltd* [1977] 2 EGLR 177.

to use the property, *rebus sic stantibus*, should be considered in deciding what the hypothetical tenant would pay by way of rent.

The definition assumes a tenancy from "year to year". This has been taken to mean a tenancy by the hypothetical tenant who can occupy the hereditament indefinitely, having a tenancy which will probably continue but which can be determined by notice. There is therefore a reasonable prospect of the tenancy continuing and no implied lack of security of tenure which would adversely affect the expected rent.

"Usual tenants' rates and taxes" include non-domestic rates, water rates and occupier's drainage rate but not Schedule A Income Tax and owner's drainage rate, which are landlords' taxes.

The cost of repairs, insurance and other expenses necessary to maintain the hereditament are deemed to be the responsibility of the tenant under the statutory definition of rateable value.

It is usual in setting up local property tax systems around the world to require the assessing body to disregard any normal lack of repair. This makes the system easier and less costly to administer as the assessing body needs only to consider disrepair when it is exceptional. For non-domestic rates, the definition places the responsibility for repairs during the life of the hypothetical tenancy on the tenant. In effect the hypothetical tenancy is an FRI lease. It is also assumed that the hereditament is in a state of reasonable repair prior to the commencement of the hypothetical tenancy, subject to the important assumption that this only applies to repairs that a reasonable landlord would, if the property were in disrepair, consider economic to remedy before letting the property. In most circumstances a landlord would consider it economic to repair and therefore the wording achieves the object of disregarding the actual state of repair and assuming that hereditaments are, for purposes of valuation, in a generally reasonable state of repair. Difficulties do arise, however, when the disrepair is substantial in deciding at what point would a reasonable landlord decide not to do repairs, and therefore the hereditament has to be valued as it is.

As has already been stated, the hereditament has to be valued having regard to the level of rental values established as at the antecedent valuation date – 1 April 2008 in respect of a 2010 rating list and 1 April 2013 in respect of the 2015 lists. However, the physical circumstances of the hereditament are taken as at the date when the new rating list comes into force (1 April 2010 or 1 April 2015 respectively), or at any other later "material day" for an alteration applying after the list came into force. These dates are set down in the Non-Domestic Rating (Material Day for List Alterations) Regulations 1992 as amended. This separation of dates is important. Valuations always take place in the AVD world. To undertake a rating valuation, the physical state of the property and its surroundings is taken as at the material day and these factors are, in effect, transported back and replace the physical factors at the AVD for purposes of valuation. So, if a new shopping centre is built in a town and is completed in June 2016, then, in considering whether this has had any depressive effect on rating assessments of other shops in the town, the valuer needs to ask whether, had the shopping centre already been built as at the AVD, 1 April 2013, the rents paid would have been lower than they actually were at the AVD. The physical circumstance of there being no shopping centre at the AVD is replaced by the physical circumstances at the later material day. What physical changes can be taken into account are listed in Schedule 6 paragraph 2(7) of the Local Government Finance Act 1988.

In general, the rating concept of rateable value is almost identical to the valuation practitioner's concept of a modern commercial lease.

7. Rating of unoccupied properties

Despite the absence of a beneficial occupier, owners of unoccupied properties have been made potentially liable for rates since 1966. In effect this is an additional rate levied on owners of vacant properties.

Under the provisions of the Non-Domestic Rating (Unoccupied Property) Regulations 2008 owners of vacant property are liable to pay an empty property rate at 100% of the occupied charge, provided the hereditament comprises a relevant non-domestic hereditament defined as consisting of, or of part of, any building, together with any land ordinarily used or intended for use for the purposes of the building or part.

The definition deals only with buildings or buildings with land. It follows that the owner of a hereditament consisting solely of land or land with ancillary buildings, e.g. a car park or playing fields with clubhouse, will not be liable for unoccupied property rating.

There are a number of exemptions from the charge:

- all properties are exempt from empty property rates for the first three months of vacancy and "qualifying industrial hereditaments", i.e. factories and warehouses for the first six months of their vacancy;
- listed buildings;
- hereditaments where occupation is prohibited by law;
- hereditaments with a rateable value below defined levels, currently £2,600; and
- certain other categories of hereditament.

Partly occupied property can also receive relief from the occupied charge under the provisions of Section 44A, Local Government Finance Act 1988, where the part not occupied is only vacant in the short term. This statutory provision is applied at the discretion of the billing authority, which seeks a certificate from the valuation officer showing the rateable value apportioned to the occupied and unoccupied parts.

8. Methods of assessment

There are different methods of assessment, but it is vital to bear in mind that, whichever valuation method may be used, the end product must be the rent on the basis of the statutory definition of rateable value. The principal valuation methods of assessment used are:

(a) by reference to rents paid, which is usually known as the rental method;
(b) the contractor's method or test; and
(c) the receipts and expenditure or profits basis.

The general rule is that if the rental method can be used, it should be used in preference to the contractor's method or the profits method. However, all these methods are means to the same end, namely, to arrive at a rental value.

Their application to particular types of property is dealt with in section 9 of this chapter, but a brief description of the first three methods is given here.

(a) Valuation by reference to rents paid

In many cases the rent paid for the property being valued may be the best guide to a rating assessment. However, rateable value can only correspond with rents actually paid, provided that (i) the rent represents the fair annual value of the premises at the relevant date, and (ii) the terms on which the property is let are the same as those assumed in the statutory definition. The rent paid for any particular property will almost certainly not be the true rental value in cases where property has been let for some time and values in the district have since changed.

Again, properties are often let at a low rent in consideration of the lessee paying a premium on entry, or surrendering the unexpired term of an existing lease, or undertaking improvements or alterations to the premises. In the case of premiums or such equivalent sums, the rent must be adjusted for assessment purposes by adding to it the annual equivalent of the consideration given by the tenant for the lease. This would not, however, apply if the premium was paid in respect of furniture or goodwill. A rent of business premises may also be low if a tenant's voluntary improvement under an earlier lease has been ignored under Section 34 of the Landlord and Tenant Act 1954, as amended. If the tenant has carried out non-contractual improvements, the rental value of these improvements must also be added to the rent paid.

By contrast, a rent may be high where a tenant has received inducements to take the lease, such as a reverse premium or a long rent-free period.

The rent paid may not be equal to the rental value because there is a relationship between the lessor and the lessee. A typical example is where the property is occupied by a limited liability company on lease from the freeholder who is a director of that company. Rents reserved in respect of lettings between associated companies are often not equal to the proper rental value. Sale and leaseback rents may also differ from rental value.

Adjustments may be necessary in cases where the terms of the tenancy differ from those of the statutory definition of rateable value. A typical example is that of a shop let on a lease, under which the lessee is responsible for internal repairs only. In such a case, the rent will be higher than it would have been had the tenant been responsible for all repairs. A deduction in respect of the external repairs must therefore be made to the rent to obtain the rateable value. As a rule of thumb, it is common practice to deduct 5% from the reserved rent where the tenant has the responsibility for internal repairs only, and 10% where there is no responsibility for repairs. However, such a general practice may be considered too arbitrary and should not be adopted if other figures can be justified. Ideally, actual repair costs should be used.

The relationship between rents in accordance with the statutory definition and lease rents has been a subject of contention. The general conclusion is that rents agreed on leases with a five-year review pattern can be regarded as being no different from the annual rent of the rating hypothesis.

In *Dawkins* v *Ash Brothers & Heaton Ltd*,[14] the House of Lords held by a 3–2 majority decision that the prospect of the demolition of part of a hereditament within a year was a factor which a hypothetical tenant would consider in making his rental bid. This resulted in a lower assessment but involved complex argument as to the point at which actual facts displaced the hypothetical circumstances to be considered in rating law.

Where the rent actually paid is considered to be above or below the prevailing rental value, or where a property is owner-occupied, a valuation must be made by comparing the property being valued with other similar properties let at true rental values. Sometimes a proper rent can be fixed by direct comparison with other similar properties. In other cases it may be necessary to compare their relative size and to use some convenient unit of comparison, such as the rent per square metre.

With certain special types of property a unit of accommodation rather than of measurement may be used, such as a figure per seat for theatres and cinemas, or per car parking space for car parks.

When using the rents paid for other similar properties as a basis of assessment, regard must be had to the terms on which those properties are let, and to the need to make any necessary adjustments to those rents so that they conform with the statutory definition.

Mention may also be made of the principle of "equation of rents". This is a theory based on the assumption that a tenant would pay only a fixed sum by way of rent and rates. If the rates burden decreases substantially, as it did with the introduction of small business relief, then the rent should rise because prospective tenants now have the spare money from the lower rates to attempt to outbid other prospective tenants. Similarly, a marked rise in rates, as happened to offices in Westminster at the 2010 revaluation due to a doubling of rental values since the previous revaluation, should have a depressive effect on rents.

---

14 See Humber Ltd v Jones (VO) & Rugby Rural District Council (1960) 53 R&IT 293 and Baker Britt & Co Ltd v Hampsher (VO) [1976] 2 EGLR 87.

(b) The contractor's method or test

This method is used in cases where rental evidence, either direct or indirect, is not available or is inconclusive. The theory behind the contractor's test is that the owners, as potential hypothetical tenants, in arriving at their bid would have regard to the possible yearly cost to them if they were, as an alternative, to acquire the ownership of the hereditament by taking a rate of interest on capital cost or capital value. Examples of properties for which the contractor's basis is appropriate are public libraries, town halls, sewage disposal works, fire stations, colleges, municipal baths, schools, specialised industrial properties and most rateable plant and machinery. The method has also been used in the case of a motor racing track where no licence was required and there was no quasi-monopoly. If a quasi-monopoly exists, as for example with a licensed hotel, then the profits basis, based on available accounts, would be more appropriate.

The contractor's method was described in outline in Chapter 2 and is applied by taking a percentage of the effective capital value of the land and buildings to arrive at the annual value.

There are two possible methods of determining the effective capital value of buildings for this purpose:

(i) by reference to the known or estimated cost, inclusive of fees, of reconstructing the existing building, known as the replacement cost approach. In the case of a new building, the effective capital value might well be the actual cost of construction unless there was some form of surplusage, such as excess capacity, or other disability which would justify a reduction in that cost. With an older property, however, the cost of reconstruction would have to be written down to take account of disabilities of the existing building, i.e. reflecting its age and obsolescence. Many old cases state that excessive embellishment and ornamentation would normally have to be ignored, as it is not something for which the hypothetical tenant would pay more rent. However, these cases often date from a time when such embellishment was not fashionable or well regarded and very simple functional buildings were preferred. It is to be noted that it is not unusual for buildings of importance to be constructed today of expensive materials and with features which aim at prestige rather than utility.

(ii) by reference to the cost, including fees, of constructing a simple modern building capable of performing the functions of the existing building. This is a more refined approach known as the "simple substitute building" solution to the same problem of valuing obsolete or older properties.

In either case, the effective capital value that is required is the cost of a modern building less the necessary deductions in respect of disabilities, so that the final figure obtained represents as nearly as practicable the value of the old building with all its disadvantages. To this figure must be added the value of the land, the value being assessed having regard to the existing buildings *rebus sic stantibus* and ignoring any prospective development value. The land value is an important factor as, whilst building costs do vary across the country, it is land value which can vary significantly and results in the contractor's valuation producing different and higher valuations in what are normally regarded as high-value areas and lower rateable values in areas where values generally are lower. Not taking full account of land value can have the surprising result that the valuation of similar buildings does not significantly vary across the country when they are valued on the contractor's method.

The next stage in the valuation is the conversion of effective capital value into a rental value by the application of an appropriate percentage which, theoretically, could be argued to be the market borrowing rate for funds. Over the years the Lands Tribunal provided some judicial guidance on this point and a range of decapitalisation rates was built up.

In *Coppin (VO)* v *East Midlands Airport Joint Committee*,[15] concerning the assessment of the East Midlands Airport, Castle Donington, the Lands Tribunal in their decision referred to the fact that both valuers agreed that 5% was the generally accepted rate to be applied to the effective capital value to arrive at net annual value in the normal type of case. In *Shrewsbury School (Governors)* v *Hudd (VO)*,[16] concerning the rating of a public school, 3.5% of the cost of a substitute building, fees and site, less 70% for age and obsolescence, formed the basis of the Lands Tribunal's decision in arriving at gross value, while in a cemetery and crematorium case, *Gudgion (VO)* v *Croydon London Borough Council*,[17] the Lands Tribunal adopted 3% on replacement cost, less surplusage and disabilities, to arrive at net annual value.

Given these variations in practice, it is perhaps not surprising that statutory decapitalisation rates were introduced for the 1990 Revaluation onwards. The Non-Domestic Rating (Miscellaneous Provisions) (No. 2) Regulations 1989 set down rates of 4% for hospitals and educational establishments and 6% for other properties, thus effectively putting an end to the debate on the appropriate rate to use. These are reconsidered afresh at each revaluation. For the 2010 Revaluation the rates were 3.33% and 5% respectively (2.97% and 4.5% in Wales). The 3.33%, in addition, now applies to defence hereditaments.

To reduce areas of contention further, the 1990 Lists led, for the first time, to guidance notes being agreed between the Valuation Office Agency and many of the various bodies that generally occupy property to which the contractor's method of valuation applies.

A typical contractor's method of valuation of a fire station in a 2010 list might be as shown in the following example.

15 (1970) RA 503.
16 (1966) RA 439.
17 (1970) RA 341.

Extension	1980	125	£1,050	25%	£98,438
Smoke house	2000	155	£750	5%	£110,438
Stores	1965	100	£500	40%	£30,000
Canopy	1980	125	£175	25%	£16,406
					£515,156
Steel tower, 7m high	1980		£15,000	10%	£13,500
Total structures					**£528,656**

External works, including surfacing, fencing, tanks and CCTV	say	330000	34.5%	£216,150

Total adjusted build costs (excl professional fees)		**£744,806.25**
factor adjustment (Outer London Borough)		1.2
Total build costs at local level		**£893,768**
Add for professional fees	11.00%	£98,314
Total adjusted build costs (inc fees)		**£992,082**

		Land value per hectare		
Site area (hectares)	0.25	£2,500,000	34.5%	£409,375

Total adjusted replacement cost	**£1,401,457**
Decapitalisation at 5% (prescribed)	£70,073
Stage 5 allowance	none appropriate
Rateable Value	**say £70,000**

The adopted obsolescence allowance for external works and the land is, in this example, directly proportionate to that adopted overall for the buildings.

It is possible to incorporate an end allowance in a contractor's method valuation to reflect a disability not otherwise dealt with. For example, an end allowance for piecemeal development, i.e. when the buildings are spread apart, thus being inefficient to operate, would be justified. However, an allowance for poor natural light within classrooms would not be justified, as this should be reflected in the obsolescence allowance.

Finally, it should not be overlooked that the hypothetical tenant's ability to pay the resultant figure of rent arrived at by the contractor's method valuation is a material factor. Once this mathematical valuation exercise has been completed, it is always necessary to "stand back" and consider whether or not the resultant answer looks sensible.

The method is thus usually described as having five stages:

- Stage 1 Estimation of cost construction
- Stage 2 Deductions from cost to arrive at effective capital value
- Stage 3 Estimation of land value
- Stage 4 Application of the appropriate decapitalisation rate
- Stage 5 "Stand back and look".

(c) The receipts and expenditure or profits basis

This method is also used where rental evidence is absent or inconclusive. The theory is that a hypothetical tenant's rental bid would relate to the profits likely to made from the business the tenant would conduct in the hereditament. It must be emphasised that the profits themselves are not rateable but that they are used as a basis for estimating the rent that a tenant would pay for the premises vacant and to let. Examples of properties to which the receipts and expenditure basis may be applied are racecourses, licensed premises, caravan sites and leisure centres, including pleasure piers. The general rule is that the method is particularly applicable where there is some degree of monopoly, either statutory, as in the case of public utility undertakings, or factual, as in the foregoing examples.

The method involves the use of the accounts of the actual occupier, but if the actual occupier's management skill is below or above the normal level of competence assumed for the hypothetical tenant, then consequential adjustment to those accounts will be necessary. From the gross receipts, purchases are deducted, after they have been adjusted for variation of stock in hand at the beginning and end of the year, thus leaving the gross profit. Working expenses are then deducted. The residue, known as the divisible balance, is divided between the tenant, for his remuneration and interest on his capital invested in the business, and the landlord, who receives the remainder by way of rent. The tenant's share must be sufficient to induce the likely tenant to take the tenancy, irrespective of the balance for rent.

Traditionally, the actual rates were not deducted as an outgoing in making the initial valuation since the year in which the valuation was being prepared was not necessarily the statutory date of valuation. However, as a result of the 1990 rating lists, and subsequent rating lists having an antecedent valuation date, it is now accepted that the actual rates payable in that year should be deducted to arrive directly at the rateable value.

A profits valuation for the assessment of a licensed hotel is set out in outline below:

Outline profits valuation

	£
	Receipts
Less	Purchases (adjusted for stock position)
	Gross profit
less Working expenses, from the trading figures, which would be appropriate to a tenant from year to year occupying under the definition of rateable value. The working expenses deducted may be adjusted, if necessary, to reflect those which the hypothetical tenant would allow and need not necessarily be those actually incurred.	
	Net profit/Divisible balance

less
Tenant's share consisting of:

(i)   interest on his capital invested in
      furniture, equipment, stock and
      working capital;
(ii) remuneration for running the business,
      including risk, etc., on capital (an
      assessed figure) *or* a percentage of
      tenant's capital (or of the divisible
      balance)

_____
Rateable value

Note: It is sometimes said that a shortened version of the profits test is undertaken by taking a percentage of gross turnover or receipts to arrive at rateable value. Whilst this is a commonly used approach to valuing properties that could be valued on a receipts and expenditure method, it is properly a rentals method as the percentage is usually derived from a comparison of rent to gross receipts. The rent is analysed by reference to gross receipts rather than area per square metre.

9. Assessment of various types of property

(a) Agricultural properties

Agricultural land and buildings are wholly exempt from rating under Schedule 5, Local Government Finance Act 1988. "Agricultural land" is defined as:

> land used as arable, meadow or pasture ground only; land used for a plantation or a wood or for the growth of saleable underwood; land exceeding 0.10 hectare and used for the purpose of poultry farming; a market garden, nursery ground, orchard or allotment, including allotment gardens within the meaning of the Allotments Act 1922.

However, the definition does not include "land occupied together with a house as a park; gardens (other than as aforesaid); pleasure grounds; land used mainly or exclusively for purposes of sport or recreation; land used as a racecourse".

"Agricultural buildings" are also defined in Schedule 5 to the Act. A building is an agricultural building if it is not a dwelling and:

(a) it is occupied together with agricultural land and is used solely in connection with agricultural operations on the land, or
(b) it is, or forms part of, a market garden and is used solely in connection with agricultural operations at the market garden.

The test of whether a building is occupied with land is not a geographic test but rather a question of whether or not the buildings are occupied at the same time as the land; see *Handley (VO)* v *Bernard Matthews plc.*[18]

_____
18 [1988] RA 222.

Bee-keeping and fish farming, excluding ornamental fish, are similarly exempted under Schedule 5. Certain buildings occupied by farmers' co-operatives are also entitled to exemption under Schedule 5.

Buildings used for extensive animal rearing, e.g. indoor pig or chicken units, are also treated as agricultural buildings.

(b) Offices

The analysis of actual rents paid, checked by comparison with rents paid for other similar premises, will usually provide the basis of valuation. Most properties of this type will be let under a lease which makes the tenant responsible for insurance, all repairs, etc. and so may well equate to the definition of rateable value. However, some leases will provide for the tenant to pay towards these costs through a fixed or variable service charge. Whatever the individual arrangements, the actual terms must be adjusted as appropriate to accord with the definition of rateable value. This is particularly important if an analysis of the rent of the property in question is to be used as the basis of valuation for other properties. Office rents are usually analysed to a price per square metre and the basis of measurement tends to be net internal area, as defined in the RICS Code of Measuring Practice (Sixth Edition, 2007). Factors that influence the value of office accommodation are:

(i) external factors such as geographic location, namely central to the perceived office location core; proximity to transport connections; proximity to staff facilities, e.g. shopping, lunch and leisure opportunities.
(ii) internal factors such as feature reception area, lifts, air conditioning, raised floors, double glazing.

(c) Shops

The assessment of shops is governed mainly by consideration of the rents actually paid. In principle, if a property is let at a market rent at the antecedent valuation date, then that rent, adjusted if necessary to conform to the definition of rateable value, should form the basis of the assessment. Indeed, as shops are usually let on full repairing and insuring leases, little or no further rental adjustment is usually necessary. An addition to the rent must be made for any tenant's expenditure such as the installation of a shop front, particularly in the case of a new shop where the tenant leases and pays rent for only a "shell" and covenants to complete and fit out the shop. This is obtained by amortising the capital expenditure at a suitable rate over the shorter of the two periods until either the improvements are of no value or their value does not have to be disregarded on review or renewal. A good example of the amortising of improvements is the case of *Edma Jewellers* v *Moore*[19] {1975} RA 343. Any necessary adjustment for repairs not undertaken by

19 (1975) RA 343.

the tenant must be applied to the rent after it has been increased by allowing for tenant's expenditure. The final figure obtained, by whatever approach, should be the rent which the hypothetical tenant would pay for the completed shop on the basis of the statutory definition.

Shops only a short distance apart can vary considerably in rental value because of vital differences in their positions. Any factor influencing pedestrian flow past the shop has its effect on value, such as proximity to "multiples", adequate and easy car parking or public transport, the position of shopping "breaks" such as intersecting roads, bus stops and traffic lights, and widths of pavements and streets.

When the time comes either to analyse adjusted rents or to value a particular shop, certain recognised methods of valuation are used with the object of promoting uniformity and facilitating comparison between different properties.

A method in general use for valuing the ground floor of a shop, which is usually a very much more valuable part than the rest of the premises, is the "zoning" method. The front area of the shop is the most valuable part, as it provides the most prominent selling space, and the value per unit of area decreases as the distance away from the front of the shop increases until a point is reached beyond which any further reduction would not be sensible. The zoning method allows for this progressive decrease in value from the front to the rear.

The first step in the analysis of a rent paid for premises including a ground floor shop is to deduct the rent attributable to all ancillary accommodation, leaving that paid for the ground floor selling space only. This rent is then broken down on the assumption that the front part of the shop, back to a certain depth, is worth the maximum figure, say £x per unit of area, the next portion £$x/2$ per unit of area, and so on. This process, known as "halving back", is used most extensively and can be varied, but usually there will be up to three zones and a remainder.

The depth of each zone varies according to circumstances and valuers have different opinions as to the most suitable depths to be used. That said, it is now common practice to divide the shop into three consistent zones of 6.1 metres (20 feet) depth.

The first zone (termed Zone A) is measured backwards from the front of the shop, which is usually the rear pavement line of the demise. This zone is clearly the most valuable and may consist entirely of shop front, arcades and show cases. Subsequent zones are measured from the end of the previous zone. The benefits of a return frontage on a corner shop can be reflected by a percentage addition to the value of the shop. As far as sales space on upper or basement floors is concerned, this may be valued as a fraction of the Zone A value in comparative terms. Other ancillary space is usually taken at a unit value applied to the area concerned.

The Lands Tribunal has indicated that, in its opinion, the number and depths of the zones to be adopted is not so important, provided that the zoning method is applied correctly and consistently according to the actual depths of the shops whose rents are being devalued and that the same method is applied subsequently when valuing.[20]

When a large shop is being valued on the basis of rental information derived from smaller shops, a "quantity allowance" is sometimes made in the form of a percentage deduction from the resulting valuation of the large shop. This procedure may be justified on the grounds that there is a limited demand for large shops and that therefore a landlord may have to accept a rather lower rent per unit of area than he would for a small shop. This may be valid when a large shop is in what is basically a small shop position, but it is probably invalid when it is in part of a shopping area where large chain stores have established themselves. For this reason, the Lands Tribunal rejected quantity allowances on two shops in Brighton in 1958, in assessments for the 1956 list.[21] Later, in a 1963 list reference concerning a new store built for Marks & Spencer in Peterborough and their old store, which they had sold to Boots, the Lands Tribunal refused quantity allowances on the ground that the limited demand and limited supply of large shops was in balance.[22] Furthermore, in *Trevail (VO)* v *C & A Modes Ltd* and *Trevail (VO)* v *Marks & Spencer Ltd*,[23] the Lands Tribunal expressed doubts as to whether the zoning method was not being stretched beyond its capabilities in the valuation of these large walk-round stores. The problems of valuing large stores for rating purposes was comprehensively considered in the Lands Tribunal's decision on the John Lewis store in Oxford Street, London.[24]

It is usual today to value such large stores on an overall pricing method, supported by analysis of reliable rental evidence which has been tested in the market.

Apart from any possible quantity allowance, a disability allowance may be an appropriate deduction in the case of shops with disadvantages such as obsolete layout, inconvenient steps, changes in floor level, excessive or inadequate height, lack of rear access, inadequacy of toilets or large and obstructive structural columns, if the valuation basis has been derived from rental evidence on shops without such disabilities.

In order to be able to check an assessment of a particular shop to see if it is fair by comparison with the assessments of other neighbouring shops, it is essential to be at least reasonably conversant with the shopping locality.

The following example illustrates the method of analysing the rent of premises comprising a ground floor shop and two upper floors occupied together with the shop for residential purposes.

20 See *Marks & Spencer* v *Collier (VO)* (1966) 197 EG 1015.
21 See *British Home Stores Ltd* v *Brighton Corporation and Burton*; *Fine Fare Ltd* v *Burton and Brighton Corporation* [1958] 51 R&IT 665.
22 See *Marks & Spencer Ltd* v *Collier (VO)* [1966] RA 107.
23 (1967) 202 EG 1175.
24 See *John Lewis & Co Ltd* v *Goodwin (VO) and Westminster City Council* (1979) 252 EG 499.

The analysis is made to ascertain the rental value per unit of area of the front zone of the shop so that it can be compared with the rents of other shops. It is not, in this case, considered likely that there would be any doubt about the correctness of the figure used for valuing the upper floors.

Example 20–1
Shop premises comprising a ground floor "shell" and two upper floors, situated in a good position in a market town, were let in April 2008 at £50,000 p.a. on a full repairing and insuring lease for 15 years with a review in the fifth and tenth years. The tenant spent £45,000 on a shop front and other finishings. Analyse this rent for the purpose of rating assessments. The shop has a frontage of 6 m and a depth of 30 m. The net floor areas of the residential first and second floors are 50 m² each.

Analysis

Rent reserved	£50,000
Add	
Annual equivalent of cost of shop front and finishings provided by the tenant	

$$\frac{£45,000}{\text{YP 15 yrs at 6\% and 3\%}} = \frac{45,000}{8.7899} \qquad = \qquad 5120$$

Full rental value in terms of RV	£55,120
Deduct	
Rental value of upper floors	
100 m² at £70 per m²	£7000
	£48,120

Therefore, leaving for shop rental value in terms of RV

Measurement shows that the effective net area of the shop is as follows:

Zone A	(first 6.1 m)	36.6 m²
Zone B	(second 6.1 m)	36.6 m²
Zone C	(third 6.1 m)	36.6 m²
Remainder	(remaining depth)	70.2 m²

Let £x per square metre represent the rental value of the first zone. Adopting "halving back":

$$\frac{36.6x}{1} + \frac{36.6x}{2} + \frac{36.6x}{4} + \frac{70.2x}{8} = £48,120$$

72.82x	$= £48,120$
x	$= £660.8$

Zone A rental value is approximately £660 per square metre in terms of rateable value.

The above analysed figures can then be compared with other Zone A rents based on the letting of similar shops in the vicinity to assist in obtaining a rent that represents a fair and consistent basis for the assessment of the shops under consideration.

A more convenient approach is that of reducing areas to "equivalent Zone A" areas. Using the above facts, the total area of 180 m^2 is expressed as:

Zone A	36.6	
Zone B	18.3	$\left(\dfrac{36.6}{4}\right)$
Zone C	9.15	$\left(\dfrac{36.6}{4}\right)$
Remainder	8.775	$\left(\dfrac{70.2}{8}\right)$

Equivalent Zone A area = 72.82 m^2.

Analysis then becomes $\dfrac{£48,120}{72.825} = £660$ approximately, as above.

The whole area has been converted to the area in terms of Zone A, and the figure of 72.825 m^2 would be expressed as "72.82 square metres ITZA".

It must be emphasised that the zoning method, although used very commonly in practice, is merely a means to an end in order to find the true rental value. The fixing of zone depths and the allocation of values to the zones are arbitrary processes, and strict adherence to an arbitrary pattern may result in absurd answers in particular cases.

On the other hand, the method is based on a sound practical principle, namely that the front portion of a shop, including windows, is the part that attracts the most customers and is therefore the most valuable part as selling space. Consequently, a shop with moderate but adequate depth is likely to have a higher overall rental per unit of floor space than one with the same frontage but a much greater depth. Similarly, a shop with a frontage of 7 m and a depth of 15 m may be worth £6,000 p.a., but an adjoining shop with the same frontage of 7 m and otherwise identical, but with double the depth, i.e. 30 m, will be found to be worth considerably less than £12,000 p.a.

Reference may usefully be made to *WH Smith & Son Ltd* v *Clee (VO)*[25] and *Hiltons Footwear Ltd* v *Leicester City Council and Culverwell (VO)*.[26]

25 (1977) 243 EG 677.
26 (1977) 242 EG 213.

(d) Factories and warehouses

Both factories and warehouses are usually assessed by reference to gross internal area, the value per square metre being fixed by comparison with other similar properties in the neighbourhood. In rare cases a flat rate per unit of floor space may be used throughout the building, but usually it is more appropriate to vary the rate applied to each floor and/or possible use for the available floor space.

Properties of this type differ considerably in such matters as construction, accommodation, planning and situation. Any special advantages or disadvantages attaching to the property under consideration must be very carefully examined when applying rents of other properties. Particular factors affecting rental values are the proximity of motorways or rail facilities, availability of labour, internal layout, clear height of the main warehouse area, nature of access to upper floors, loading and unloading facilities, natural lighting, heating, ventilation, adequacy of toilets, nature of ancillary offices, canteen facilities, yard areas, availability of public transport and car parking for staff.

There may be no general demand for certain industrial properties of highly specialised types in some areas and therefore no evidence of rental values that could be used as a basis for comparison, particularly since most are owner-occupied in any event. With such cases as, for example, oil refineries, chemical plants, cement works and steelworks, it will probably be necessary to resort to the contractor's method by taking an appropriate percentage of the effective capital value of the land and buildings as a method of obtaining the rateable value.

Plant and machinery are only to be taken into account in the rating assessment if they are deemed to be a part of the hereditament by regulation. The Valuation for Rating (Plant and Machinery) Regulations 2000, as amended, specify four classes of plant and machinery that are deemed to be a part of the hereditament and therefore rateable. These are as follows:

Class 1   relates to the generation, storage, primary transformation or main transmission of power in the hereditament.

Class 2   relates to "services" to the hereditament, i.e. heating; cooling; ventilation; lighting; drainage; water supply; protection from trespass, criminal damage, theft, fire or other hazard.
    This class excludes items used to provide services that are in the hereditament mainly as part of the manufacturing or trade processes.

Class 3   comprises a variety of items in various categories, including railway lines, etc.; lifts, elevators, etc.; cables, pylons, etc., used for transmission, etc. of electricity which are part of an electricity hereditament; cables, masts, etc. used for the transmission of communications; pipelines.

Class 4   comprises structural process plant. It treats as rateable plant and machinery named in Tables 3 and 4 to the regulations except –

    (a)   any such item which is not, and is not in the nature of, a building or structure;

(b) any part of any such item which does not form an integral part of such item as a building or structure or as being in the nature of a building or structure;

(c) so much of any refractory or other lining forming part of any plant or machinery as is customarily renewed by reason of normal use at intervals of less than fifty weeks;

(d) any item in Table 4 the total cubic capacity of which (measured externally and excluding foundations, settings, supports and anything which is not an integral part of the item) does not exceed four hundred cubic metres and which is readily capable of being moved from one site and re-erected in its original state on another without the substantial demolition of any surrounding structure.'

The effect of these provisions is that the rent to be ascertained is that which a tenant would pay for the land and buildings, including that plant and machinery which is rateable in accordance with the statutory definition.

There may be no difficulty in assessing the rental value, on a unit of floor space basis, of premises including certain rateable plant and machinery such as sprinkler and heating systems. Other rateable plant and machinery may have to be valued separately from the land and buildings on the contractor's method, applying the statutory decapitalisation rate to its effective capital value to arrive at the rateable value.

Rateable plant and machinery is not restricted to that found on industrial hereditaments; for example, a passenger lift in a department store is rateable and so are the lighting and heating in an office building.

The valuation officer, if asked to do so by the occupier, must provide written particulars of what items of plant and machinery are deemed to be part of the hereditament and have been included in the valuation officer's valuation for the assessment.

(e) Cinemas and theatres

A method of valuation which has been approved by the Lands Tribunal involves finding the gross receipts from the sale of seats plus the takings from bingo, screen advertising and the sale of ice-cream, cigarettes and sweets, etc. A percentage of this total figure is then taken to obtain the rateable value. In *Thorn EMI Cinemas Ltd v Harrison (VO)*,[27] a percentage of 8.5% was applied to gross receipts (adjusted to 1973 levels as the 1973 Valuation List was in force). An alternative method of making the assessment is to relate it to a price per seat, but very wide variations can be found in analysing assessments by this method.

27 [1986] 2 EGLR 223.

(f) Licensed premises

Public houses are valued on a rentals approach by reference to rents, as many pubs are rented. Rents are not normally analysed on an area basis because the size of public house is not something a prospective tenant would primarily be interested in when thinking of renting a pub: what the hypothetical tenant would need to know is the likely trade that could be achieved in running the pub. Rents are therefore analysed by reference to the "fair maintainable trade" likely to be achieved by a reasonably competent publican at the AVD.

The analysis produced is a percentage of gross receipts, which will show a variation depending largely on the level of receipts. Usually the analysis is made separately for drinks and gaming machines, which are treated as liquor or wet trade, and the food and letting trade, known as the dry trade. The percentages for the different income streams may vary due to their differing costs and profitability.

(g) Caravan sites

A caravan pitch used as a sole or main residence will be liable for Council Tax (see Section B below). In the case of other caravans and sites, Statute[28] provides that the whole geographic unit comprising caravan pitches, together with any caravans thereon, and common facilities shall be treated as a single assessment. The site operator is made liable for payment of rates in respect of any such assessment.

In the absence of good rental evidence, caravan sites are normally assessed by a profits test. A Lands Tribunal decision on a 1963 List reference, *Gorton v Hunter (VO)*,[29] established that various methods of valuation were admissible but that it was up to the then Local Valuation Court (now Valuation Tribunal) to decide the most appropriate.

(h) Garages and service stations

These are normally valued by reference to throughput. The rental value depends primarily on throughput, being the quantity of petrol sold. A rental value for a certain throughput (per thousand litres) is taken to represent the tenant's bid in terms of rateable value. The figure varies on a sliding scale which increases with throughput, as the tenant's overheads would be proportionately less as the trade increases. The scale also has regard to varying revenue from credit card sales, petrol accounts, opening hours, incentives offered, etc., since these all affect profitability. The rateable value thus arrived at allows for the petrol sales area and is taken to include any small sales kiosk but not a shop as such. It also assumes a canopy over the pump sales area, and an allowance will be appropriate if there is no canopy. In recent years the size and importance of shops with petrol filling stations has

28 [1986] 2 EGLR 223.
29 Non-Domestic Rating (Caravan Sites) Regulations 1990, SI 1990/673.

increased to such an extent that the main profitability of a site can lie with the shop rather than the fuel sales. Additions to the valuation would need to be made for other buildings and land such as car washes, storerooms, showrooms, lubrication bays, workshops and lock-up garages.

(i) Minerals

Mineral hereditaments are valued according to their component parts. The mineral itself is valued on a royalty basis applied to the volume extracted. Plant, machinery and buildings are valued on an estimated capital value basis. Surface land may be valued by reference to its value to the overall operation. Fifty per cent relief is applied to the mineral value and the value placed on any land used for actual quarrying operations. Any value for waste disposal purposes should be taken into account.

Mineral hereditament valuations are subject to annual review.

SECTION B: DOMESTIC PROPERTIES

1. General

Prior to 1 April 1990, domestic properties were liable for local rates.

On 1 April 1990 the Community Charge was introduced. This sought to raise local revenue by taxing persons resident in the local authority's area – a poll tax. However, this proved difficult both to administer and to collect, and the government quickly sought a replacement. It returned to the idea of a property tax, effectively bringing back a form of rating, but the new Council Tax was based not on rental values but on capital values allocated into broad value bands.

2. Procedures

The Local Government Finance Act 1992 provided for the compiling and then maintenance of a valuation list for each billing authority. Billing authorities are the same authorities as for non-domestic rates.

Valuation lists show every dwelling in the billing authority area, together with a band of value applicable to that dwelling. A dwelling is defined as any property which would have been a hereditament for the purposes of the General Rate Act 1967, is not included in a non-domestic rating list and is not exempt from local non-domestic rating.

Responsibility for valuation for banding lies with the listing officers appointed by the Commissioners of HM Revenue and Customs. Like the valuation officers, they are supported by the Valuation Office Agency.

For England the band values are set at 1 April 1991 levels of value, but in Wales, where there was a revaluation in 2005, they are set at 1 April 2003 levels of value.

The bands for England are as follows:

- Band A – Value not over £40,000
- Band B – Value over £40,000 but not over £52,000
- Band C – Value over £52,000 but not over £68,000
- Band D – Value over £68,000 but not over £88,000
- Band E – Value over £88,000 but not over £120,000
- Band F – Value over £120,000 but not over £160,000
- Bank G – Value over £160,000 but not over £320,000
- Band H – Value over £320,000.

The bands for Wales are as follows:

- Band A – up to 44,000
- Band B – 44,001–65,000
- Band C – 65,001–91,000
- Band D – 91,001–123,000
- Band E – 123,001–162,000
- Band F – 162,001–223,000
- Band G – 223,001–324,000
- Band H – 324,001– 424,000
- Band I – 424,001 upwards.

3. Basis of assessment

The appropriate banding is decided by determining the value the property might reasonably have been expected to realise if it had been sold on the open market by a willing vendor on 1 April 1991 (1 April 2003 in Wales), subject to the following assumptions:

(a) that the sale was with vacant possession;
(b) that the interest sold was the freehold or, in the case of a flat, a lease for 99 years at a nominal rent;
(c) that the size, character and layout of the property and the physical state of its locality were the same as at the time the valuation of the property was made; and
(d) that the property was in a state of reasonable repair.

It should be noted that in England the valuation date is over twenty years ago and, in deciding on the band for a newly built house, the valuer has to establish what it would have sold for had it been built back in 1991. This may be straightforward in existing residential locations but is challenging in the case of new developments in estates which did not exist at all back in 1991.

The actual liability to Council Tax is determined by the banding, i.e. value, of the property and also by the number of adults living in it. The legislation provides for appeals to be made by way of making a proposal under certain circumstances where it is considered that the banding is incorrect. Appeals are determined by the Valuation Tribunal, whose decisions can be appealed to the High Court on points of law only.

Whilst the hereditament forms the basis for the "dwelling" for Council Tax, the regulations require a hereditament to be treated as several "dwellings", each with its own band, where there ais more than one self-contained unit, as defined, within the hereditament. Self contained is defined as "a building or part of a building which has been constructed or adapted for use as separate living accommodation". There have been a number of cases decided by the High Court to clarify what this means. In particular, it has been confirmed that being "self contained" does not depend on the actual use made of the part but whether it was in fact constructed or adapted to be capable of use as separate living accommodation.

4. Compilation and maintenance of new lists

A revaluation coming into force in 2005 was undertaken in Wales, but the planned revaluation for England in 2007 was cancelled, leaving the tax based on the old relativities and values from 1991. It is not known when the government will order a revaluation to update the tax base.

Chapter 21

Taxation

Capital gains tax

1. Taxation of capital gains – capital assets and stock in trade

The Finance Act 1965 introduced a comprehensive system for the taxation of long-term capital gains, by which is meant gains realised when a capital asset is disposed of and the proceeds exceed the costs incurred in acquiring the asset.

These provisions apply to assets of a capital nature. In the case of property these would be the land and buildings owned by a person or company which they occupy or let as an investment. It is important to note that in some instances such property will not be regarded as capital but as stock in trade, which is not subject to the provisions applicable to capital gains.

The determination of whether or not property is held as stock in trade is not always clear. Broadly, stock in trade refers to those instances where a property was purchased with the purpose of making a profit. A clear example is land purchased by a company engaged in house building. The company buys land on which to build houses and the land is as much a raw material as the bricks and timber that go into the houses themselves. Such land is clearly not a capital asset since the company will sell the houses with the land as soon as possible and so move on to the next site. Thus, the cost of buying the land is one of the general costs of the business, whilst the proceeds of sale from the disposal of the houses constitute its income. Any gain in respect of the land is therefore a part of the company's profits, to be treated as income. On the other hand, if the company owns an office building wherein its staff are housed, such a building is a capital asset since the offices were not purchased with a view to selling them on for a profit. If the company does choose to sell its offices at any time, any gain realised would be subject to capital gains tax legislation. This legislation is examined below.

2. Taxation of chargeable gains act (TCGA) 1992

(a) Occasions of charge

Capital gains tax (CGT) is payable on all chargeable gains accruing after 6 April 1965, and on the disposal of all forms of assets including freehold and leasehold property, options and incorporeal hereditaments.

An asset may be disposed of by sale, exchange or gift. Disposal also includes part disposals and circumstances where a capital sum is derived from an asset. The transfer of ownership following a death is not a disposal.

A part disposal arises not only when part of a property is sold but also when a lease is granted at a premium. By accepting a premium the landlord is selling a part of the interest and the lessee is purchasing a profit rent. Premiums may be liable to both income tax under the Income and Corporation Taxes Act 1988, which deals with taxation of premiums, and capital gains tax under the 1992 Act. The Act contains provisions for avoiding double taxation on any part of the premium.

A capital sum is derived from an asset when the owner obtains a capital sum such as a payment received to release another owner from a restrictive covenant or to release him from the burden of an easement, and would also include compensation received for injurious affection on compulsory purchase.

It also includes capital sums received under a fire insurance policy, although if the money is wholly or substantially applied to restore the asset no liability will arise.

The transfer of an asset out of fixed assets into trading stock is deemed to be a disposal. This would arise, for example, if an owner-occupier of commercial premises decided to move elsewhere and carry out a redevelopment with a view to selling the completed new property.

Where a disposal is preceded by a contract to make the disposal, the disposal is taken to occur when the contract becomes binding. This typically occurs on the sale of property where the parties enter into a contract with completion at a later date. If the contract is conditional, for example subject to grant of planning permission, it becomes binding on the date when the planning permission is granted, which will be the date of disposal.

(b) Computation of gains

The general rule applicable to the disposal of non-wasting assets, which includes freeholds, and leases with more than 50 years unexpired, is that in arriving at the net chargeable gain certain items of allowable expenditure are deductible from the consideration received on disposal. These items are:

1. Expenditure wholly and exclusively incurred in disposing of the asset. This includes legal and agents' fees, advertising costs and any valuation fees

incurred in preparing a valuation required to assess the gain. It does not include fees incurred for a valuer to negotiate and agree the values with the district valuer.[1]

2.　The price paid for the asset, together with incidental costs wholly and exclusively incurred for the acquisition, for example legal and agents' fees and stamp duty.

3.　Additional expenditure to enhance the value of the asset and reflected in the asset at the disposal date, including expenditure on establishing, preserving or defending title. A typical example would be improvements to the property carried out by the taxpayer, such as an extension. There is an exception where premises above a shop is converted into more than one flat. In this case the expenditure can be charged against income and not added to the capital cost.

Any item allowed for revenue taxation cannot be allowed again, and thus only items of a capital nature are deductible. Special rules apply where expenditure has been met out of public funds.

Certain of the above-mentioned items will need to be apportioned in the case of a part disposal, and further modifications will be required to meet special circumstances, as described later.

Example 21–1
A purchased a freehold interest in a shop in 1982 for £200,000. In 2008 A sold his interest for £400,000.

Proceeds of disposal			£400,000
Less			
Agents' fees on sale, say		£4,000	
Advertising for sale, say		£800	
Legal fees on sale, say		£2,000	£6,800
Net proceeds of disposal			£393,200
Less			
Acquisition price		£200,000	
Add			
Agents' fees on purchase, say	£2,000		
Legal fees on purchase, say	£1,000		
Stamp duty	£2,000	£5,000	£205,000
Gain			£188,200

---

1 *Couch (Inspector of Taxes)* v *Administrators of the Estate of Caton*, The Times, 16 July 1997; *Sub nom. Administrators of the Estate of Caton* v *Couch (Inspector of Taxes)* [1997] STC 970.

(c) Disposals after 5 April 1988

The Finance Act 1988 introduced a significant reform of the capital gains tax rules in respect of disposals on or after 6 April 1988. The general provisions are noted elsewhere, but one change requires specific attention. This concerns cases where the asset, the subject of the disposal, was acquired before 1 April 1982. In such cases the taxpayer may elect (irrevocably) for capital gains and losses on all assets held at 31 March 1982 to be calculated by reference to values at that date. This relieves those who make the election of the need to maintain records going back beyond that date.

The gain is determined by deducting from the net proceeds of disposal the market value of the asset as at 31 March 1982 in lieu of the actual price paid at the actual time of acquisition. The effect of this is to charge to capital gains tax only the gain arising after 31 March 1982. The Act states that the taxpayer is deemed to have sold the asset and immediately to have re-acquired it at market value. No allowance is given for the notional fees that would have been incurred on the notional purchase. Further, any capital expenditure incurred before 31 March 1982 is excluded from the calculation of the gain since it took place before the asset was (notionally) purchased.

(d) Disposals before 6 April 1998

As already stated, properties acquired before 31 March 1982 are revalued as at 1 April 1982 to give the capital value base for the assessment of gain. Until 5 April 1998 this value would be increased by what was known as indexation up to the date of sale to give a revised base figure, the extent of the indexation being the increase in the retail price index (RPI) between the date of purchase (or value as at 1 April 1985) until the date of sale. For indexation allowance, see CG17207.

(e) Disposals after 6 April 1998

Indexation was frozen as at 6 April 1998 and replaced with taper relief until 5 April 2008, when both were abolished for individuals.

The capital tax regime was then simplified for individuals and a single rate of CGT is now applied for disposals on or after 6 April 2008. As part of the CGT simplification, frozen indexation and taper relief are not now available for such disposals by individuals and the single rate of CGT has been replaced by a new rate structure for gains accruing on or after 23 June 2010; see CG21200.

Companies and other concerns within the charge to corporation tax are not affected by these changes. They pay corporation tax on their chargeable gains, and continue to be entitled to indexation allowance. (For indexation allowance and taper relief see HM Revenue and Customs (HMRC) customer advisory notes CG17207 and CG17895 respectively.)

(f) Rates of tax from 23 June 2010 for individuals

Taxable and chargeable taxes act 1992/S4, TCGA92/S4A, F(No. 2)10/SCH1

The first part of a gain is exempt for individuals, with half that amount for trustees. This varies from year to year. For example, in the tax years 1994/1995 to 1999/2000 the exempt amount rose from £5,800 to £7,100. For 2007/2008 it was £9,200 and for 2011/2012 it will be £10,600. There is no exempt amount for companies.

The rates at which gains are taxed have been changed for chargeable gains that accrue on or after 23 June 2010. From that date gains are taxed at either 28% or 18%, according to the extent that an individual has any unused part of the basic rate band. However, gains that accrue on or after 23 June 2010 may qualify for entrepreneurs' relief, and these are taxed at 10%. There is detailed guidance on entrepreneurs' relief in CG63950. Where an individual has gains that qualify for entrepreneurs' relief those gains use up the basic rate band in priority over other gains (TCGA92/S4(6)).

The amount of gain that is charged to capital gains tax at 18% may be modified in certain special cases that affect the amount of income that is treated as using up an individual's basic rate band; see CG21230. An individual's basic rate band may be extended to permit higher rate relief on certain payments and this may affect the amount of gain that is taxed at 18%; see CG21220.

2010–11 is the only year in which the rates of capital gains tax have been changed part of the way through a tax year. Gains that accrue in 2010–11 up to and including 22 June 2010 are taxed at 18%; see CG21100. There are transitional provisions that determine when in 2010–11 a gain is to be deemed to accrue in cases where the relevant legislation does not provide for a precise time of accrual. These transitional provisions are described in CG21240. In other cases the time of accrual depends on the facts and no transitional rule is needed, just as no transitional rule is needed when rates change from one tax year to the next.

For companies, indexation still applies by virtue of the Finance Act 1982/S86 – FA82/S89 and FA82/SCH13, which introduced the deduction, known as the "indexation allowance", to be made in computing gains. For assets acquired before March 1982 these are revalued as at 1 April 1982 and for these, and for assets acquired after March 1982, the chargeable gain becomes approximately the "real" gain, that is, the gain after making an allowance for inflation. But relief was severely restricted (see CG17540) by three particular provisions.

FA94/S93 and FA94/SCH12 introduced restrictions on the extent to which indexation allowance can create or increase a loss. These rules operate in respect of disposals of assets on or after 30 November 1993; see CG17700. However, there are special rules for dealing with indexation following no gain/no loss transfers of an asset (see CG12705) before 30 November 1993.

Deferred gains

Freezing of indexation allowance does not change the treatment of any gain that is computed on a disposal or a deemed disposal before 6 April 1998 but postponed so that it is treated as accruing on or after 6 April 1998. Indexation allowance will be computed to the date of the disposal or deemed disposal and not to the date of accrual.

Where the date of the disposal or deemed disposal is on or after 6April 1998 indexation allowance will be computed to April 1998, except in the case of companies within the charge to corporation tax, where the existing treatment will continue.

The most common examples of postponed gains include:

* reorganisations involving qualifying corporate bonds, see CG53820+; and
* business asset roll-over relief where the replacement asset is a depreciating asset, see CG60370+.

The withdrawal of indexation allowance with effect from 6 April 2008 does not change the treatment of any gain on an asset owned before April 1998 that is computed on a disposal or a deemed disposal before 6 April 2008 but postponed or deferred so that it is treated as accruing on or after that date. Indexation allowance will be computed to 5 April 1998 (or the date of disposal).

CG63950 – Entrepreneurs' relief: scope and layout of guidance

The Finance Act 2008 introduced entrepreneurs' relief for gains arising from a material disposal of business assets by individuals and the trustees of certain settlements and disposals of other business assets associated with a material disposal of business assets where certain conditions are met.

Entrepreneurs' relief was first announced in January 2008, with draft legislation being published shortly after. Entrepreneurs' relief was introduced by FA2008/S7 and Sch3, which inserted the provisions into TCGA 1992 from S169H to S169S. FA2008/Sch3/paragraph5 contains the commencement provisions for entrepreneurs' relief. It applies to disposals made on or after 6 April 2008.

Subsequent amendments have been made by FA2010/S4 and F(No.2)A2010/Sch1/paragraphs 4 to 9.

The guidance is organised so that for each aspect there is an introductory paragraph giving a general overview of the topic, followed by more detailed information. The introductory paragraphs are shown in the table below. For convenience, the statutory references in TCGA92 are used throughout this guidance. They cover the following areas.

The maximum gain to which relief can apply is £2 million up to 22 June 2010, £5 million from 23 June 2010 and £10 million from 6 April 2011. This is a lifetime limit, so an individual making a number of disposals of businesses during his lifetime may exceed the limit. The value of the relief from 6 April 2011 is therefore £10 million $\times$ 18% (the reduction in tax) = £1,800,000.

These taxation rules are very complicated and therefore the above paragraphs are given for general guidance. The advice of tax experts should always be sought when computing any tax liability.

(g) Exemptions and reliefs

1. Exempted bodies

Certain bodies are exempt from liability, including charities, local authorities, friendly societies, scientific research associations and pension funds, as are non-resident owners.

2. Owner-occupied houses

Exemption from capital gains tax is given to an owner in respect of a gain accruing from the disposal of a dwelling-house (including a flat) which has been the owner's only or main residence, together with garden and grounds up to 0.5 ha or such larger area as may be appropriate to the particular house ("required for the reasonable enjoyment of the house").

Where, in addition to the main building, there are outbuildings such as garages, greenhouses, sheds, and even staff accommodation in the grounds of the house, the first step is to determine which buildings are part of the house. The test is whether the buildings are within the curtilage of the house, which requires that they must be close to the house (*Lewis* v *Rook*).[2] In that case a gardener's cottage 190 yards from the house was not part of the house.

The test of which buildings are included with the house is settled between the taxpayer and the inspector of taxes, whilst it is the district valuer who negotiates with the taxpayer as to the area of land required for the reasonable enjoyment of the house.

Where part of the garden of a house is sold but the house is retained, the exemption applies if the land sold is within a garden of less than 0.5 hectares or, if larger, is within the area required for the reasonable enjoyment of the house. This will not be so if the house is sold first without the land in question, and the land is sold later (*Varty* v *Lynes*).[3] This can prove to be a valuable exemption if part of the garden is sold for a high development value.

The exemption does not apply, however, if the house was purchased with a view to making a gain or was not used as a home – which implies that the taxpayer must show a degree of permanence and expectation of continuity. In *Goodwin* v *Curtis*[4] the taxpayer purchased a farmhouse on 1 April 1985 and sold it on 3 May 1985 at a considerably enhanced price. It was held that the private residence relief did not apply.

The degree of exemption is proportionate to the period of owner-occupation during ownership. Full exemption is given if the owner has lived in the house throughout the whole period of ownership. To cover the circumstances where a house is vacant after the owner has moved and is looking for a purchaser, the exemption includes the last three years of ownership. Certain other periods of absence have also to be disregarded under the Act, such as where the taxpayer has to live elsewhere temporarily as a condition of employment.

Thus, if a property is sold and was not owner-occupied for the whole period of ownership, the degree of exemption depends upon the ratio between the period of owner-occupation (including in any event the last three years of ownership) and the total period of ownership.

If, therefore, X buys a house on 1 April 2002, and lives in it for two years until 1 April 2004, when he lets it, subsequently occupying the house on 1 April 2008 and selling it on 1 April 2008, the proportion of any capital gain which is exempt from capital gains tax is:

$$\frac{2+3}{9} = \frac{5}{9}$$

2 [1992] STC 171.
3 [1976] STC 508.
4 *The Times*, 2 March 1998.

Where premises are used partly as a dwelling and partly for other purposes, the exemption applies only to the residential part and it will be necessary to apportion any capital gain on the whole property between the two portions. This apportionment should be made on a value basis. Thus, if a house is half let and half owner-occupied, the gain on disposal is not necessarily apportioned on a 50:50 basis. It is possible that more value is attributable to the owner-occupied part and that the growth in value is greater for that part. The apportionment should reflect these circumstances.

The exemption also applies to one other house which is occupied rent free by a dependent relative if the house was bought before 6 April 1988, and also where trustees dispose of a house which has been occupied by the beneficiary under the trust.[5]

3. Replacement of business assets

Where trade assets, including buildings, fixed plant and machinery, and goodwill are sold and the whole of the proceeds are devoted to the replacement of the assets with other trade assets, the trader may claim to defer any capital gains tax which ordinarily may have been payable on the sale. The trader may choose, instead of paying the tax that arises, to have the actual purchase price of the replacement written down by the amount of any chargeable gain on the sale of the original assets. This process may be repeated on subsequent similar transactions so that tax on the accumulated capital gain will not be paid until the assets are sold and not replaced. To qualify for this relief, the acquisition of the new assets must be within a period commencing one year before and ending three years after the sale of the old assets.

The Act also deals with circumstances where not all the proceeds of sale are re-invested.

This is known as "rollover relief" and is also available to an investor who is forced to sell assets by virtue of a compulsory purchase order or grants a lease extension or sells a freehold under the Leasehold Reform legislation.

Example 21–2

X purchased a freehold shop in June 1991 for £15,000, the expenses of acquisition being £500, and occupied and commenced trading as a grocer. In September 1997 he sold this shop for £40,000, his expenses of sale being £900. He immediately purchased a new shop for £45,000, incurring expenses of £1,000, and continues in business in the same trade.

X claimed relief from the payment of tax in 1997 on the sale for £40,000, so the cost of acquisition of the new shop will be dealt with as follows:

5 For further details see CGT4, issued by the Board of Inland Revenue and available from Tax Offices free of charge.

Chargeable gain on sale in 1997

Proceeds of disposal		£40,000
Less expenses		900
Net proceeds of disposal		£39,100
Less acquisition price	£15,000	
Add costs	£500	£15,500
Gain		£23,600
Less indexation relief		
$\dfrac{159.3 - 134.1}{134.1} \times 15,500 =$		£2,900
Chargeable gain		£20,700

The chargeable gain is deducted from the purchase price of the new shop so that, on a subsequent sale, £24,300 (£45,000 – 20,700) and not £45,000 will be deducted as the purchase price. In this way, the chargeable gain is said to be "rolled over" into the new asset – hence rollover relief.

Cost of replacement asset	£45,000
Less gain on sale of old asset	£20,700
Notional cost of replacement asset	£24,300

If the replacement shop had been sold in 2008 for £280,000, with the expenses of sale being £8,000 and the owner then ceasing to continue in business, liability to capital gains tax would then arise as follows (ignoring indexation and taper relief):

Proceeds of disposal		£280,000
Less expenses		£8,000
Net proceeds of disposal		£272,000
Less		
Notional acquisition price	£24,300	
Expenses of acquisition	£1,000	£25,300
Gain		£246,700

Thus, the actual net gain in respect of the new asset of £272,000 – (£45,000 + £1,000) = £226,000 is boosted by the previous gain of £20,700 carried (or "rolled") over to give a terminal gain of £226,000 + £20,700 = £246,700.

Taper relief, which would have accrued between 1997 and 2008, was lost as a result of the CGT simplifications applying after 6 April 2008.

Where the sum re-invested in the replacement asset is less than the net proceeds realised on the sale of the old asset, the difference is subject to capital gains tax, since it is part of the gain itself which is not being rolled over.

Where rollover relief was claimed between 31 March 1982 and 6 April 1988 and part of the gain accrued before 31 March 1982, any gain arising on the disposal of the replacement asset is halved.

It should be emphasised that rollover relief applies to properties occupied for business purposes but not to properties held as investments. The exception to this is where an interest in an investment property is acquired under compulsory purchase powers and, under the Leasehold Reform legislation, when the owner has 12 months to roll over the compensation proceeds into another investment property.

(h) Losses

A capital loss might arise as a result of a disposal, say where the asset is sold for less than the price paid for it. Such a loss is allowable if a gain on the same transaction would have been chargeable. The loss would normally be set off against gains accruing in the same year of assessment but if, in a particular year, losses exceed gains, the net loss can be carried forward and set off against gains accruing in the following or subsequent years.

(i) Part disposals

The rule for calculating any gain or loss on a part disposal is that only that proportion of the allowable expenditure which the value of the part disposed of bears at the date of disposal to the value of the whole asset can be set against the consideration received for the part. Thus, it is necessary to value the whole asset, and this necessitates a valuation of the retained part, since for these purposes the value of the whole asset is treated as being the aggregate of the values of the part disposed of and the part retained. The proportion disposed of is then:

$$\frac{A}{A+B} = \frac{\text{Consideration on part disposal (A)}}{\text{Consideration on part disposal (A)} + \text{value of retained part (B)}}$$

Example 21–3

X bought an asset in 1984 for £100,000, including costs, and later sells part of it for £600,000, no disposal costs being involved. The first step is to value the part retained, say £900,000. The proportion of the asset disposed of is:

$$\frac{£600,000}{£600,000 + £900,000} \text{ or } 40\%$$

i.e., X is realising 40% of the value of the asset. (£600,000 out of an asset worth £1,500,000.)

Allowable expenditure of costs of acquisition and enhancement is therefore taken to be 40%. Hence the gain realised, ignoring indexation, is:

Proceeds of disposal		£600,000
Less		
Acquisition price	£100,000	
$\dfrac{\times A}{A+B} = 40\% =$	0.4	
Apportioned acquisition price		£40,000
Capital gain		£560,000

Where enhancement expenditure relates solely to the part sold, it is fully allowable; where it relates solely to the part retained, then none is allowable on the part disposal. In other cases it is similarly apportioned by $A/(A + B)$.

Indexation relief is not applied to the apportioned acquisition price (unless X was a company).

An alternative basis of calculating the cost of a part disposal of an estate for capital gains tax purposes avoids having to value the unsold part of the estate. Under this alternative basis the part disposed of will be treated as a separate asset and any fair and reasonable method of apportioning part of the total cost to it will be accepted, e.g. a reasonable valuation of that part at the acquisition date. Where the market value at 31 March 1982 is to be taken as the cost of acquisition, a reasonable valuation of the part at that date will be accepted.[6]

The cost of the part disposed of will be deducted from the total cost of the estate (or from the balance) to determine the cost of the part retained. This avoids the total of the separate parts ever exceeding the whole. Thus, in the above example, the total costs were £100,000. Of this, £40,000 was treated as the cost of the part disposed; the remaining £60,000 was carried forward as the cost of the retained interest.

The taxpayer can always require the general rule to be applied, except in cases already settled on the alternative basis. However, if he does so, it will normally be necessary to adhere to the choice for subsequent part disposals.

Where a part of a holding is disposed of for £20,000 or less, such consideration being not more than 20% of the value of the whole holding, it is termed a small part disposal[7] and the taxpayer can elect to pay no capital gain on such a part disposal. This is effected by deducting the consideration from the acquisition costs. Thus, in the above example, if the value of the whole is taken to be £150,000 at the time of disposal, a disposal of part at £20,000 would be less than 20% of the whole. No tax would arise, but the acquisition cost of the retained land would be £100,000 − £20,000 = £80,000. Where other land has been sold in the same tax year the aggregate consideration must not exceed £20,000.

6 This alternative basis is an extra-statutory concession introduced by the Inland Revenue; see Statement of Practice D1, published on 22 April 1971.
7 Taxation of Chargeble Gains Act (TCGA) 1992, section 242.

(j) The market value rule on certain disposals

In a normal case the actual sales price will be treated as the consideration received on disposal. On certain occasions, however, the Act provides for a deemed sale at market value at the date of disposal. This rule on market value operates in the case of gifts. It also applies in the case of other dispositions that are not by bargain at arm's length, such as transfers between closely related or associated persons. It applies as well where an asset is transferred from being a fixed asset into trading stock. The rule applies to both transferor and transferee, so that if X gives a property to his son worth £100,000, he will be deemed to have sold the property for that sum and will be assessed for capital gains tax accordingly. The son would then adopt £100,000 as his cost of acquisition on a subsequent disposition.

Market value is the price which the asset "might reasonably be expected to fetch on a sale in the open market".[8]

(k) Sale of leasehold interests

A long lease, that is a lease with more than 50 years unexpired, is not treated as a wasting asset and the whole of the original acquisition cost and other expenditure can be set against the consideration received on the sale of the lease.

A lease with 50 years or less to run, however, is treated for capital gains tax purposes as a wasting asset and, on the sale of such an interest, the whole of the acquisition cost and other expenditure cannot be deducted. Instead, it is deemed to waste away at a rate which is shown in a table of percentages in Schedule 8 to the Taxation of Chargeable Gains Act (TCGA) 1992. Only the residue of the expenditure that remains at the date of disposition can be set against the consideration received.

The table of percentages is derived from the years' purchase (YP) 6% single rate table. The YP for 50 years is taken to be 100. Thus, against 40 years the figure is 95.457 and against 25 years, 81.100. This implies that the value of the lease with 40 years to run should be 95.457% of its value when it had 50 years to run, and 81.100% with 25 years to run. Similarly, the value of the lease with 25 years to run should be 81.100/95.457% of the value when it had 40 years to run.

The Act provides a formula for writing down acquisition costs whereby there is deducted P1 – P3/P1 from the costs, where P1 = figure from the table in Schedule 8 relating to years to run at time of purchase and P3 = figure from the table relating to years to run at time of sale.

It is suggested that a more convenient approach is to multiply the acquisition cost by P3/P1.

---

8 TCGA 1992, section 272(1).

Example 21–4

A lease with 40 years to run was purchased for £50,000 and sold for £55,000 when it had 25 years to run. The gain, ignoring costs and indexation, is:

Proceeds of disposal		£55,000
Less		
Acquisition price	£50,000	
$\times \dfrac{P3}{P1} = \dfrac{81.100}{95.457} =$	0.850	£42,500
Gain		£12,500

If a lease with more than 50 years to run is purchased, the appropriate percentage at acquisition will be 100 and it will commence to be a wasting asset when it has less than 50 years unexpired.

Where an election is made to adopt market value at 31 March 1982, it should be remembered that this is a deemed acquisition cost which is also subject to writing down by applying P3/P1. Indexation relief is applied to the written down value.

If any value of P3 or P1 is required, it can easily be calculated by dividing YP at 6% for the years unexpired by YP 50 yrs at 6% and multiplying by 100.

e.g., P3 when 25 years to run

$$= \frac{\text{YP 25 yrs at 6\%}}{\text{YP 50 yrs at 6\%}} \times 100 = \frac{12.783}{15.762} \times 100 = 81.100$$

(l) Premiums for leases granted out of freeholds or long leases

The receipt of a premium on the grant of a lease is treated as a part disposal of the larger interest out of which the lease is granted. Against the premium may be set the appropriate part of the price paid for the larger interest and any other allowable expenditure as determined by the formula applicable to part disposals.[9]

Where a lease for a period of 50 years or less is granted at a premium, a part, at least, of the premium will be liable for income taxation under Schedule A. To avoid double taxation, the part of a premium for a lease which is thus chargeable to income tax is excluded from liability to capital gains tax.

Example 21–5

X purchased a freehold shop for £50,000 (inclusive of costs) in 1986. In 1999, he grants a 25-year lease at £6,000 per annum net, taking a premium of £75,000.

Step 1

Calculate the amount taxable under Schedule A.

9 See Section 2(j) of this chapter.

The amount taxable is the whole premium less 2% of the premium for each complete year of the term except the first year. Hence:

Whole premium	£75,000
Less 2(25 – 1)% of £75,000	£36,000
Part subject to Schedule A	£39,000

£39,000 is excluded from any capital gains tax liability, leaving £36,000 as the amount of premium on which to base capital gains tax.

Step 2

Allowable expenditure is then to be considered in accordance with the part disposal formula[10] but there is to be excluded from the consideration in the numerator of the fraction (but not in the denominator) that part of the premium taxable under Schedule A.

The application of the formula involves a valuation of the interest retained, which includes the capitalised value of the rent reserved under the lease:

Rent reserved	£6,000 p.a.		
YP 25 years at 6%	12.78	£76,680	
Reversion to estimated full rental value (net)	£12,000 p.a.		
YP in perp. deferred 25 years at 6%	3.88	£46,560	£123,240
Value of retained interest, say			£123,000

Using the modified part disposal formula, the gain (ignoring costs and indexation) is:

$$Gain = £36,000 - \left\{ £50,000 \times \frac{£36,000}{£75,000 + £123,000} \right\}$$

$$= £36,000 - £9,090$$

$$= £26,910$$

(m) Premiums for subleases granted out of short leases

As the larger interest is a wasting asset, the normal part disposal formula does not apply. Only that part of the expenditure on the larger interest that will waste away over the period of the sub-lease in accordance with the table of percentages[11] can be set against the premium received.

Thus, if a person acquired a 40-year lease for £5,000 and, when the lease had 30 years to run, granted a sublease for seven years at the head lease rent, taking a premium of £6,000, the following percentages from the table are required:

10 See Section 2(j) of this chapter.
11 See Section 2(k) of this chapter.

40 years, 95.457 (percentage when interest acquired);
30 years, 87.330 (percentage on grant of sub-lease);
23 years, 78.055 (percentage on expiration of sub-lease).
The percentage for the seven years of sub-lease is:

$$87.330 - 78.055 = 9.275$$

and the expenditure which is allowable against the premium of £6,000 is:

$$£5,000 \times \frac{9.275}{95,457} = £486$$

The capital gain is therefore £6,000 – £486 = £5,514.

(*Note*: The allowable expenditure has to be written down if the sublease rent is higher than the head lease rent.)

In this case $2 \times (7 - 1)\% = 12\%$ of the premium $= 12\%$ of £6,000 = £720, leaving £5,280 liable to income tax under Schedule A. The method of avoiding double taxation is different in this case. The amount of the premium chargeable to income tax is deducted from the capital gain of £5,514 and only the balance of £5,514 – £5,280 = £234 is subject to capital gains tax. If the income tax proportion exceeds the capital gain, then no capital gains tax is payable.

INHERITANCE TAX

I. General

The Finance Act 1975 repealed the long-established provisions for taxing the value of a deceased's estate at death by the imposition of estate duty. The Act replaced estate duty with a new tax, capital transfer tax. The provisions relating to this tax were incorporated in the Capital Transfer Tax Act 1984. The Finance Act 1986 then abolished capital transfer tax and replaced it with inheritance tax (IHT). Since many of the features of inheritance tax are the same as capital transfer tax, the 1984 Act was renamed the Inheritance Tax Act 1984. There is a single tax rate of 40% on the taxable estate.

2. Taxable transfers

IHT applies when there is a transfer of value. This will arise where people make a gift in their lifetime or their estate passes in its entirety on their death. A gift may be an absolute gift, whereby an asset is passed to another without any charge, or one where an asset is sold to another at an "under value", by which is meant the sale price is deliberately low so as to confer an element of gift in the price, i.e. there is an intention to confer a gratuitous benefit.

3. Basis of assessment

Since IHT relates to the transfer of value, the value on which the tax is assessed is not the value of the asset itself but the diminution in the value of the donor's estate caused by the transfer. This will commonly be the same figure as the value, but not necessarily so. For example, suppose that a person owns a home which is vacant and which they let to a son at a nominal rent of £100 p.a. for 20 years. The grant of the lease at a low rent, an under value, is a gift to the son. The value of the house before the gift, with vacant possession, is, say, £300,000, but after the gift when subject to the lease it is, say, £80,000. The value of the lease is, say, £32,000. Although the subject of the gift, the lease, is worth £32,000, the person has actually made a transfer of value of £300,000 (pre-gift) − £80,000 (post-gift) = £220,000, which will be subject to IHT (if the donor does not live 7 years).

Sums that may be set against the transfer of value are any incidental costs incurred in making the transfer and any capital gains tax payable on the gift. Note that it is the capital gains tax itself and not the whole gain, so that there is an element of double taxation.

Since it is the transfer of value which is to be taxed, it is important to establish who is paying the tax, as IHT may be paid by the donor or donee in the case of a lifetime gift. If the donor pays, then the IHT is assessed on the value transferred. If, for example, the value transferred was £32,000 and the tax rate was 28%, the donor would pay 28% of £32,000 = £4,800. On the other hand, if the donor pays, the transfer value is not only the asset but may also be the IHT payable out of the estate. Hence it is necessary to gross up the value transferred before determining the tax so that the grossed-up value less IHT equals the value transferred.

These are very complicated tax calculations and therefore they are best left to specialist tax advisers, who should also be consulted before such transactions are carried out.

4. Rates of tax

Prior to 1988 there were several rates of tax on bands of value, the rates and bands changing yearly. The Finance Act 1988 introduced a greatly simplified system whereby the first £110,000 of value was exempt from IHT, a rate of 40% being the rate on all value in excess of this sum. The exempt band has been increased since then by changes made in successive Finance Acts. For example, the band for the tax year 2010/2011 was £325,000 and this is intended to remain at that level until 2015. Since October 2007 married couples and registered civil partners can claim double the threshold when the second partner dies and the first partner has not used their allowance (or all of it). Lifetime gifts are subject to capital gains tax calculated on the basis that they are transferred at their full value. There is no capital gains tax on death.

5. Exemptions and reliefs

(a) General

There are several exemptions and reliefs of a general nature, such as exemptions for transfers between husband and wife, gifts to charities and some other bodies, lifetime gifts up to £3,000 per annum, wedding gifts up to a certain amount, and small gifts of £250 to as many individuals as the donor likes.

(b) Potentially exempt transfers

Where a gift is made by an individual to another individual or into certain trusts such as an accumulation and maintenance trust, no IHT liability will arise if the donor lives for seven years after making the gift. In such cases the gift is an exempt transfer. However, if the donor dies within seven years IHT becomes payable, although there is tapering relief whereby a proportion only of the tax is charged. The tapering relief is:

Years between gift and death	Percentage of full IHT payable
0–3	100
3–4	80
4–5	60
5–6	40
6–7	20

IHT is not charged at the rate of tax applicable at the time of death but on the value at the time of making the gift. Thus, if an asset was given to another individual in June 2006 with a value of £100,000 and the donor dies in September 2011 when the asset has a value of £150,000, the IHT liability will be 60% of 40% of £100,000. The executors can elect to adopt the value at death if this is less than the value at the time of the gift.

Because of these provisions, only gifts made within the period of seven years prior to death normally attract IHT liability.

All lifetime gifts are chargeable to capital gains tax.

(c) Business assets

Where a person owns a business as the sole proprietor, or otherwise has a controlling interest (husband and wife being treated as one) on the transfer of relevant business assets, including shareholdings in unlisted companies or in listed companies where the person has a controlling interest, the relief is 100%. In the case of land and buildings used by a partnership of which the person was a partner, or by a company which he controlled, the relief is 50%. Business does not include dealing in land or investment, so that property held for such purposes is not included. On the other hand, development is a business for this purpose.

(d) Agricultural property

Where people own or occupy property for agricultural purposes and are engaged in agriculture – the "working farmer" – they are entitled to relief of 100% of the agricultural value of the property on most transfers of that property.

Agricultural value is the value assuming it will always be used for agricultural purposes, so any value attributable to non-agricultural use is ignored, particularly development value or hope value. Hence, assume two hectares of agricultural land which had a possibility of residential development were bequeathed by a farmer to his son. The residential value is £1,000,000 but the value as farmland is £20,000. The residential value would be taken into account in arriving at the value of the estate, but the relief would be 100% of £20,000. Similarly, when farm cottages are the subject of the transfer, the agricultural value is determined by assuming that they could only be occupied by farm workers, and so ignoring any enhanced value if they were to be sold as, say, "weekend cottages".

(e) Woodlands

Where, on death, the deceased's estate includes woodlands, an election may be made to exclude the value of the timber from the value of the estate. Thus, no IHT will be payable on the value of the timber at that time. If there is a subsequent disposal or transfer of the timber in the new owner's lifetime, then IHT becomes payable at the deceased's marginal rate of IHT on the disposal proceeds. If the new owner retains the timber to death, then no IHT will be payable on the original owner's estate at death in respect of the timber.

In determining the value of the deceased's estate where this relief is claimed, the value of the land on which the woodlands grow is included at what is often termed "prairie value". Since it is land covered with trees, the value will be relatively low.

Where a person owns and runs woodlands as a commercial enterprise, the business relief referred to above may be claimed. Where the woodlands are ancillary to an agricultural holding, then agricultural relief may be claimed. Both of these reliefs are more favourable.

6. Valuation

The general principle for determining the value of a person's estate when it is the subject of a transfer is to assume a sale at market value, this being the price which might reasonably be expected to be fetched if sold in the open market at the time of the transfer. No reduction is to be made because such a sale would have "flooded the market". Further, it is to be assumed that the assets would be sold to achieve the best price, which means that if a higher overall price would be achieved by selling parts rather than as a single lot, then it is to be assumed that the sale would follow that approach.[12]

INCOME TAX ON PROPERTY

1. Schedule A

Income tax is charged, under Schedule A of the Income and Corporation Taxes Act 1988, on income from rents and certain other receipts from property after deductions have been made for allowable expenses.

Liability under Schedule A extends to rent under leases of land and certain other receipts such as rent charges. Premiums receivable on the grant of a lease for a term not exceeding 50 years are also subject to taxation under Schedule A. The amount of the premium taxable is the whole premium less 2% for each complete year of the lease, except for the first year[13] ($2(n-1)$% of the premium, where n is the number of complete years of the lease).

Deductions from the gross rents receivable may be made in respect of allowable expenses incurred, including cost of maintenance, repairs, insurance, management, services, rates and any rent payable to a superior landlord.

Rents receivable from furnished lettings are similarly chargeable to income tax on the gross rents less allowable expenses. The whole of the rent will be charged under Case VI of Schedule D, but the landlord may elect for the profit arising from the property itself to be assessed under Schedule A, with the profit from the furniture continuing to be assessed under Case VI. A writing-down allowance may be claimed in respect of furniture and any other items provided. As an alternative to this, the taxpayer may choose to deduct the expenses of renewing these items.

2. Capital allowances

Items of a capital nature, such as expenditure on improvements, additions or alterations, cannot be deducted except for writing-down allowances of the cost of certain specific categories. These capital allowances include expenditure on such items as the provision or improvement of agricultural or industrial buildings and the provision of certain plant and machinery. These allowances generally take the form of annual allowances such as 4% of the cost of industrial or agricultural buildings, and 25% of plant and machinery on a reducing cost basis. 100% allowances are available in Enterprise Zones. The provisions that govern the availability and calculation of capital allowances are contained in the Capital Allowances Act 1990,

---

12 *Duke of Buccleuch* v *Inland Revenue Commissioners* [1967] 1 AC 506.
13 See Part 1 of this chapter for capital gains tax provisions concerning premiums.

and specialist advice should be taken in respect of them. Their allowance will usually have a value to a purchase of a new building as they reduce the taxable income from that property for investors.

Capital allowances can be sold by a developer with the investment or can be retained by the developer and used again in turn from other properties if the development company is converted into an investment company. They cannot be sold independent of both the building to which they apply and the company which developed the building (i.e. one or the other). They are very valuable to a high taxpayer.

VALUE ADDED TAX

I. VAT and property

Value added tax (VAT) was introduced in 1973 when the UK joined the European Community (EC). The rules for membership of the EC are contained in the Treaty of Rome and associated Directives, and include a rule that member states must operate a VAT system of taxation. VAT is thus a eurotax and the detailed provisions of VAT legislation are consequently capable of challenge in the European Court. That this is so is well illustrated by the fact that VAT in respect of property was radically changed in 1989 following a decision of the European Court in 1988.

Prior to 1989, most property activities were effectively outside the VAT net but, since the Finance Act 1989, many activities are now subject to VAT and the number is growing. The main legislation is contained in the Value Added Tax Act 1994.

2. VAT – the general scheme

(a) Added value

VAT is chargeable on added value when a taxable supply is made by a registered person in the course or furtherance of a business. Some definitions are required before this statement can be understood.

(i) Supply

This refers to the sale of goods or the provision of services. They are expressed as being the supply of goods or the supply of services.

(ii) Registered person

Where a company, individual, partnership or other body makes taxable supplies above a prescribed level over a prescribed period, that supplier is required to register for VAT. The result of that is that each supplier is allotted a unique VAT registration number which is used in respect of all VAT activities. VAT returns are then made at the end of successive three-monthly periods when the supplier provides the details required by Her Majesty's Revenue and Customs (HMRC), which administers the tax.

(iii) In the course or furtherance of business

There is no detailed definition of "business". It "includes any trade, profession or vocation" and also the provision of facilities to subscribing members of a club and admission of persons to any premises for a consideration.[14] The meaning of business is fairly widely interpreted but clearly excludes activities by individuals acting in their personal capacity, such as when they sell their own car or house, or assist organisations in a voluntary capacity.

(iv) Added value

The VAT system operates by applying tax to the value added to goods as they pass through the business chain.

Example 21–6

A produces raw materials which are sold to B for £100. B converts the raw materials into parts which are sold to C for £240. C assembles the parts to produce an item which is sold to D, a person who will use the item in his home, for £400.

Transaction	Price paid (a)	Sale price (b)	Value added (b) − (a)
A to B	–	£100	£100
B to C	£100	£240	£140
C to D	£240	£400	£160
			£400

Assuming the rate of tax is 20% then the tax payable to HMRC will be:

Transaction	Value added	Tax (at 10%)
A to B	£100	20
B to C	£140	28
C to D	£160	32
		80

In practice, this result is achieved by each party in the chain acting in the course or furtherance of a business, charging tax on the whole sale proceeds and deducting any tax paid on the purchase costs.

---

14 VAT Act 1994, section 94.

Party	Tax charged on sale (x)	Tax paid on purchase (y)	Tax paid to HMRC (x − y)
A	100 @ 20% = 12	–	20
B	240 @ 20% = 48	20	28
C	400 @ 20% = 40	48	32
			80

As can be seen, the one person who bears the tax, since it is a real irrecoverable cost, is D, the final consumer, whereas it is collected and passed over to HMRC by the parties in the chain of production as it passes down the chain.

Tax charged on a sale price is termed output VAT and that paid on the purchase price is termed input VAT. The quarterly tax returns referred to above require the taxpayer to total the output and input taxes over the quarter and to pay the excess of output VAT over input VAT to HMRC. If input VAT exceeds output VAT, then HMRC reimburses the difference to the taxpayer.

(b) Taxable supplies

The VAT scheme applies to persons making taxable supplies in the course or furtherance of a business. As has been shown, the scheme operates by taxpayers making VAT returns whereby they pay to HMRC the excess of output VAT which they have collected from the input VAT which they have paid out. This is described as the recovery of input VAT. The effect is that VAT is a neutral tax for such persons (apart from the costs of managing their VAT records).

However, the VAT legislation divides supplies of goods or services into three categories – two of which are of taxable supplies and one is not. Hence, where a person makes supplies in the non-taxable supply category, the scheme does not apply as described so far for such a person, who is treated as a consumer; any input VAT paid out is not recoverable.

The two categories of taxable supplies are standard-rated supplies and zero-rated supplies.

(i) Standard-rated supplies

Standard-rated supplies are all supplies that are not found in the detailed list of supplies coming within the other two categories. They are known as standard-rated supplies since VAT is applied at the prevailing standard rate of VAT – 20%. This rate is subject to change from time to time according to general economic policy. Hence, if goods are sold for £100 and constitute a standard-rated supply, then VAT of £20.00 is added to the sale price. (The supply of domestic fuel has an imposed rate below the standard rate, but this may be more by accident than by design.)

(ii) Zero-rated supplies

These supplies are also taxable supplies. However, the rate of VAT charged is zero, so that no VAT is actually added to the price. Zero-rated goods sold for £100 will keep the sale price of £100 with nothing added. Nonetheless, the supplier of zero-rated goods or services is entitled to recover input VAT paid out on purchases. If the supplier makes wholly zero-rated supplies, the quarterly return would show output VAT of nil so that any input VAT paid out would be fully refunded by HMRC.

Zero rating exists within the VAT scheme to keep down the price of goods and services to consumers where it is considered socially desirable to do so. This is demonstrated by the types of supplies that are zero-rated, including food, children's clothing, most public transport, and products designed for use by disabled persons. A full and detailed list of zero-rated supplies is contained in Schedule 8 to the VAT Act 1994. Those relating to property are considered below.

(c) Non-taxable supplies

The third category of supplies of goods or services is termed "exempt supplies". Where a person makes an exempt supply, no VAT is added to the sale price. To this extent it is like a zero-rated supply. However, wholly unlike a zero-rated supply, the goods or services are not subject to any VAT charge at all, neither 20% nor 0%. A person making wholly exempt supplies does not make taxable supplies, does not register for VAT, makes no VAT returns and does not recover any input VAT paid out on purchases of goods or services. For such a person any VAT paid out on purchases of goods or services is a charge which increases the cost of purchases, as it does for consumers generally. Where a person makes some taxable supplies and some exempt supplies ("makes partially exempt supplies"), any input VAT incurred on goods or services attributable to an onward exempt supply by that person cannot be recovered, notwithstanding that the person is registered for VAT.

Those items that comprise exempt supplies are fully set out in Schedule 9 to the VAT Act 1994. They include supplies by the Post Office of postal services, financial services provided by banks, products of insurance companies, the betting industry, education provided by schools and universities, and health services provided by doctors, dentists and opticians. It is important to be familiar with those items constituting exempt supplies since exempt suppliers are likely to be VAT-averse and this could condition their response to proposals put to them. Those supplies related to property are considered below.

3. VAT and the property market

In order to consider the impact of VAT on the property market, it is essential first to identify the VAT treatment of the various property activities.

(a) Residential property

Zero-rated supplies are mainly confined to the provision of new residential accommodation. For this purpose, residential encompasses the provision of accommodation with a view to long-term residence, such as care homes for the aged or infirm, or children's homes, student hostels, hospices, even nunneries. It does not include hotels or prisons.

Thus, the construction of new houses or flats, including most of the building materials, involves zero-rated supplies, as does the final sale or letting, other than when the buildings are let for less than 21 years. This applies also to similar sales or lettings following the conversion of non-residential property, and to certain types of significant work to listed buildings so as to produce residential property which is to be sold or let for not less than 21 years.

It follows that the sale of a new house or flat has no actual VAT added to the sale price, but is subject to the notional VAT at nil percentage. All other transactions in relation to residential property are either exempt supplies or are supplies not in the course or furtherance of a business – which is the situation for the great majority of transactions. As a result, no VAT is chargeable on any transactions involving residential property.

On the other hand, work carried out to residential property such as repairs or improvements is a standard-rated supply. Since the supply is in respect of property for which no taxable supply can be made, VAT increases the cost of such work to the property owner. Hence the valuation of a residential property where a deduction is made for repairs requires the cost to be included gross of VAT (unless the work is of a small scale which may be carried out by a small builder – see *(b)(ii)* below).

There is, however, a concession for a scheme known as an "an urban regeneration scheme", where the VAT rate is 5%. This applies when converting a non-residential building into a residential building, converting flats or a house into bed-sits, refurbishing residential property that has been empty for three years and changing the number of household dwellings. Details of this scheme are available in VAT Notice 708.

(b) Non-residential property

Consideration of the VAT implications of activities in respect of non-residential properties will be considered in respect of sales and lettings, and development and building works.

(i) Sales and lettings

All sales and lettings are generally exempt supplies. However, there are specific exceptions to this general rule. Most important are:

- The sale of the freehold interest in a new or partly completed building. "New" means a building completed less than three years ago. Such a sale is a standard-rated supply, so that VAT must be added to the sale price.

- The sale of the freehold interest in a new or partly completed civil engineering work. This is also a standard-rated supply. In practice, this will usually relate to infrastructure. For example, if an estate road and some services have been built into a development site and completed within the past three years, then, on a sale of the freehold interest in the site, VAT is added to the sale price, or that part of the price reflecting the infrastructure. This could require an apportionment of the price between the standard-rated supply, on which VAT is charged, and the balance, which is an exempt supply where no VAT arises.
- The provision of space for car parking is a standard-rated supply. This applies to licences, a fact familiar to all car drivers who pay VAT on the charge for parking their car.

There are various other supplies set out in Group 1 of Schedule 9 which are also standard-rated, many of them related to sports and leisure activities.

(ii) Development and building works

The supply of building works of all kinds is a standard-rated supply. Hence, demolition, new construction, refurbishment and repairs are all standard-rated supplies on which VAT is added at 20% to the cost of the work.

The exception would be if a builder has a low turnover below the VAT registration threshold. Such "small builders" are not required to register and so do not add VAT to their charges.

4. Option to tax

As is clear from the above, most transactions are exempt supplies, whereas physical works are standard-rated. This creates a VAT burden where the owner of a property carries out work to the property on the cost of which VAT will be charged. Where the owner will make exempt supplies if the property is subsequently let or sold, the VAT on the cost of the work is irrecoverable.

However, the Act provides that an owner of commercial property may elect to have all supplies of the property treated as standard-rated supplies – the option to tax. The incentive to elect is that, because supplies become taxable supplies, input VAT is recoverable. So an owner carrying out work to a property will be able to recover the input VAT on the cost of the work if the election is made. If the cost of the work is significant, with a large amount of input VAT payable, the consequence of an election could be highly beneficial.

Example 21–7

X owns the freehold interest in office premises which have recently been vacated on the expiry of a lease. The property can be let in its present condition for £100,000 p.a. If X spends a total sum of £300,000 on refurbishment, the property will be let at £120,000 p.a. and the yield will fall from 9% to 8%.

(a) Value without refurbishment

Rental value	£100,000	p.a.
YP in perp. at 9%	11.1	
Capital value	£1,110,000	

(b) Value with refurbishment (no election to tax)

Rental value		£120,000	p.a.
YP in perp. @ 8%		12.5	
Less		£1,500,000	
Refurbishment cost	£300,000		
Add			
VAT @ 20.0%	£60,000	£360,000	
Capital value		£1,140,000	

Increase in capital
 value £1,140,000 – £1,110,000 = £30,000
Return on costs = 8.33%

(c) Value with refurbishment (X elects to tax)

Rental value	£120,000	p.a.
YP in perp. @ 8%	12.5	
	£1,500,000	
Less		
Refurbishment cost	300,000	
Capital value	£1,200,000	

Increase in capital
 value £1,200,000 – £1,110,000 = £90,000
Return on costs = 30.00%

Note

- As a result of the election, the return on the additional investment is 30%, which would justify carrying out the work. If no election is made, the return of 10.64% would render the additional investment of marginal benefit.
- VAT would be payable on the rent by the tenant, but this would be "passed on" to HMRC so is ignored.
- If the offices are in an area where potential tenants making exempt supplies form a dominant part of the market, the yield might need to be increased to, say, 8.25% if X elects. This would still justify an election.

The effect of opting to tax a property is that all supplies become standard-rated so that, if the property is let, VAT at 20.0% is added to the rent, or, if sold, to the sale price.

If the tenant or purchaser makes taxable supplies, then input VAT on the rent or price is of little or no consequence since they would recover the input VAT. However, if they make exempt supplies, then the imposition of input VAT would be a deterrent to them in taking a lease or purchasing the property. This might have an adverse effect when marketing the property, particularly if exempt suppliers are the most likely tenants or purchasers. This is a factor that an owner would need to take into account when deciding whether to elect.

When an election is made, if the property is a building then the election will apply to the whole building (but does not apply to owners of other interests in the building). In the case of land, the owner can define the area of land to which the option shall apply.

If a building is part residential, the option will not apply to the residential part. Hence, if a shop with a flat over is let in a single lease at £20,000 p.a. and the landlord opts to tax, the rent would have to be apportioned between the shop and the flat (assume £18,000 p.a. and £2,000 p.a., respectively), when VAT will be payable on £18,000 p.a. only.

An election is irrevocable, apart from an initial three-month period or after 20 years. The election ceases to be of effect if a building to which it applies is demolished. It also ceases when ownership ceases following a sale. It does not bind a purchaser. However, since the purchaser will have paid VAT on the purchase price there will be a strong incentive on the purchaser to elect in order to recover the input VAT. This will not be the case if the property is vacant and the purchaser proposes to occupy it for business purposes, making taxable supplies: in these circumstances the purchaser can recover the input VAT on the purchase price as normal input VAT of the business.

The option to tax may be ineffective if the building is one designed to be occupied by a business making wholly or mainly exempt supplies. Hence, a bank building offices for its own occupation may not opt to tax the building. However, if a developer built the offices and let them to the bank, with the bank making no financial contribution to the development, then an option would be effective.

5. Transfer of a going concern

Under the VAT legislation, the transfer of a going concern (TOGC) is not a supply for VAT purposes. Hence no VAT charge can arise when a transaction is a TOGC.

HMRC takes the view that the letting of property can be a going concern. If so, the change of ownership of a property investment may be treated as a TOGC. Whether this is the case is of no importance if the owner of such a property has not opted to tax, since the sale of the landlord's interest will be either an exempt supply or a TOGC. In either case, no VAT charge will arise. An exception to this would be if the landlord's interest is the freehold and the building is less than three years old, as the sale of the freehold interest would be of a new building and so standard-rated if not a TOGC.

Whether a sale of the landlord's interest is a TOGC or not is of crucial importance where the landlord has opted to tax (or if it is the sale of the freehold interest in a new building), since VAT is chargeable if it is not a TOGC but is not chargeable if it is a TOGC. The position affects both vendor and purchaser.

The view of HMRC is that, if the purchaser elects to tax the building prior to completion of the purchase, the sale will be a TOGC but not otherwise. The reasoning behind this is that all circumstances, including the VAT position, will be the same both before and after the change of ownership and VAT will thus be payable on the rents from the lettings without a break.

6. Value and VAT

In nearly all cases, VAT is charged on the consideration passing. In some cases the consideration may not consist wholly or partly of money, in which case the value may need to be determined. This can also be the case where there is a transaction between connected persons at less than open market value.

Valuers may be called upon to provide the value required. There is no clear guide in the legislation as to valuation assumptions that the valuer may or may not make, save that it must be assumed that the parties are, effectively, at arm's length.

7. Valuation and VAT

As a general principle, the inclusion of VAT in the elements of a valuation will arise where VAT is a cost to the interest being valued because it is not recoverable.

For example, in preparing a valuation where an allowance for expenditure has to be made and VAT is payable on that expenditure, the valuation will allow for the gross of VAT cost if the input VAT cannot be recovered. So, if the valuation is of a residential property, any allowance for expenditure on matters such as repairs or improvements will be included gross of VAT. An exception would be the valuation of a listed building where the expenditure is of a nature which is within the zero-rated provisions of Schedule 8 to the VAT Act 1994.

In the case of residual valuations, building costs and fees will normally be included net of VAT. This is so because, if the development is residential, it will probably result in a zero-rated supply; if not residential, it is probable that the developer (if not selling the freehold interest of the new building on completion – a standard-rated supply) will opt to tax in order to recover VAT on the development costs and so will make a standard-rated supply of the building.

An exception would be the residual valuation of a building to be developed and used to make exempt supplies. For example, the residual valuation of a hospital or doctor's clinic development, or of a university teaching block, would include costs and fees gross of VAT.

The valuation of an occupier's leasehold interest in a building where the landlord has opted to tax and the occupier makes exempt supplies, and the use under the lease is limited to the making of exempt supplies, is a further example where expenditure, in this case the rents payable, should be included gross of VAT.

As is clear, the inclusion of VAT in a valuation is something which turns on the facts of the case, with the test being: can the input VAT be recovered?

8. Special cases

(a) Incentives

When the market is weak, property owners are prone to offer incentives to tenants to take a lease of their premises. These take various forms, with the most common being long rent-free periods and capital payments to the tenant (reverse premiums).

In the case of rent-free periods, this is not treated as a supply by the tenant, so that no VAT is chargeable. The exception is where the rent-free period is in return for some act by the tenant which benefits the landlord when it is, in effect, a reward or payment to the tenant. In such cases, the value of the rent-free period has to be determined and the tenant charges the landlord VAT on that value. A typical example is where a tenant takes a lease of property which requires substantial work to it which the tenant agrees to carry out in return for the rent-free period.

As to reverse premiums, it was held in the *Neville Russell* case[15] that a payment to the tenant must be in return for services by the tenant which are a standard-rated supply. Hence the tenant should charge the landlord VAT on the amount of the premium. Failure to do so brings into play the general rule that, if VAT should have been charged but no explicit addition for VAT has been made, the payment is deemed to include the VAT element. So, for example, a tenant receiving £100,000 as a reverse premium should charge the landlord £20,000 VAT (at standard rate of 20%). Failure to do so would result in HMCE treating the payment as £85,106.33 plus VAT (at 20%) of £16,667, a total of £100,000. The VAT element would be payable by the tenant to HMCE.

The *Neville Russell* decision is contradicted by the VAT treatment of other incentives and similar situations and it would be no surprise if it were overturned. It would appear to remain in effect, however, at the time of writing.

VAT is a very specialist subject and experts should be consulted when any special situations arise.

(b) Statutory and other compensation

Payments to agricultural or business tenants under the relevant statutes following the termination of their leases are not a supply for VAT purposes. Indeed, such payments are tax-free for the tenants.

Compensation payable on the settlement of a claim under section 18(1) of the Landlord and Tenant Act 1927 is not subject to VAT. However, the claim may include a VAT element where landlords are not able to recover VAT on any expenditure which they have made, or may make, in order to justify the claim. Where tenants can recover VAT on building/repairs expenditure but landlords cannot,

15 *Neville Russell* v *HMCE* (1987) VAT TR 194.

tenants are often advised to carry out the work themselves and recover the VAT and not allow landlords to do the work after the expiration of the lease.

(c) Mixed supplies

Where a mixed supply is made of two or more elements which would be different types of supply if made separately, the type of supply is determined by the dominant purpose if one supply is ancillary to the other. For example, a letting of offices is an exempt supply (unless the landlord has opted to tax), whereas a separate letting of car parking spaces in the building to the same tenant would be a standard-rated supply. However, a letting of offices with car spaces in a single lease would be an exempt supply since the dominant purpose is the letting of the offices to which the car spaces are an ancillary matter.

Similarly, if a building is opted to tax and is compulsorily purchased, VAT would be payable by the acquiring authority on the purchase price of the building and also on any payment of compensation for disturbance, which is an ancillary matter. However, no VAT is payable on compensation for injurious affection since no supply of the land affected is being made.

Where each supply is complementary to the other, then any consideration would have to be apportioned between them. So, the letting of a shop with a flat above in a single lease would probably be treated as two separate supplies.

STAMP DUTY LAND TAX

I. Purchases

Stamp duty is payable on certain deeds. In the case of property, a sale is effected by a deed. So stamp duty is payable on a sale by the purchaser, the amount of duty being dependent on the price. The rate of duty remained fixed for many years but was increased sharply in successive years from 1997. For non-residential property/land and mixed use, the rate fixed for sales is currently nil on prices up to £125,000 (£150,000 in disadvantaged areas), 1% up to £250,000, 3% above £250,000 up to £500,000, 4% above £500,000 up to £1m, 5% up to £2m and thereafter at 7%. Note that the duty is charged on the whole price so that, for example, a purchaser paying £151,000 pays 1% of £151,000 = £1,510, whereas no duty is payable on a price of £150,000.

The budget of 2012 introduced a measure to combat SDLP avoidance whereby a new charge of 15% will apply to transfers of residential property worth over £2m into a company. Avoidance will still be possible thereafter, but it is hoped that the high upfront tax will act as a disincentive to the practice of using SVPs as ownership vehicles. It is also proposed to charge CGT on overseas companies on gains on disposal of UK residential properties and on shares and interests in such companies.

2. Leases

Further stamp duty is payable by lessees when taking a lease. The amount of duty depends on the length of the lease and the amount of rent. The calculation of the stamp duty payable is extremely complicated and HMRC provides online facilities for filing the relevant returns and calculating the tax. This is the province of solicitors and no further comment will be provided here.

Reference

UKGN3 (Red Book 2012) *Valuations for capital gains tax, inheritance tax and stamp duty land tax* deals with the statutory basis of "market value" for CGT, IHT and SDLT.

Chapter 22

Principles of the law of compulsory purchase and compensation

1. Legal basis of compulsory purchase

There are numerous Acts of Parliament under which government departments, local or public authorities or statutory undertakings may carry out schemes for the general benefit of the community involving the acquisition of land or interference with owners' proprietary rights.

An owner whose property is taken under statutory powers is entitled to compensation as of right, unless the Act that authorises the acquisition expressly provides otherwise. Where no interest in land is taken but a property is depreciated in value by the exercise of statutory powers, the owner's right to compensation depends on the terms of the Act under which these powers are exercised.

This chapter sets out a brief outline of the principles involved in compulsory purchase valuations, which are considered in more detail in the subsequent chapters.

Compulsory purchase of land normally brings into play four main sets of statutory provisions, as follows. First, there is the authorising Act, normally a public general Act authorising a public body or class of public bodies (e.g. county councils) and former public bodies, which are now private following privatisation, to carry out some specified function, and going on to state:

- whether such a body may acquire land for the purpose;
- whether the body may buy it compulsorily;
- whether the body may obtain power to do this by compulsory purchase order (CPO) specifying the land required; and
- if so, what procedure is to be followed when making the CPO.

There is now a standardised procedure laid down by the Acquisition of Land Act 1981, though alternative procedures are occasionally specified. Second, therefore, is the Act of 1981 (or such alternative code as the authorising Act may prescribe), which governs the making of the CPO: it may be said that the great majority of acquisitions are made under that Act. Third is the Compulsory Purchase Act 1965, which has to all intents and purposes replaced the Lands Clauses

Consolidation Act of 1845 and governs the actual procedure for acquisition after the CPO has sanctioned it. The Act is supplemented by provisions in the Compulsory Purchase (Vesting Declarations) Act 1981 and in the Land Compensation Act 1973. Fourth is the Land Compensation Act 1961, which contains the current rules for assessing compensation in so far as it relates to land. These rules, too, are supplemented by provisions in the Land Compensation Act 1973 and Part 8 of the Planning and Compulsory Purchase Act 2004 (see Chapter 23 below).

Disputes over compulsory purchase fall broadly into two main classes, depending on whether or not they relate to the assessment of compensation. If they do relate to compensation, they must be brought before the Upper Tribunal (Lands Chamber). Prior to its name change in 2009 this was known as the Lands Tribunal, and in common parlance it continues to be known by its previous, more familiar, name. The Lands Tribunal, as we shall continue to call it, is a specialised body staffed by valuers and lawyers (the President of the Tribunal is always a lawyer, and is normally a QC). Appeal lies to the Court of Appeal on a point of law only, by way of case stated, and within six weeks of the Tribunal's decision.

2. Compulsory purchase procedure

Any acquiring authority that is empowered by an appropriate authorising Act to select and acquire compulsorily the particular land it needs by a CPO must normally make the CPO in accordance with the procedure laid down in the Acquisition of Land Act 1981. This involves making the order in draft and submitting it to a "confirming authority", which will be the appropriate Minister or Secretary of State – unless of course the Minister him- or herself is acquiring the land. There must be prior press publicity and notification to the owners and occupiers of the land, and the hearing of objections by an inspector from the Ministry or Department concerned. The order may be confirmed, with or without modifications, or rejected. If confirmed, it takes effect when the acquiring authority publishes a notice in similar manner to the notice of the draft order and serves it on the owners and occupiers concerned. The order is subject to the standard procedure for appeal to the High Court within six weeks on the ground of *ultra vires* or a procedural defect that substantially prejudices the appellant.

Section 100 of the Planning and Compulsory Purchase Act 2004 amends the Acquisition of Land Act 1981 in relation to the making and confirmation of compulsory purchase orders by acquiring authorities other than Ministers. The "confirming authority" (i.e. the relevant government department) can confirm any such CPO if there are no objections, or if they are withdrawn, with or without modifications. If there are objections the "confirming authority" must either hold a public inquiry or grant a hearing to objectors, unless it prefers to proceed on the basis of "written representations" (which it can choose to do if (i) the CPO is not subject to special parliamentary procedure and (ii) the objectors consent). Confirmation in stages is also possible. Notices of confirmation of a CPO must be served on the relevant owners, etc., and also published and displayed, in a prescribed form. Section 101 makes broadly similar provision for the making of CPOs by ministers. Section 102 empowers acquiring authorities that are not ministers to confirm their own CPOs if the relevant government department, as "confirming authority", so allows (e.g. where ownership of the land being acquired is unknown), but such confirmation cannot apply if the CPO is to be modified, or confirmed in part.

The CPO will lapse, in relation to any of the land comprised in it, unless it is acted on within three years. When the authority wishes to act on the order, it must serve a notice to treat on the persons with interests in the land to be acquired, requiring them to submit details of their interests and their claims for compensation. When the compensation is agreed in each case, it and the notice to treat together amount to an enforceable contract for the sale of the land. This is then subject to completion by the execution of a registered or unregistered conveyance in the same way as a private land transaction.[1]

There is, however, a more modern alternative procedure at the authority's option whereby the notice to treat (or more than one) and the conveyance are combined in a "general vesting declaration". The authority must notify the owners and occupiers concerned, in the same notice as that which states that the CPO is in force (or a separate and later notice), that they intend to proceed in this manner by making a vesting declaration not less than two months ahead. This, when made, will by unilateral action vest the title to the land in the authority at a date not less than 28 days after notification to the owners concerned. It will, by and large, have the same consequences as if one or more notices to treat had been served, but fixes the valuation date as well as giving the acquiring authority legal title.

Freeholds and leaseholds are capable of compulsory acquisition. Leasehold tenancies with a year or less to run, including periodic tenancies, are not subject to acquisition and compensation and are normally allowed to expire, after due service of notice to quit, though an authority needing possession quickly can take it compulsorily under the CPO, subject to payment of compensation under s.20, Compulsory Purchase Act 1965.

An authority cannot normally, without clear statutory authorisation, take rights over land in the limited form of an easement or other right less than full possession (even a stratum above or beneath the surface). For example, in *Sovmots Ltd* v *Secretary of State for the Environment*[2] the House of Lords quashed a compulsory purchase order for acquisition of a lease and a sub-lease (not the freehold) of certain property, together with new easements of access and support which would have been required because only the upper part of a building was to be taken, on the ground that appropriate statutory authority was lacking. For local authorities the necessary statutory authority is provided by the Local Government (Miscellaneous Provisions) Act 1976, section 13 and Schedule 1. However, an authority acquiring

1 Compulsory Purchase Act 1965, sections 4, 5 and 23.
2 [1979] AC 144; [1977] 2 All ER 385.

a dominant tenement will also acquire the easements appertaining to it, as in private conveyancing, and an authority acquiring a servient tenement will either allow the easements and other servitudes over it to subsist without interference or else pay compensation for "injurious affection" to the dominant land if it is necessary to interfere.

If part only of an owner's land is to be acquired, the owner of "any house, building or manufactory" or of "a park or garden belonging to a house" can require the authority to take all or none; but the authority can counter this by saying that to take part only will not cause any "material detriment" or that the part can be taken without seriously affecting the amenity or convenience of the house (s.8, Compulsory Purchase Act 1965). Any such dispute has to be settled by the Lands Tribunal. Similar rules apply to farms.[3]

Unjustifiable delay by the authority after service of a notice to treat may, in an extreme case, amount to abandonment of the acquisition.[4] The Planning and Compensation Act 1991, section 67, amplifying the Compulsory Purchase Act 1965, section 5, provides that a notice to treat expires at the end of three years unless it has been acted on (e.g. by entry, or settlement of compensation, or reference to the Lands Tribunal, or substitution of a general vesting declaration). A claimant who fails to act may lose the right to claim after six years (*Hillingdon London Borough Council v ARC Ltd* [1998] 3 EGLR 18), although the acquiring authority may be estopped from relying on the expiration of the time-limit (*Hillingdon London Borough Council v ARC Ltd (No. 2)* [1999] 3 EGLR 125), which can be waived (*Chester-le-Street DC v Co-operative Wholesale Society Ltd* [1998] 38 EG 153). Entry before payment of compensation entitles an owner to receive interest on the compensation to be paid, and to an advance payment of compensation.

Many acquisitions by authorities are made by agreement under the shadow of available compulsory powers and consequently involve the same rules of compensation. An owner must not increase the authority's liability to compensation by creating new tenancies and other rights in the land, or by carrying out works on it after service of the notice to treat, at any rate if such action "was not reasonably necessary and was undertaken with a view to obtaining compensation or increased compensation".[5]

3. Compulsory purchase compensation

The detailed rules of compensation are discussed in Chapters 23–25. It will, however, be convenient to consider them briefly in outline here in order to demonstrate the legal basis on which they rest.

3 Compulsory Purchase Act 1965, section 8; Land Compensation Act 1973, sections 53–58.
4 *Grice v Dudley Corporation* [1957] 2 All ER 673.
5 Acquisition of Land Act 1981, section 4.

The acquiring authority must compensate the expropriated owner for the land taken, by way of purchase price, and for any depreciation of land retained by the owner, as well as for "all damage directly consequent on the taking".[6]

The basis of compensation for the taking or depreciation of land is "market value", namely "the amount which the land if sold in the open market by a willing seller might be expected to realise". "Special suitability or adaptability" of the land which depends solely on "a purpose to which it could be applied only in pursuance of statutory powers, or for which there is no market apart from the special needs of the requirements of any authority possessing compulsory purchase powers", must be disregarded. There must be no addition to nor deduction from market value purely on the ground that the purchase is compulsory, nor any addition or reduction solely and specifically on account of the project to be carried out (on the claimant's land or any other land) by the acquiring authority. An increase in the value of adjoining land of the owner not taken by the authority, if it is attributable solely to the acquiring authority's project, must be "set off" against compensation.[7]

If the property has been developed and used for a purpose that has no effective market value, such as a church or a museum, then the Lands Tribunal may order that compensation "be assessed on the basis of the reasonable cost of equivalent reinstatement", if "satisfied that reinstatement in some other place is bona fide intended".[8]

These intricate legal rules are intended for the guidance of valuers rather than lawyers. Valuers engaged in the assessment of the compensation are required, subject to such guidance, to reach a figure that will put the expropriated owner in a position as near as reasonably possible to the position if there had been no compulsory acquisition and the land had been sold in an ordinary private sale.

In addition to purchase price compensation, there is compensation for "severance and injurious affection", i.e. depreciation in the value of land retained. This is "severance" if the value of the land retained plus the value of the land taken is less than the value of the combined holding prior to the two parts being severed. Depreciation caused by what is done on the land taken by the acquiring authority is injurious affection and is closely analogous to damages in tort for private nuisance, though it may well include loss not compensable in tort.[9] If the harm done goes beyond what is authorised by the statutory powers of the acquiring authority, then it will in any case be unlawful and so compensable (if at all) in tort and not as "injurious affection".

It is also possible to obtain compensation for "injurious affection" when no land has been acquired from the claimant under the Compulsory Purchase Act 1965, section 10 and/or Part I of the Land Compensation Act 1973.

6 *Harvey* v *Crawley Development Corporation* [1957] 1 All ER 504, *per* Denning LJ ("Crawley costs").
7 Land Compensation Act 1961, sections 5–9 and Schedule 1, as amended by the Planning and Compensation Act 1991.
8 Land Compensation Act 1961, section 5 (rule 5).
9 *Buccleuch (Duke of)* v *Metropolitan Board of Works* (1872) LR 5 HL 418. On this, see Chapter 24.

Another head of compensation is "disturbance", which is not strictly land value but "must ... refer to the fact of having to vacate the premises".[10] Thus it may include the loss of business profits and goodwill, removal expenses and unavoidable loss incurred in acquiring new premises. Disturbance compensation is (somewhat inconsistently as it is a separate head of claim) regarded in law as an integral part of land value. It is therefore not payable where the acquiring body, having expropriated and compensated the landlord, does not expropriate a short-term tenant but instead displaces that tenant by notice to quit. In such cases the Land Compensation Act 1973 (sections 37–38) provides for "disturbance payments" by the acquiring body to the tenant (except an agricultural tenant, for whom separate compensation provisions exist). The 1973 Act also authorises the payment of "home loss payments" to displaced occupants of dwellings who have lived there for one year (sections 29–33). The 2004 Act has now added "basic loss" and "occupier's loss" payments. For further details see Chapter 25 below.

A claimant "must once and for all make one claim for all damages which can be reasonably foreseen".[11] The date of the notice to treat fixes the interests that may be acquired, but does not govern the valuation date which, as the House of Lords held in the case of *Birmingham Corporation* v *West Midland Baptist (Trust) Association (Inc)*,[12] must be assessed as at the time of making the assessment of compensation, or of taking possession of the land by virtue of a notice to treat (if earlier), or of the beginning of "equivalent reinstatement". The 2004 Act, section 103, has now endorsed this rule.

Where a General Vesting Declaration is used as an alternative to Notice to Treat, the valuation date is the vesting date or the date compensation is assessed, whichever is the earlier.

---

10 *Lee* v *Minister of Transport* [1965] 2 All ER 986, per Davies LJ.
11 *Chamberlain* v *West End of London etc. Rail Co.* (1863) 2 B&S 617, per Erle CJ.
12 [1970] AC 874; [1969] 3 All ER 172; (1969) RVR 484.

Chapter 23

Compulsory purchase compensation I

Compensation for land taken

I. Introduction

The basis of compensation for land compulsorily acquired was, up to 1919, regarded as a matter of valuation alone and not law. The statutes, private and public alike, prescribed the procedural steps to be taken but left the question of payment to be dealt with by valuers.

This separation from law was illusory because no valuer can compel agreement. In the absence of agreement on compensation, arbitration is essential, whether by privately appointed arbitrators or official arbitrators prescribed under the Acquisition of Land (Assessment of Compensation) Act 1919 (and earlier statutes) or, later, the Lands Tribunal (Lands Tribunal Act 1949) and, since 2009, the Upper Tribunal (Lands Chamber). Arbitration is a process regulated by law, because an allegation that any arbitrator's decision is faulty is a matter that can be made the subject of litigation. This has led to judicial decisions giving rise to case law and then to statutes governing principles of assessment applicable to compensation.

The modern compensation code is, therefore, based on a complex interaction of statute and case law, the latter being constantly reviewed. It should be noted that in *Melwood Units Property Ltd* v *Commissioner of Main Roads*[1] (a case taken on appeal from Australia) the Privy Council stated emphatically that the wrongful application of valuation principles is not merely a valuation issue but an issue of law.

The first statute to prescribe rules of assessment was the Acquisition of Land (Assessment of Compensation) Act 1919. Section 2 of this Act prescribed six rules governing the compensation payable for interests in land acquired compulsorily by any government department, local or public authority or statutory undertaking. The general basis prescribed was the price the land might be expected to realise if sold in the open market by a willing seller, known in practice as "open market value". The Act also replaced the various earlier procedures for arbitration. From

1 [1979] AC 426.

1950, appeal from the Lands Tribunal has been by way of "case stated" (on points of law only) direct to the Court of Appeal.

The substantive principles of assessment of compensation, as distinct from procedural matters, are to be found in the Land Compensation Act 1961, as amended. The Act applies to all cases where land is authorised to be acquired compulsorily, including acquisitions made by agreement but under compulsory powers (which in compensation terms count as compulsory acquisitions).

2. General principles of compensation

The Land Clauses Consolidation Act 1845 was intended to deal with procedural rather than compensation matters and seemed to assume that valuers would be able to resolve any issues as to quantum of compensation through negotiation. This misplaced confidence resulted in disputes which had to be settled by the courts. The resulting body of case law created the rules of compensation that apply to all cases of compulsory purchase and acquisition by agreement under compulsory powers, unless expressly or impliedly excluded by Statute. This compensation code may be briefly summarised as follows:

1. Service of the acquiring body's notice to treat fixes the property to be taken and the nature and extent of the owner's interest in it. This is because the notice is a semi-contractual document and a step towards the eventual conveyance to the acquiring body of a particular owner's leasehold or freehold interest. Indeed, the courts have said that notice to treat plus compensation (when settled) does in fact constitute a specifically enforceable legal contract.[2] Any interest in respect of which notice to treat is *not* served therefore remains unacquired. The effect of section 20 of the Compulsory Purchase Act 1965 is that holders of yearly tenancies and lesser interests are not entitled to receive notices to treat. The assumption is that the acquiring authority acquires the reversion to such an interest and the interest is then terminated at common law (e.g. by notice to quit), though compensation will have to be paid if it is terminated sooner.

    Where general vesting declarations are used in place of notices to treat and conveyances of the normal kind, a notice to treat is deemed to have been served, so that what has been said applies in those cases equally.

2. The old leading case *Penny* v *Penny* (1868)[3] was assumed to have established a rule that the value of the property acquired must be assessed as at the date of notice to treat. In *Birmingham Corporation* v *West Midland Baptist (Trust) Association*[4] the House of Lords held that this supposed "rule" was without

2 As stated by Upjohn J, in *Grice* v *Dudley Corporation* [1957] 2 All ER 673.
3 (1868) LRS Eq 227.
4 [1969] 3 All ER 172; [1969] RVR 484.

foundation, and expressed the view that the appropriate date for assessing the value of the land is either:

(a) the date when compensation is assessed, or
(b) the date when possession is taken (if this is the earlier), or
(c) the date when "equivalent reinstatement" can reasonably be started, if that mode of compensation is to be applied (as it was in the *West Midland Baptist Trust* case itself).

This new rule has now been confirmed by section 103 of the Planning and Compulsory Purchase Act 2004, which introduced a statutory date of valuation into s5A Land Compensation Act 1961. Where a General Vesting Declaration is used, however, the valuation date is the earlier of the vesting date and the date compensation is assessed – s5A(4). This has the advantage of creating certainty to the acquiring authority, which will know levels of value at the time of vesting. It can, however, be problematic for claimants if their actual relocation date is some years later, when the cost of a replacement property may have risen significantly. Where the compensation is assessed by the Lands Tribunal the valuation date is the last day of the hearing if possession has not been taken by then.

3. Compensation must be claimed "once and for all … for all damages which can be reasonably foreseen".[5] But this can be varied if the parties agree; and under the Land Compensation Act 1973, sections 52 and 52A (inserted by the Planning and Compensation Act 1991), claimants are entitled to an advance payment of 90% of the compensation, as estimated (in default of agreement) by the acquiring authority, plus payment of statutory interest, at rates set by the Treasury, on amounts of compensation outstanding after entry on to the land. Further advance payments can be made if the acquiring authority is satisfied that their earlier estimate was too low.

Special provisions set out in section 104 of the Planning and Compulsory Purchase Act 2004 apply to advance payments of compensation in respect of land that is mortgaged.

4. Compensation must be based on the value of the land in the hands of the owner, not its value to the acquiring body.[6] It is the former value which is "market value" because it is the vendor, not the purchaser, who is at liberty (in a sale by agreement) to market the land or not. Any increase or decrease in value solely attributable to the scheme underlying the acquisition is not a market factor and must be disregarded.[7]

---

5 *Chamberlain* v *West End of London and Crystal Palace Rly Co* [1863] 2 B&S 617, per Erle CJ.
6 *Cedar Rapids Manufacturing and Power Co* v *Lacoste* [1914] AC 569; *Re Lucas and Chesterfield Gas and Water Board* [1909] 1 KB 16.
7 The *Pointe Gourde* principle or "no-scheme world" rule discussed below.

Establishing what comprises the scheme is often difficult. In *Waters* v *Welsh Development Agency*[8] the House of Lords held, upholding the decisions of the Court of Appeal and the Lands Tribunal, that when land (including water) was compulsorily purchased under the Cardiff Bay Barrage Act 1993, the inclusion in that project of adjacent land required for a new nature reserve, to compensate for interference with an existing wild life habitat, was to be treated as part of the scheme ("compensatory wetlands provision") even though it was 10 miles distant along the coast. "When assessing compensation payable for the claimant's land, the authority's need to acquire land as a palliative measure, necessary as a result of the environmental consequences of the Cardiff Bay Barrage, is to be disregarded" (per Lord Nicholls of Birkenhead).

5. Covenants, easements etc. already in existence and affecting the land, whether by way of benefit or of burden, must be considered in assessing compensation.[9] For instance, the property may enjoy the benefit of a covenant restricting building or other works on adjoining land, or it may itself be subject to the burden of such a covenant and be less valuable in consequence.

Similarly, it may be a "dominant tenement" with the benefit of an easement of way, support, light, etc. over adjoining property; or, on the other hand, it may be a "servient tenement" subject to such a right which benefits adjoining property. Also, the possibility of the removal or modification of restrictive covenants under section 84 of the Law of Property Act 1925, as amended, is a factor which might properly be taken into account.

6. Where lessees have a contractual or statutory right to the renewal of their lease, that right will form part of the value of their leasehold interest; but the mere possibility of a lease being renewed is not a legal right existing at the date of notice to treat and cannot be the subject of compensation.[10]

7. Owners are entitled to compensation not only for the value of the land taken but also for all other loss they may suffer in consequence of its acquisition.[11] For example, the occupier of a private house compulsorily acquired will be put to the expense of moving to other premises and will suffer loss in connection with the fixtures. An occupier of trade premises will suffer similar losses and in addition may be able to claim for loss on sale of the stock or for injury to the goodwill of the business. It will be convenient to discuss the "disturbance" compensation payable under these heads in Chapter 25. But it is important at this point to recognise them as part of

8 [2004] 2 All ER 915.
9 *Corrie* v *MacDermott* [1914] AC 1056.
10 See *Re Rowton Houses' Leases, Square Grip Reinforcement Co (London) Ltd* v *Rowton Houses* [1966] 3 All ER 996.
11 *Horn* v *Sunderland Corporation* [1941] 2 KB 26; *Venables* v *Department of Agriculture for Scotland* [1932] SC 573.

the compensation to be paid for the compulsory taking of the owner's interest in the land.

8. Where part only of an owner's land is taken, section 7 of the Compulsory Purchase Act 1965 makes it quite clear that the owner is entitled not only to the value of the land taken, but also to compensation for severance or injurious affection of other land, previously held with the land taken, which the owner retained after the acquisition.

9. There are acts, for example the Local Government (Miscellaneous Provisions) Act 1976, which expressly give specified bodies the power to acquire "new rights" (e.g. easements, leases, etc. not currently in existence), and the appropriate payment for those (in principle) will presumably be the prevailing market figure for the grant of such a right.

The underlying principle throughout is that a normal compulsory purchase is, in truth, a compulsory assignment of an existing freehold or leasehold right in land; and consequently the fundamental question which valuers should ask themselves is: what capital sum would a purchaser of this interest expect to pay to acquire it in the open market? In this chapter we concentrate upon the question of compensation for land actually taken, with that principle in mind.

3. Market value

(a) Six basic rules of valuation

Section 5 of the Land Compensation Act 1961, amended as to Rule 3 by section 70 of the Planning and Compensation Act 1991, sets the framework of the compensation code by prescribing six rules for assessing compensation in respect of land. Five of these rules relate to the valuation of land and interests in land; the sixth preserves the owner's right to compensation for disturbance and any other matter not relating to land value, which is suffered in consequence of the land being taken from the owner.

Rule 1. No allowance shall be made on account of the acquisition being compulsory

In assessing compensation under the Land Clauses Consolidation Act 1845 it had become customary to add 10% to the estimated value of the land on account of the acquisition being compulsory. This addition was no doubt to reflect the fact that the "willing seller" assumption is a falsehood, and therefore to compensate the seller for being forced to give up his interest in land for public works. There was nothing in the 1845 Act which expressly authorised such an allowance and, since this rule was introduced in the 1919 Acquisition of Land Act, it is now expressly excluded (though the principle is now acknowledged by the range of loss payments referred to at the end of this chapter).

Rule 2. The value of land shall, subject as hereinafter
provided, be taken to be the amount which the land
if sold in the open market by a willing seller might
be expected to realise

The purpose of this rule is to indicate the true basis of compensation as being open market value, which is what the courts meant in earlier cases by stressing "value to the owner". It takes for granted that there will be a "willing buyer", i.e. that there exists a demand for the land – presumably because of the compulsory nature of the acquisition. Yet whether in the open market there would be a willing buyer at a price acceptable to a willing seller is a question which may well cause problems in certain cases (some of which may be solved by applying Rule 5 below).

Where interests in land are purely of an investment nature, there is probably no difference in any valuer's mind between value to owner and value in the open market. In other cases the rule should put it beyond doubt that there must be excluded from the valuation any element which would have no effect on the price obtainable for the property under normal conditions of sale and purchase, i.e. anything that distorts a proper calculation of "market value".

The meaning of the words "amount which the land if sold in the open market by a willing seller might be expected to realise" was fully examined by the Court of Appeal in the case of *Inland Revenue Commissioners* v *Clay & Buchanan*,[12] as follows:

1. "In the open market" implies that the land is offered under conditions enabling every person desirous of purchasing to come in and make an offer, proper steps being taken to advertise the property and to let all likely purchasers know that it is in the market for sale.
2. "A willing seller" does not mean a person who will sell without reserve for any price he can obtain. It means a person who is selling as a free agent, as distinct from one who is forced to sell under compulsory powers.
3. "Might be expected to realise" refers to the expectations of properly qualified persons who are informed of all the particulars ascertainable about the property and its capabilities, the demand for it and likely buyers.

In assessing compensation, then, it must be assumed that the owners are offering the property for sale of their own free will, but taking all reasonable measures to ensure a sale under the most favourable conditions. The compensation payable will be *that price which a properly qualified person, acquainted with all the essential facts relevant to the property and to the existing state of the market, would expect it to realise under such circumstances.*

---

12 [1914] 3 KB 466. (In that case judgment was directed to identical words used in the Finance (1909–1910) Act 1910, with regard to taxation.) The decision of the Privy Council in the "Indian case" *(Raja Vyricherla Narayana Gajapatiraju* v *Revenue Divisional Officer, Vizagapatam).*

An estimate of market value on this basis will take into account all the potentialities of the land, including not only its present use but also any more profitable use to which, subject to the requisite planning permission, it might be put in the future, i.e. prospective development value as well as existing use value, subject to planning control.

For example, if, in the absence of a scheme, land at present used as agricultural land is reasonably likely (having regard to demand) to become available for building in the future, the prospective building value may properly be taken into account under Rule 2, provided that it is deferred for an appropriate number of years. Again, if buildings on a well-situated site have become obsolete or old-fashioned, so that the rental value of the property could be greatly increased by capital expenditure on improvements and alterations (again, having regard to demand), compensation may properly be based on the estimated improved rental value, provided that the cost of the necessary works is deducted from the valuation.

It should be emphasised, however, that in the case of land capable of development, the price which it might be expected to realise under present-day conditions of strict planning control depends very largely on the kind of development for which planning permission has already been obtained, or is reasonably likely to be given having due regard to the development plan for the area. But it is essential to bear in mind the words used by Lord Denning MR, in *Viscount Camrose* v *Basingstoke Corporation*:[13]

> Even though (land may be) having planning permission, it does not follow that there would be a demand for it. It is not planning permission by itself which increases value. It is planning permission coupled with demand.

In that case land had to be assumed to have planning permission for the acquiring authority's own use – housing – but in the no-scheme world (no new town of Basingstoke) this was planning permission which had no value because any development would have no infrastructure around it. Thus, existing use value rests on demand, whereas development value rests on demand plus planning permission; in principle both go to make up "market value" of land.

In practice, prospective purchasers in the open market would probably obtain permission for their proposed development before deciding the price they are prepared to pay. This factor is normally absent in compulsory purchase cases, and the Land Compensation Act 1961, as amended by the Localism Act 2011, therefore provides that, besides taking into account any existing planning consents, certain assumptions shall be made as to the kinds of development for which planning permission might reasonably have been expected to be granted but for the compulsory acquisition. These assumptions are considered in detail later in this chapter.

---

13 [1966] 3 All ER 161.

In the case of leasehold interests, their value should be determined by reference to the length of the unexpired lease at the date the valuation is made (plus any additional security of tenure).

Rule 3 (as amended). The special suitability or adaptability of the land for any purpose shall not be taken into account if that purpose is a purpose to which it could be applied only in pursuance of statutory powers, or for which there is no market apart from the requirements of any authority possessing compulsory purchase powers

Until it was amended by the 1991 Planning and Compensation Act, Rule 3 also excluded any value to a "special purchaser". In its amended form Rule 3 is rarely used, except to exclude any value attributable to the acquiring authority's proposed use, where that use requires statutory powers. Therefore, where land is being acquired to build a motorway, then to the extent that the proposed motorway use enhances the value of the land (which in itself is doubtful) that enhancement is to be disregarded.

Rule 3 excludes the consideration of "special suitability or adaptability" of the land, provided that this relates to its use for a "purpose" and not merely to the position of any party to the transfer.[14] In relation to any "purpose", the fact that its application depends solely on compulsory purchase or the existence of statutory powers suffices to bring the rule into operation.

It is clear that "purpose" in Rule 3 means some prospective physical use of the land itself and does not extend to a purpose connected with the use of the products of the land, e.g. minerals, nor to a factor such as the special attraction of a landlord's reversion to a "sitting tenant" (i.e., to the latter the purchase of the reversion means an enlargement of the tenant's interest; to anyone else it means merely the acquisition of an investment).[15] If, therefore, "special suitability" of the land for "any purpose" is not established, Rule 3 does not come into play in any case. It follows that "ransom value" is payable as being genuine market value; this is endorsed by the House of Lords decision in *Hertfordshire County Council v Ozanne*[16] and the Court of Appeal decision in *Wards Construction (Medway) Ltd v Barclays Bank plc.*[17]

---

14 In *Inland Revenue Commissioners* v *Clay & Buchanan* [1914] 3 KB 466, it was held that the special price already offered and paid for a house by the adjoining owner, who wanted the premises for the extension of his nursing home, was properly taken into account in assessing market value for the purposes of the Finance (1909–10) Act 1910.
15 See *Pointe Gourde Quarrying and Transport Co Ltd* v *Sub-Intendent of Crown Lands* [1947] AC 565; and *Lambe* v *Secretary of State for War* [1955] 2 QB 612.
16 [1991] 1 All ER 769.
17 [1994] 2 EGLR 32.

Rule 4. Where the value of the land is increased by reason of the use thereof or of any premises thereon in a manner which could be restrained by any Court, or is contrary to law, or is detrimental to the health of the inmates of the premises or to the public health, the amount of that increase shall not be taken into account

The general purpose of this rule seems clear, and it is often known as the "illegal and immoral user" clause. In *Hughes* v *Doncaster Metropolitan Borough Council*[18] a scrapyard had a certificate of established use, which for all practical purposes was as good as a written planning permission. The Council, however, argued that technically the use was not legal, merely "illegal but unenforceable", so should be left out of account under Rule 4. The House of Lords held that the rule excludes value attributable to a use carried on in breach of planning control, but does not exclude such a use after it is no longer susceptible to enforcement proceedings because of lapse of time. This attempt to exclude a valuable use on a technicality therefore failed. In any event, the following year Certificates of Lawful Use replaced Certificates of Established Use, and it would be impossible to argue that a use with a new certificate remained unlawful. In *Taff* v *Highways Agency* [2009] UKUT 128 (LC), ACQ/23/2007, PLSCS 279, the lack of a waste management licence prevented a claim for disturbance due to the application of Rule 4.

Rule 5. Where land is, and but for the compulsory acquisition would continue to be, devoted to a purpose of such a nature that there is no general demand or market for land for that purpose, the compensation may, if the Lands Tribunal is satisfied that reinstatement in some other place is bona fide intended, be assessed on the basis of the reasonable cost of equivalent reinstatement

There are certain types of property which do not normally come onto the market and whose value cannot readily be assessed by ordinary methods of valuation. Such properties include churches, schools, hospitals, public buildings and certain classes of business premises where the business can only be carried on under special conditions.[19]

Rule 5 gives statutory authority to assessing compensation on the basis of the cost of providing the owner, so far as is reasonably possible, with an equally suitable site and equally suitable buildings elsewhere – as an alternative to assessment on the basis of market value – provided that:

18 [1991] 1 EGLR 31.
19 See *Festiniog Rail Co* v *Central Electricity Generating Board* (1962) 13 P&CR 248.

- the land is devoted to a purpose;
- there is no general demand or market for land for that purpose;
- compensation is limited to the "reasonable cost" of equivalent reinstatement; and
- the Lands Tribunal is satisfied that reinstatement in some other place is "bona fide" intended.

"Equivalent reinstatement" would seem to imply putting claimants in the same position (or in an equally advantageous position) as that which they had when their land was acquired.

In certain cases the only practical method of reinstatement may be the provision of a new site and new buildings. But where, for instance, claimants are using an old building which has been adapted to their purposes, the term "equivalent reinstatement" might cover the cost of acquiring another similar property, if that is possible, together with the expenses of any necessary adaptations.

Bearing in mind that there is no reason in strict principle why one prospective purchaser should not constitute a "market", it will be apparent that the application of Rule 5 depends not on there being "no market" but on there being "no general market". This was endorsed by the House of Lords in *Harrison and Hetherington* v *Cumbria County Council* [1985] 2 EGLR 37, in which the claimants were held to be entitled to equivalent reinstatement compensation for the site of the old cattle market in the centre of Carlisle. The demand for land for cattle markets normally extends only to one such site at a time in any given town, and therefore that demand, though genuine, cannot be described as "general".

What therefore matters is that the Lands Tribunal should be satisfied that the facts of the case require Rule 5 to be applied. It should also be noted that it is the purpose which is to be reinstated, not the building – the question to be asked is how large the new building needs to be to meet that purpose, not: how large is the existing building?

It is important to be aware that a qualifying claimant does not have to have compensation assessed under Rule 5 if a Rule 2 approach is in his best interests. If the existing use of land under Rule 2 points to a market value figure of £100,000 and the reasonable cost of equivalent reinstatement is assessed at £200,000, it is obviously reasonable for the owner to claim the latter figure. If, however, that land has development value assessed at £300,000, it is reasonable to claim this figure instead. It is necessary, therefore, to always undertake two valuations in a Rule 5 case: one based on market value, and one on equivalent reinstatement.

Worked example – Rule 5

Frenchwood Village Hall is to be acquired by the highways authority under compulsory purchase for a road-widening scheme. It is owned by the local parish council and used regularly by local community organisations. It is over 100 years old and in poor condition, and it is considerably larger than is currently required. Having regard to the following information, and making whatever additional assumptions you consider necessary, formulate a compensation claim on behalf of the council.

It is agreed that the property is devoted to a purpose for which there is no general market or demand, and that there is a bona fide intention to reinstate.

The site is approximately 1,200 m^2.

The building has a gross external floor area of 400 m^2.

The property is within the development boundary of the village, where planning policies favour residential development. Individual house plots sell for around £100,000.

Build costs for similar structures are currently around £1,250 per m^2 including fees, services and infrastructure works.

Land which could be used for building a new village hall is available on the edge of a nearby business park for £75,000.

The claimant qualifies for equivalent reinstatement under Rule 5. However, market value should also be considered in case it may be financially beneficial to the claimant.

Market value – based on development value
Assume the site is large enough for two good-sized house plots.

Two house plots at £100,000 each   £200,000 (less demolition costs)

Plus basic and occupiers' loss payments totalling 10% of market value. No disturbance compensation is payable as, in the absence of the scheme, they would have had to incur those costs in order to achieve development value.

Equivalent reinstatement
Assuming a building around half the size would be adequate to reinstate the purpose:

Site purchase cost		£75,000
Rebuilding cost		
200 m^2 @ £1,250 per m^2	£250,000	
less, say, 25% for age and condition		£187,500
add for fees, finance, contingencies, etc., say		£27,500
Plus disturbance – removal costs, etc., say,		£20,000
Total, say		£310,000

On the basis of this calculation the claimant should be advised to opt for compensation under Rule 5, rather than market value.

Rule 6. The provision of Rule 2 shall not affect the assessment of compensation for disturbance or any other matter not directly based on the value of the land

This rule merely preserves the owners' right (established under the Lands Clauses Acts) to be compensated not only for the value of their land but for any other loss they suffer through the land being taken from them, chiefly "disturbance". This is discussed in Chapter 25.

4. Additional rules of assessment

Sections 6 to 9 of the Land Compensation Act 1961 prescribe four additional rules to be applied in assessing market value for compensation purposes.

1. Ignore notional increases or decreases in value due solely to development under the acquiring body's scheme

Section 6 of and Schedule 1 to the 1961 Act give statutory authority to a principle that certain increases and decreases in value due to the development of the scheme underlying the acquisition must be ignored. Section 6 has been much criticised by the courts. While many of the rules of compulsory purchase are open to criticism for being vague and uncertain, section 6 has the problem of being too precise. By setting out in the schedule a detailed list of the types of scheme which should be left out of account, it incurs the risk that schemes of an unusual nature, or of a type not anticipated at the time the schedule was written, may not feature and will therefore be allowed to affect the compensation payable. As a result the courts tend to prefer the *Pointe Gourde* principle as a "no-scheme rule" rather than sections 6 and 9 which were intended to replace it. They will, however, have regard to the statutory rules in preference to the common law principle where they consider this produces a fairer outcome.

Under s.6 it is made clear, by the use of the words "land authorised to be acquired other than the land to be valued" and "other land", that what is in question is a modification of the value assessed for the claimant which is made because of what has happened, or is thought likely to happen, on neighbouring land also being acquired.

In principle, there is, of course, no objection to modifying an assessment of value for reasons of this kind. Section 6 and Schedule 1 therefore exclude such modifications only in the following circumstances:

- the "other land" is being acquired in order to be developed for the same purposes as the claimant's land because it is included in the same compulsory purchase order or in the same area of comprehensive development;
- an increase or decrease in value of the claimant's land is being envisaged on the hypothesis that it is not being acquired for those purposes – contrary to the reality of the situation, since its value is being assessed precisely because it is in fact being so acquired; and
- such an increase or decrease in value cannot be justified independently in the way that it would be justified if the development were "likely to be carried out" in other circumstances (that is to say, if there had been a market demand for such development and a likelihood of its being permitted).

In other words, what is being ruled out is an unreal modification of value, dependent on the blatant contradiction that the land is assumed not to be acquired when the price for its acquisition is being assessed.

In *Davy* v *Leeds Corporation*[20] the claimant's land was acquired (at "site value", because it comprised houses unfit for habitation) as part of a clearance area. Davy claimed that the prospect of clearance and redevelopment of the adjoining land in the area would increase the value of his own land if it were not being acquired: as a cleared site among standing houses there would be no development potential, but if it could be assumed that the surrounding houses would also be cleared there would be potential to combine sites and create a developable plot of land. Section 6 did not rule out this notional increase because it was not justifiable to say that the clearance "would not have been likely to be carried out except for the compulsory acquisition". This contention depended on the possibility that the clearance might have been brought about in circumstances other than compulsory purchase by the local housing authority. The House of Lords rejected this argument, on the ground that the facts made it inconceivable that the clearance would have come in any other way. Section 6 therefore applied, and the argument for an increase in value failed.

The application of these rules to assessing the value of the claimant's land requires the valuer to decide what would have happened in the absence of the scheme for which the compulsory purchase is taking place, assuming that no authority would have been given compulsory powers for an alternative scheme.

So, for example, when valuing an area of land which is part of a new town and is surrounded by houses built as part of the new town scheme, and which is itself to be developed with new houses, the valuer must decide what would have happened if there had been no new town scheme. Although valuers can assume that planning permission would be granted for the residential development (see below), they cannot assume that the infrastructure to enable the development to go ahead exists (as it actually does), unless they can show that it would have been provided even if there had been no new town scheme.

It is obvious that any exercise that seeks to establish what would have happened in the absence of the scheme is fraught with difficulties. This was recognised by Lord Denning MR, in *Myers* v *Milton Keynes Development Corporation*,[21] who stated that:

> It was apparent, therefore, that the valuation in the present case has to be done in an imaginary state of affairs in which there was no scheme. The valuer must cast aside his knowledge of what had in fact happened in the last eight years due to the scheme. He must ignore the developments which would in all probability take place in the future ten years owing to the scheme. Instead, he

20 [1965] 1 All ER 753. "Site value" applied (then) to slum properties.
21 (1974) 230 EG 1275.

must let his imagination take flight to the clouds. He must conjure up a land of make believe, where there had not been, nor would be, a brave new town.

The scheme can affect values in either a positive or a negative way; for example, a bypass may enhance the value of land by creating access and therefore development potential, but reduce the value of houses due to increased noise and visual intrusion. Both increases and decreases must be left out of account if the "no-scheme rule" is to operate as intended. Because of the shortcomings of s.6 it is important to consider in more detail the general principle arising from *Pointe Gourde etc. Transport Co Ltd* v *Sub Intendant of Crown Lands,* commonly referred to as the "Pointe Gourde principle".[22]

The "Pointe Gourde principle"

In the *Pointe Gourde* case, land in Trinidad (containing a stone quarry) had been acquired for a naval base. The owners claimed, in addition to land value and disturbance, a sum apparently representing savings to the Government by reason of having stone from the quarry conveniently on hand to build the base installations instead of needing to transport it there from afar. This item was rejected, being (in the words of the Privy Council) "an increase in value which is entirely due to the scheme underlying the acquisition". Compared to s.6, this principle is seductively simple, and although the original decision was authority only to disallow any increase in value resulting from the nature of the scheme, it has been relied on by the courts to disallow decreases as well as increases in value, whether arising from the nature of the scheme or the threat of compulsory acquisition. The *Pointe Gourde* principle is not, however, without problems.

In *Wilson* v *Liverpool Corporation*[23] it deprived the claimant of some of the development value which, on the facts, clearly accrued to his land. In *Jelson Ltd* v *Blaby District Council,*[24] recently approved of in *Spirerose* v *Transport for London (2009)* 4 All ER 810, it gave the claimant development value which, on the facts, his land did not possess. It concerned the compulsory purchase of a narrow strip of land which, many years before the compulsory purchase, had been reserved for a new road while the surrounding land was developed for housing. At the valuation date the land had no development potential, but *Pointe Gourde* was used to justify valuing the land as if no road scheme had ever existed and the strip had been sold with the adjoining land for residential development. Spirerose went further: the land was to be valued not as if sold in its actual physical condition at the valuation date but as if sold at such time, and in combination with such adjoining land, as it was likely to have been sold in the no-scheme world. In *Trocette Property Co Ltd*

22 [1947] AC 565.
23 [1971] 1 WLR 302.
24 [1977] 2 EGLR 14.

v *Greater London Council*,[25] the Jelson approach gave the claimant's leasehold interest in a cinema substantial "marriage value" in respect of redevelopment prospects which did not in fact exist because no such redevelopment was permitted (whether it could be assumed to be permitted, under sections 14–16 of the Land Compensation Act 1961, might have been considered). In *Birmingham City District Council* v *Morris & Jacombs Ltd*[26] the rule was not applied when valuing a piece of access land which could in fact have been used for additional housing development had this use not been prohibited. Moreover, the two last-mentioned cases all arose on acquisitions under purchase notices, where there can be no relevant "scheme" at all because the authority is acquiring against its will and therefore without any kind of functional purpose in view.

In each of these cases the compensation awarded for land was different from its actual market value at the valuation date because that value had been influenced by the scheme itself. The *Pointe Gourde* principle requires the valuer to estimate what that land value would have been in a no-scheme world and, while this is an inexact science, it is essential if the claimant is not to suffer, or benefit, financially from the scheme itself. The task is made somewhat simpler by s.232 of the Localism Act 2011, which makes it clear that the scheme should be ignored by assuming it has been cancelled at the date the CPO is made. This approach has the advantage of clarity and relative simplicity compared to the alternative of ignoring the scheme by imagining it had never been thought of, though it is less effective at leaving out the scheme, because many of the planning policies and surrounding developments in effect at the time the CPO is made may well have been influenced by the knowledge that the scheme is to take place.

2. Set-off increases in value of adjacent or contiguous land in the same ownership

Where land is acquired compulsorily, it may well be that other adjoining land belonging to the same owner is increased in value by the carrying out of the acquiring body's undertaking on land taken. It is arguable that, in fairness to the acquiring authority, any benefit to the claimant created by the scheme should be taken into account when assessing the amount of compensation to which he is entitled.

The Land Compensation Act 1961 applies the principle of "set-off" to all cases of compulsory acquisition, as follows.

The effect of section 7 of the 1961 Act is that where, at the date of notice to treat, the owner has an interest in land contiguous or adjacent to the land acquired, any increase in the value of that interest in the land retained due to development under

25 (1974) 28 P&CR 408.
26 [1976] 2 EGLR 143.

the acquiring body's scheme is to be deducted from the compensation payable for the interest in the land acquired.[27]

As under section 6, "development" refers to either actual or prospective development under the acquiring body's scheme which would not be likely to be carried out but for the compulsory acquisition. In this case, however, the prospect of development on the land acquired must be considered, as well as development on other land taken under the same compulsory purchase order, or special Act, or included in the same area of development.

While this may seem fair from the viewpoint of acquiring authorities, the set-off provisions are often criticised as being unfair to claimants who have their land acquired. For example, when a new bypass is built, creating development potential in a previously rural area, why should a claimant who has some of his land acquired be required to give up some or all of his windfall on his retained land, while his neighbour who is fortunate enough to have no land within the CPO is allowed to keep all his newly created development value?

3. Subsequent acquisition of other land of the owner

Section 8 of the 1961 Act provides as follows:

- If, either on a compulsory purchase or a sale by agreement, the purchase price is reduced by setting off (under section 7 or any corresponding enactment) the increase in the value of adjacent or contiguous land due to the acquiring body's scheme, then if the same interest in such adjacent land is subsequently acquired that increase in value (which has served to reduce the compensation previously paid) will be taken into account in assessing compensation and not ignored as it otherwise would have been under section 6.
- Similarly, if a diminution in the value of other land of the same owner due to the acquiring body's scheme has been added to the compensation payable for land taken, then, if the same interest in that other land is *subsequently* acquired, that depreciation in value (for which compensation has already been paid) will be taken into account in assessing compensation and not ignored as it otherwise would have been under section 6.

If in either of the above cases part only of the adjoining land is subsequently acquired, a proportionate part of the set-off for injurious affection will be taken into account.

The underlying principle in sections 6–8 can be perhaps best expressed by saying that sections 7 and 8 require genuine market increases and decreases in

---

27 In *Leicester City Council* v *Leicestershire County Council* [1995] 2 EGLR 169 it was held that the compensation could be reduced to nil if the increase in value equalled or exceeded the compensation otherwise payable.

value to be taken properly into account, whereas section 6 requires that increases and decreases which are not genuine market calculations be disregarded.

4. Ignore loss of value due to prospect of acquisition

Section 9 of the Land Compensation Act 1961 provides that, in assessing compensation, no account should be taken of any depreciation in value due to any proposals involving the acquisition of the claimant's interest, whether the proposals are indicated in the development plan, by allocation or other particulars in the plan, or in some other way. For instance, no account should be taken of depreciation in the value of the land due to its inclusion in a compulsory purchase order which has been publicised under the Acquisition of Land Act 1981, or any comparable provisions. The effect of s.9 is that any depreciation caused by the blighting effect of the CPO will be left out of account, though, as with s.6, the courts often prefer to use the *Pointe Gourde* rule to achieve the same objective. Section 9 refers only to depreciation in land value as it is inconceivable that the threat of compulsory purchase could cause an increase in value.

5. Development value

(a) Assumptions as to planning permission

Sections 14 to 17 of the Land Compensation Act 1961 prescribe certain assumptions as to the grant of planning permission which are to be made in assessing the market value of the owner's interest in the land to be acquired. These assumptions were drafted based on the planning system in place at the time of the Act, and have caused increasing issues as they have become less and less relevant to the current planning system. To address this issue, sections 14–17 have been completely replaced by amendments contained in s.232 of the Localism Act. The following section sets out the new legislation, but also explains the problems the reforms are intended to address and, because many current claims will have a valuation date before the reforms took effect in April 2012, a summary of the old legislation.

> **Section 14** – In assessing compensation for land taken by compulsory purchase under rule 2 of s.5 1961 Land Compensation Act, regard shall be had to any planning permission in force at the valuation date, on the relevant land or other land. Regard shall also be had to the prospect of planning permission being granted on such land, in the same way as it would be in the open market, subject to the following assumptions:
>
> (a) that the scheme of development underlying the acquisition had been cancelled on the launch date;
> (b) that no action has been taken (including acquisition of any land, and any development or works) by the acquiring authority wholly or mainly for the purposes of the scheme;

(c) that there is no prospect of the same scheme, or any other project to meet the same or substantially the same need, being carried out in the exercise of a statutory function or by the exercise of compulsory purchase powers; and

(d) if the scheme was for use of the relevant land for or in connection with the construction of a highway ("the scheme highway"), that no highway will be constructed to meet the same or substantially the same need as the scheme highway would have been constructed to meet.

The requirement to take into account existing planning consents is unchanged from the original provisions, and is arguably superfluous as it requires the valuer to do something he would clearly do in any event. What is new is that this applies to other land, not just the relevant land. This removes the option for either the acquiring authority or the claimant to argue that, although planning permission is in place on adjoining land, such planning permission would not have been granted in a no-scheme world. For example, land is being acquired for a bypass. The landowner claims development value, arguing that all the surrounding land has planning permission for business park use, so it is reasonable to assume his land would have obtained a similar consent, had it not been included in the CPO. The acquiring authority argues that the local plan has been drafted in the knowledge that a bypass is to be built, and that in a true no-scheme world, with no prospect of a bypass, the land would have been in the green belt and none of the adjacent land would have planning permission. This is no longer an argument that will be available.

This is reinforced by assumption (a), which resolves the long-standing argument as to whether the scheme should be left out of account by assuming it has just been cancelled (this approach has the advantage of clarity and simplicity) or by assuming no scheme has ever been thought of and planning policies and surrounding land use have developed over many years in a no-scheme world (technically this is the only true way to leave the scheme out of the account, but how land uses and planning policies would have developed is highly speculative).

Assumption (b) makes it clear that, unlike other developments that have taken place or have planning permission, any land acquisitions and developments on the subject or other land which are directly associated with the scheme must be assumed not to have taken place.

Assumptions (c) and (d) address the possible scenario in which an assumption that the scheme has been cancelled could still leave open the argument that another similar scheme is inevitable. For example, agricultural land is being bought for a bypass. The bypass will open up the area for development, so the owner claims for compensation reflecting development potential. This is excluded as the development potential is entirely due to the scheme underlying the acquisition, which must be disregarded. The claimant then argues that, even disregarding the specific scheme, the land is in an area where there is a desperate need for a bypass, which will inevitably need to be built in the near future. The valuation of the land should still, therefore, reflect an element of hope value for development. Assumptions (c) and (d) render this argument invalid.

A second element to the new section 14 is the taking into account of the prospect of planning permission being granted on or after the valuation date for development of the relevant land or other land. This replaces the old section 16 which is repealed, and effectively requires the valuer to do exactly what he would do in the real world – assess the prospect of development value, having regard to the Local Plan and all other relevant considerations, to the extent that such development value is likely to be reflected in a market transaction.

Section 16 had often been criticised for being too prescriptive in setting out exactly what planning assumptions have to me made and, being 50 years old, for setting out those requirements in a way which is hard to relate to the modern planning system (*Essex County Showground Group Ltd* v *Essex County Council* (2006) ACQ 120 2004). It has now been replaced by a much more common-sense and practical approach.

Section 15

Planning permission is to be assumed for the acquiring authority's proposals. This is little changed from the original section 15, which requires that the possibility of carrying out the type of development for which the land is acquired may be reflected in the compensation payable. This is logical, because the local authority will require planning permission for its scheme, and if it can get planning permission it is reasonable to assume that any private developer would be able to obtain planning consent for a similar development. For instance, if the local authority acquires land for housing, it will be assumed that planning permission is available for the kind of housing development which the local authority proposes to carry out.

There is an apparent conflict between s.15 requiring an assumption of planning permission for the acquiring authority's scheme and Rule 3 of section 5 of the Act, which requires any value attributable to a scheme requiring statutory powers to be left out of account. In fact the two provisions complement each other and ensure that compensation reflects the acquiring authority's scheme, except where the proposed development is for a purpose to which the land could only be applied in pursuance of statutory powers.

The main change to section 15 is that there is no longer an assumption that planning permission will be granted for development falling within schedule 3 of the Town and Country Planning Act 1990. It previously required an assumption that planning permission would be granted for any form of development specified in schedule 3 (section 15(3) and (4)). This, broadly, assumes permission to rebuild a building that stood on the site prior to the 1947 Town and Country Planning Act and has been demolished, either during the war or since 1948. It also assumes permission to divide a residential property into small residential units, e.g. conversion of a house into flats.

Schedule 3 is an anachronistic provision which only has relevance in the context of its original purpose to protect existing use values at the time of the first Town and Country Planning Act in 1947. Its repeal is only a matter of time, but until the recent amendment its existence forced the Lands Tribunal into decisions which at times can only be described as bizarre. In *Greenweb* v *Wandsworth* (2007), for example, the claimant served a purchase notice in respect of an area of public open space which had a value of only around £15,000 in the real world. The Lands Tribunal was forced to assume planning permission for rebuilding nine terraced houses which had stood on the site prior to 1947, and therefore to award compensation of £1.6 million. In response to the complaints of the acquiring authority that this result was clearly inequitable, the tribunal calmly pointed out that "there is simply no authority that even begins to suggest that the application of the assumptions in sections 15 to 16 is discretionary". This decision was subsequently approved by an equally frustrated Court of Appeal [2008] All ER (D) 420. The courts will be thankful for the removal of this antiquated planning assumption.

Of course, the fact that, under the Act, land may be assumed to have the benefit of planning permission does not necessarily imply a demand for that land for those purposes. It is important always to remember the words of Lord Denning MR, referred to earlier in this chapter, about the elements necessary for development value: "It is not the planning permission by itself which increases value; it is planning permission coupled with demand."[28] This has also been expressed in the words: "planning permission does not create development value, it unlocks development value". In some cases normal demand – apart from the acquiring authority's scheme – might be so far distant as to warrant only a "hope" value for development.

6. Certificates of appropriate alternative development

Where an interest in land is proposed to be acquired by an authority possessing compulsory powers, Part III of the Land Compensation Act 1961, as amended by s.232 Localism Act 2011, allows the claimant to apply to the local planning authority for a Certificate of Appropriate Alternative Development stating whether or not there is development which would have been permitted on the relevant land in the absence of the scheme.

The application must state the type of development that is considered by the applicant to be appropriate, or that in the applicant's opinion there is no such development, and give reasons for that opinion.

The certificate must state whether such development would have been permitted, and must also identify every description of development which, in the opinion of

28 *Viscount Camrose* v *Basingstoke Corporation* [1966] 3 All ER 161.

the planning authority, is appropriate alternative development. It must also give a general indication of:

(a) the conditions to which a planning permission would have been subject,
(b) when the planning permission could be expected if it is one that could only be expected some time after the valuation date, and
(c) any preconditions to the granting of permission, e.g. a section 106 agreement, that could reasonably be expected to be attached to any consent.

A Certificate of Appropriate Alternative Development is a useful tool for both claimant and acquiring authority where they are unable to agree on what planning permissions would have been permitted on land subject to a CPO. In the real world such disputes can be resolved simply by making a planning application. This is not helpful in the compulsory purchase world as any such application would be refused on the grounds that the land is required for a scheme. A certificate is effectively a "virtual" planning application asking the planning authority: "in the absence of a the scheme, would you have granted planning permission for a specific development, and if not, what development would you have permitted?".

The main problem with these certificates has been that they provide information relevant to the date of application, and by the valuation date (8 years after the date of the certificate in the Spirerose case) planning policies and nearby developments may suggest that a very different form of development would be granted. This problem will hopefully be rectified by an amendment in the Localism Act which requires the certificate to relate to the valuation date, though it is hard to see how this will work in practice as the parties may require planning guidance well before compensation is agreed, or the authority takes physical possession.

A major change introduced by the Localism Act 2011 is that any appeal is to the Lands Tribunal, rather than to the Secretary of State (who would deal with any planning appeal). Section 18 of the 1961 Act provides that either the owner or the acquiring authority may appeal against a certificate, or against a failure to provide a certificate within the requisite time period, and the Lands Tribunal must, in response:

(a) confirm the certificate, or;
(b) vary it, or;
(c) cancel it and issue a different certificate in its place.

It will be interesting to see how this transfer of responsibilities will work in practice. The Lands Tribunal will concentrate on issues affecting value, so is less likely than the Secretary of State to get bogged down in detailed planning issues. However, where those planning details are critical to the valuation the Secretary of State can be expected to have a higher level of planning expertise.

Since 1991, compensation payable to a claimant includes the reasonable costs incurred in connection with the issue of a certificate, including costs incurred in an appeal under section 18 if any issues of the appeal are determined in the claimant's favour.[29]

7. Special compensation rules in particular cases

(a) Reinvestment by investors

The Land Compensation Act 1961, section 10A (inserted by the Planning and Compensation Act 1991, section 70), provides that the payment of compensation by the acquiring authority to a landowner not in occupation, e.g. a landlord, shall include reimbursement for the costs of acquiring "an interest in other land in the United Kingdom". The costs must have been incurred within one year from the actual date of entry. This provision goes some small way to addressing the inequity under which claimants who are not in occupation of their property do not qualify for disturbance compensation. It provides them with some recompense for consequential losses, though it is not true disturbance compensation, either technically or in financial terms.

(b) Special provisions in Private Acts

Where compulsory purchase powers are derived from Private Acts there may be provisions for a particular basis of compensation unique to purchases under that Act. These Acts are normally passed to enable a particular scheme to be carried out, and include powers to acquire the land required. Petitioners against the measure sometimes obtain concessions on compensation.

As an example, the Croydon Tramlink Act 1994 contains provisions for the compensation payable to the owners of houses who have a negative equity in the house (i.e. whose outstanding mortgage debt is greater than the value of the house) to be on a more generous basis than would be the case where the compensation basis is the general basis described above.

It follows that advisors dealing with cases under Private Acts need to determine whether any special provisions exist and how they should be construed.

8. Summary of general basis of compensation

It may be useful at this point to summarise briefly and in very general terms the basis of compensation for land taken as prescribed by the 1961 Act.

Essentially, it is the best price that the owner's interest might be expected to realise if voluntarily offered for sale in the open market by a hypothetical willing seller. In many cases it may be necessary to make two or more estimates of the value

29 Planning and Compensation Act 1991, section 65.

of the property, each based on different assumptions permitted by the Act in order to determine what that "best price" is and (in particular) whether in addition to the existing use of the land the assessment should also take account of the prospect of development (having regard to "planning permission coupled with demand", in the words of Lord Denning MR, quoted earlier in this chapter).

The possible bases for these estimates of value may be formulated from the valuation standpoint as follows:

1. The value of the property as it stands, taking into account also the terms of any planning permission already existing at the date of notice to treat, but not yet fully implemented by development, and any realistic "hope value" for development. For the majority of properties, both in built-up and in rural areas, this will represent the highest price obtainable in the open market.
2. The value of the land, as in 1, but with the assumption that planning permission would be given for the development which the acquiring body proposes to carry out. This may or may not give a higher value than 1 according to whether permission to carry out this kind of development would be of value (a) to purchasers generally or (b) only to bodies armed with statutory powers.
3. The value of the land subject to the assumption that planning permission would be given for one or more classes of development specified in a "certificate of appropriate alternative development".

In all the above cases, the valuation will exclude any increase or decrease in the value of the land in so far as it is attributable to actual or prospective development under the acquiring body's scheme, and also any decrease in value due to the threat of acquisition.

Where the owner has an interest in other land held with that taken, additional compensation for severance and injurious affection may be payable under the provisions of the Compulsory Purchase Act 1965.

Where the owner has an interest in land adjacent or contiguous to that taken, any increase in the value of such land due to actual or prospective development under the acquiring body's scheme, either on the land taken or on other land, is to be deducted from the compensation.

Where development properties are acquired under the compensation provisions of the Land Compensation Act 1961, the problem is mainly one of the selecting the basis most favourable to the claimant.

9. Arbitration of compensation disputes

If disputes over the assessment of compensation cannot be resolved by agreement, then they must be decided by arbitration, either privately or (failing that) by reference to the Upper Tribunal (Lands Chamber) – the Lands Tribunal.

Of particular importance is the formal "sealed offer" procedure, under the 1961 Act, section 4(1), (3). Either side may offer unconditionally to pay or accept a specified sum. If the offer is not accepted it may be made a "sealed offer", submitted but not disclosed to the Lands Tribunal. If after the Tribunal's award the "sealed offer" turns out to be as favourable to the side that rejected it, or more favourable, then the side that made the offer must (unless there are special reasons to the contrary) be awarded costs against the rejecting side, in so far as these are costs incurred after the "sealed offer" was made. "Sealed offers" resemble "payments into Court" in civil actions for damages, or "Calderbank letters" (i.e. "without prejudice" except as to costs)[30] in rent review arbitrations.

This does not alter the general rule that "the cost of litigation should fall on him who caused it", as stated by the Lord President Hope in a leading case in Scotland, *Emslie & Simpson Ltd* v *Aberdeen City District Council*,[31] in the Court of Session. In a civil action this means that normally the loser pays the winner's taxed costs. In the Lands Tribunal the acquiring authority is regarded as having "caused the proceedings". In the *Aberdeen* case there was an estimate of compensation, and the claimants obtained an advance payment of 90% of the estimate. They then referred their claim to the Lands Tribunal for Scotland. At the hearing of the reference the claim figure was very high and the authority's offer figure was very low, but neither side made a "sealed offer". The award was *higher* than the offer but *lower* than the claimed sum. It was also lower than the advance payment, and the claimants had to repay the difference. However, the authority, having made a low offer, had thereby "caused the proceedings" and therefore still had to pay the claimants' costs.

10. Valuation

In general, the approach to valuing an interest subject to compulsory purchase is no different from that adopted in preparing a normal market valuation. There are, of course, the special rules described in this chapter to be taken into account, but these do not affect the method of valuation.

In most cases the valuation will be on an existing use basis when the direct comparison, investment or profits methods will be adopted.

In preparing a valuation to be presented to the Lands Tribunal, the quality of evidence is assessed in the same way as for any arbitration. The best evidence is open market transactions. However, such transactions are made in the "scheme world": if the scheme has an adverse effect the value will be reduced, and increased if the property benefits from the scheme. If such evidence is "tainted" in this way, then an adjustment may be required to correct it to its value in the no-scheme world.

Where open market evidence is scarce, the valuer may seek to draw upon other settlements of claims as providing comparable evidence. This is a valid approach, but open market evidence will always be regarded as preferable.

---

30 *Calderbank* v *Calderbank* [1975] 3 All ER 333.
31 [1994] 1 EGLR 33.

Where development value is a factor, the residual method, or a direct capital comparison approach, is appropriate. The Lands Tribunal has, however, been reluctant to accept the residual method to arrive at the market value of land with development value. Although the use of the residual method is widespread in practice, it is true that relatively small adjustments to one or more of the assumptions can lead to a large change in the resulting land value, and the Tribunal has expressed concerns that valuers can abuse this method to produce whatever level of value suits their client. However, the almost universal acceptance of this method in the real world, coupled with the introduction of the *Practice Statement and Guidance Notes for Surveyors Acting as Expert Witnesses* published by the RICS in (which explicitly requires a valuer when presenting evidence as an expert witness to submit opinions honestly held with a view to assisting the Tribunal rather than supporting their client) may help to overcome the Tribunal's reluctance.

If evidence is scarce or non-existent, the valuer may present *ipse dixit* evidence derived from experience and knowledge and understanding of values. The Tribunal has expressed a cautionary view on such an approach. In *Marson (HMIT) v Hasler*,[32] a valuer submitted a valuation based on his man and boy experience. The Tribunal commented that:

> It had every respect for able and practical surveyors who belong to the (man and boy) 'school', but the fact should be recognised that when a member of this 'school' finds himself unable to agree values with an equally able and practical member of the 'analytical school' then, on a reference to the Lands Tribunal, the latter surveyor is apt to have the easier passage.

II. Home loss, basic loss and occupiers' loss payments

The Land Compensation Act 1973 introduced additional payments for persons affected by compulsory purchase in the form of home loss payments. These payments are intended to recognise the fact that such persons are not, in reality, willing sellers as assumed by rule 2 and represent a partial return to the time, prior to 1919, when it was customary to add 10% to the compensation assessed on a market value basis.

Home loss payments may be claimed under sections 29–33 of the 1973 Act, as amended by section 68 of the Planning and Compensation Act 1991, by a freeholder, leaseholder, statutory tenant or employee, in respect of a dwelling substantially occupied by them (in that capacity) as their main residence for 12 months prior to displacement. The claim must be made within six months of the displacement. If the claimant has been resident for 12 months but not in a capacity described above (and thus has been more a licensee for part of the time), the period

32 [1975] 1 EGLR 157.

can be made up by adding the period of occupancy of the claimant's predecessors. The payment consists of a set figure, which is 10% of the value of the interest in the dwelling, subject to a maximum limit, currently £47,000 and a minimum of £4,700, but subject to regular review.

12. Basic loss and occupier's loss payments

Sections 106–110 of the 2004 Act set out provisions for "loss payments" similar to those under the Land Compensation Act 1973, sections 29–33 and 34–36. Section 120 of the Act repealed farm loss payments, which had previously been payable under the 1973 Act.

Basic loss payments (section 106) may be claimed by a freeholder or a tenant of land of at least one year's standing at the date of entry by the acquiring authority, or (if earlier) the date when compensation is agreed (or is determined by the Lands Tribunal) "to the extent that he is not entitled to a home loss payment in respect of any part of the interest". The amount payable is 7.5% of the value of the interest, up to a ceiling of £75,000. This applies to any compulsory purchase, including purchase notices and blight notices.

Occupier's loss payments (section 107) may be claimed by anyone who qualifies for a basic loss payment and who has occupied the property for a year. If the property is agricultural, the amount payable is (i) 2.5% of the value of the claimant's interest, or (ii) the "land amount" (which is £100 per ha up to 100 ha plus £50 per ha for the next 300 ha, subject to a minimum of £300), or (iii) the "buildings amount" (£25 per m^2 of the gross external floor space of the buildings), whichever of these three amounts is the greatest. If the property is *not* agricultural the amount payable is in principle the same, except that the "land amount" (above) is £2.50 per m^2 of the area of the land, or £2,500 (but only £300 if only part of the land is taken) if greater. Once again, these figures are likely to be regularly reviewed.

Section 108 states that neither a basic loss payment nor an occupier's loss payment is obtainable if the property is not in a proper state of repair or maintenance, or is unfit for habitation, under specified provisions of planning and housing legislation. Section 109 contains supplementary provisions relating to:

* survival of any claims for loss payments in the event of death or insolvency;
* in the case of agricultural land, entitlement to either an occupier's loss payment or a payment under section 12(1) of the Agricultural (Miscellaneous Provisions) Act 1968, but not both; the greater may be claimed; and
* referral of disputed claims to the Lands Tribunal.

The introduction of basic and occupiers' loss payments is to give all claimants a similar right to additional compensation to that previously only enjoyed by residential claimants. If they are both owners and occupiers the full 10% will be paid (7.5% basic loss payment plus 2.5% occupiers' loss).

13. Interest on compensation

If the acquiring authority takes possession of the land ("entry") before compensation is assessed and paid, the vendor is entitled to interest on the capital sum that constitutes the amount agreed or awarded.[33] This applies not only to compensation under the "willing seller" principle but also to "equivalent reinstatement", as held by the Court of Appeal in *Halstead* v *Manchester City Council*.[34]

---

33 Land Compensation Act 1961, section 32. The Treasury prescribes the rate of interest to be paid. Subject to the provisions of the Land Compensation Act 1973, sections 52 and 52A, regarding advance payments (discussed in section 2(3) of this chapter), payment is, inevitably, retrospective.
34 [1988] 1 All ER 33. The "willing seller" and "equivalent reinstatement" principles are discussed in section 3 of this chapter.

Chapter 24

Compulsory purchase compensation II
Compensation for severance and injurious affection

1. General principles

The principles governing the right to compensation for injurious affection were considered in Chapter 22, where a distinction was drawn between:

(i) claims for severance and injurious affection to (i.e. depreciation in) the value of land owned by the claimant in consequence of the acquisition, under statutory powers, of adjacent land of the claimant, some or all of which was formerly "held with" it; and
(ii) claims for injurious affection to the value of land owned by the claimant in consequence of the exercise of statutory powers on adjacent land, none of which was formerly "held with" it.

The right to compensation in case (i) derives from section 7 of the Compulsory Purchase Act 1965. In respect of case (ii) the relevant sections are 10 of the 1965 Act and Part I of the Land Compensation Act 1973.

The following may help to make this distinction clear:

| A | B | C |

LAND A – Claimant (i)
LAND B – Claimant (i)
LAND C – Claimant (ii)

Land "B" is being acquired under statutory powers. Claimant (i), from whom it is being acquired, is entitled to claim compensation not only for its market value on acquisition but also for any injurious affection under section 7 of the 1965 Act to his other land "A" which, until the acquisition, had been held with it. Claimant (ii), however, is only entitled to claim compensation for any injurious affection to her land "C" under section 10 of the 1965 Act and/or Part I 1973 Act because none of land "B" has been held with it. Land "B" is being "severed" from land "A", but not from land "C".

Lands such as "A" and "B" are held with each other even if the claimant's rights are not the same in "A" as they are in "B" (as in *Oppenheimer* v *Minister of Transport*,[1] where the claimant's property right in the land taken was merely an option), and even if the lands are not actually contiguous, provided they are such that "the possession and control of each" gives "an enhanced value to the whole" (*Cowper Essex* v *Acton Local Board*[2]): this is a valuation question of obvious practical importance.

In so far as the claim for compensation is based on depreciation in the "value" of the land, as opposed to disturbance, it comes within the scope of rule (2) of section 5 of the Land Compensation Act 1961. Section 5 expressly applies in general terms to "Compensation *in respect of* any compulsory acquisition"; and the exception from its scope which is contained in rule (6) is expressly confined to "disturbance or any other matter not directly based on the value of land" – not merely the value of the land acquired.

In all cases of compulsory acquisition pursuant to a notice to treat or a General Vesting Declaration, compensation for injurious affection, in that it is based on the value of the land, will represent the depreciation in the value of that land in the open market.

It would appear, however, that the assumptions as to planning permission to be made under sections 14 and 15 of the 1961 Act, in the case of lands taken, do not apply to assessments of market value in the case of retained lands which are injuriously affected.[3]

2. Compensation where part only of the land is taken

This is "severance". Cases of this type, governed by the Compulsory Purchase Act 1965, section 7, frequently occur, particularly in connection with the construction of new roads or the widening of existing roads or town centre redevelopment. An example occurred in *Ravenseft Properties Ltd* v *Hillingdon London Borough Council*,[4] though in that case the acquiring authority was eventually compelled to buy the whole of the property in dispute, instead of severing it as they had intended by acquiring part of the garden.

It is necessary to consider the principal losses likely to be suffered in connection with the land formerly held with the land taken and the bases on which claims for compensation should be prepared. Compensation where part of a property is taken

---

1 [1942] 3 All ER 485.

2 (1889) 14 App Cas 153.

3 These sections apply to the "relevant interest" and "relevant land", which section 39(2) defines "in relation to a compulsory acquisition in pursuance of a notice to treat", as (respectively) "the *interest* acquired in pursuance of that notice" and "*land* in which the relevant interest subsists"; no mention is made of any land retained by the claimant.

4 (1968) 20 P&CR 483.

falls under three main heads: land taken, severance and injurious affection to land held therewith, and disturbance comprising other incidental losses resulting from the compulsory taking. Other matters that may have to be considered are accommodation works and apportionment of rents.

As a matter of practice these two items are the first to be addressed since they affect the amount of compensation payable. The accommodation works, if any, are reflected in the injurious affection to the land retained while the apportionment of rent has a direct bearing on both the value of the land taken and also the land retained.

(a) Accommodation works

In the case of railway undertakings, the acquiring body was obliged under section 68 of the Railways Clauses Consolidation Act 1845 to provide certain "accommodation works" for the benefit of owners of land adjoining the undertaking. These included bridges and other means of communication between severed portions of land, the adequate fencing of the works, means of drainage through or by the side of the railway and provision of watering places for cattle. These works tended to reduce the effect of severance and other injury likely to be caused by the construction of the railway and were, of course, taken into account in assessing compensation.

Under modern legislation, there is no such obligation on the acquiring body to provide accommodation works or on the owner to accept them in lieu of compensation. In practice, however, it is frequently agreed between the parties as a matter of joint convenience that the acquiring body will carry out certain "accommodation works" or "works of reinstatement", and that compensation shall be assessed on the basis of these works being provided. For instance, where a new road is constructed across agricultural land, the acquiring body will not only fence along the boundaries of the road, but will also provide new gates where convenient, or perhaps connect the severed plots. If part of the front garden of a house is taken under a street widening scheme, the acquiring body will probably agree to provide a new boundary wall and gates, to plant a hedge to screen the house, to make good the connection to the house drains and similar works.

All these are items for which compensation would otherwise have to be included in the claim. Some acquiring authorities compare the cost of accommodation works with the consequent reduction in the compensation payable and, if the cost is significantly greater than the savings to them, will decline to carry out the works. For example, if a farm is severed by a motorway, the provision of a bridge or tunnel will allow the two parts of the farm to continue to function as a single entity. The injurious affection will thereby be reduced by bridges or tunnels, which can be costly to provide. If a bridge costs £80,000 and the reduction in the injurious affection is around £80,000 or more, then it makes sense to provide a bridge; but if the reduction in the injurious affection is, say, only £5,000, then there seems little point in providing a bridge since it clearly does not have any significant impact on the use of the farm whether it is severed into two parts or not.

It is usually convenient for both parties that the acquiring body should carry out the accommodation works while the work on the scheme is still in progress. Details of the works agreed upon will therefore be included in any settlement of compensation by agreement or, in case of dispute, may be referred to the Lands Tribunal and considered by them in awarding compensation.

(b) Apportionment of rent (Compulsory Purchase Act 1965, section 19)

Where part of land subject to a lease is taken, it may be necessary to apportion the rent between the part taken and that which is left. Naturally, the lessee should not be required to continue to pay the same rent, particularly if the portion taken is considerable; but it is often difficult for the lessor and lessee to agree on the amount of the deduction to be made. The lessee can only demand a fair apportionment of the rent paid, not of the annual value of the property. Also, the lessee cannot include in the reduction of rent any figure representing reduced value to the rest of the property. That loss should be met by compensation from the acquiring body to the lessor or lessee as appropriate.

3. Setting out a claim

Having considered these two preliminary issues, the claim should be set out under three separate heads:

(a) Land taken

The area of land to be taken will usually be indicated on the plan accompanying the notice to treat.

Compensation for the land taken will be assessed in accordance with the principles prescribed by the Land Compensation Act 1961 (as amended). This sets out the underlying assumptions, though the valuation itself will be undertaken in the same way as any other valuation, using the method appropriate to the type of property.

The same item of value must not be compensated for twice over. This is known as "double counting" and is one of the pitfalls resulting from the practice of separating the claim into three heads. For example, where a property such as a petrol filling station is valued for land taken by the profits method, any claim for loss of profits under the head of disturbance should only be accepted where those profits have not already been reflected in the claim for land taken.

(b) Severance and injurious affection

Section 7 of the 1965 Compulsory Purchase Act gives claimants who have land taken an additional right to compensation for severance and injurious affection to land retained. Severance is one variety of "injurious affection" compensation. That is to say, an owner suffers compensable loss (referred to in section 7 as "damage") because the owner's "other land", which he retains, is depreciated whether by severance or some other factor. The loss by severance is loss whereby the market values of the land taken and the land retained amount to less than the total market value of the land before it was severed. A classic case is *Holt* v *Gas Light & Coke Co*[5] in which a small portion of land used as a rifle range was compulsorily purchased. The part taken was peripheral only, but so situated that the safety area behind the targets was no longer sufficient and the land retained could not in future be used as a range.

The main kinds of severance can perhaps be best understood in this way:

A	B

A	B	C

LAND A – NOT TAKEN
LAND B – TAKEN
LAND C – NOT TAKEN

Land "B" is acquired from the owner and severed from retained land "A" in the first example and from retained lands "A" and "C" in the second. The essential "severance" is that of land "B" from the other land; the severance of land "A" from land "C" is a secondary matter (though none the less important). More complicated forms of severance may, of course, cause further fragmentation of an owner's land; but the basic principles of compensation are the same. A typical modern example is where part of a farm is separated from the rest, including the farmhouse and farm buildings, by the construction of a road across the property, or where the area of a farm is significantly reduced in size by the taking of a part, even though the part remaining is not split up.

Such severance is likely to result in increased working costs. For example, if a portion of arable land is cut off by severance of intervening land there will probably be an increase in the cost of all the normal operations of ploughing, sowing, reaping, etc. as well as additional supervision necessitated by the separation of the land from the rest of the farm. If the severed portion is pasture land, extra labour and supervision may be required in driving cattle to pasture as well as possible risk of injury to cattle, as for instance in crossing a busy new road. These additional expenses are likely to involve some reduction in rental value of the land which, capitalised, will represent the injury due to severance.

The extent of loss from severance will naturally vary greatly according to the nature of the undertaking and other circumstances, and whether or not

5 (1872) 8LR QB 728.

accommodation works such as a private bridge or underpass are provided to enable the owner to cross the road or railway. It must not be forgotten that land ownership extends physically downwards and upwards from the land surface, so that viaducts and tunnels involve "severance" of land in principle just as surface works do; though it may be that the financial loss caused to an owner by having a tunnel below the land surface will in some circumstances be nil (see *Pepys* v *London Transport Executive*).[6] Ownership of property fronting a highway commonly includes half the highway, subject to the dedication of the surface. So in *Norman* v *Department of Transport*[7] the ownership of the subsoil of a highway which was indicated in a draft CPO constituted land taken and entitled the owners of the property fronting the highway to compensation (in this case under a blight notice, but this would also apply to a claim under section 7).

In whichever sense it is "severed", the property as a whole may suffer depreciation in value by reason of the fact that it can no longer be occupied and enjoyed as one compact holding.

The valuer must be careful to include in the claim for compensation any injury likely to be caused to the rest of the property by the authorised user of the works. The user must, however, be authorised. If it is not covered by the statutory authorisation conferred on the acquiring authority, they will be liable instead to damages in a civil action. Here "injurious affection" arises "otherwise" than by severance, and is akin to depreciation actionable at common law in the tort of nuisance. For example, the use, as a major road, of a strip of land across a country estate may seriously depreciate the value of the mansion by reason of noise and fumes, loss of privacy and spoiled views. Such factors (other than loss of privacy and views) are actionable nuisances if they occur to any significant degree. All of them (including loss of privacy and views) should be reflected in a claim under s.7, to the extent that they cause depreciation in the value of the land retained.

The leading authority is the decision of the House of Lords in *Buccleuch (Duke of)* v *Metropolitan Board of Works*,[8] when part of the garden of the claimant's house (in fact its entire riverside frontage) was compulsorily acquired for the construction of the Victoria Embankment at Westminster. The full depreciation of the claimant's property was held to be payable, on the basis of severance and injurious affection in combination, in so far as the insignificant purchase price or market value of the strip of garden actually taken was insufficient of itself to make up the total loss in land value which the claimant had suffered (calculated using a "before and after" valuation).

In addition to consideration of the effect on value of the retained land by the authorised use of the works, the actual carrying out of the works can also

6 [1975] 1 All ER 748 (deep level tube tunnel).
7 [1996] 1 EGLR 190.
8 (1872) LR 5 HL 418.

have an impact. The valuation of the retained land needs to reflect the fact that work is to be carried out on neighbouring land which may have a significant impact on the valuation. For example, if some garden land of a house is taken to build a motorway, the valuation of the retained land is of a house which will be close to the site of a proposed motorway where work may be going on for several years which will be noisy and unpleasant, following which there will be a motorway with heavy traffic. All of these factors should be reflected in the valuation.[9]

Every head of damage that can reasonably be anticipated should be included in the claim, since no further claim can be made later for damage which might have been foreseen at the time when the land was taken.

The claim for depreciation in value of the land retained will be based on its market value, which will take into account the benefit of any actual planning permission already given. It may be permissible, in some cases, to consider the effect on value of the possibility that planning permission might have been given, having regard to the terms of the development plan and other circumstances. However, no assumptions of planning permission can be made, as in the case of land taken, because the relevant provisions of the Land Compensation Act 1961 do not refer to land retained. The theory behind this seems to be that no planning assumptions are necessary, as there is no reason why an actual planning application could not be made on retained land. However, this overlooks that the scheme taking place on adjacent land could affect the decision.

The Land Compensation Act 1973, section 8(4A), requires that there be entered in the local land charges register details of: (a) the works for which the land taken is required, and (b) that part of the land which is not taken. This is to avoid duplication of compensation in the event of a possible claim under Part I of the Land Compensation Act 1973 (discussed below).

(c) Other incidental injury (disturbance)

In addition to the value of the land taken and injurious affection to other land held therewith, owners are entitled to compensation for all other loss and expense which they may incur in consequence of the compulsory acquisition. The generic term for these items of compensation is "disturbance", which is the subject of the next chapter. But some disturbance may in practice be closely connected with claims for severance and injurious affection compensation and can, for the sake of convenience, be touched on here. Such losses will vary with the circumstances of

---

9 Such an approach is supported by, for example, *Cuthbert* v *Secretary of State for the Environment* [1979] 2 EGLR 183 in correctly applying section 44 of the Land Compensation Act 1973 that compensation for injurious affection "shall be assessed by reference to the whole of the works and not only the part situated on the land acquired" (i.e. from the claimant). In *Budgen* v *Secretary of State for Wales* [1985] 2 EGLR 203, the Lands Tribunal adopted the approach of "looking at the property before and after the acquisition and construction of the road".

each case, so that it would be misleading to attempt to suggest an exhaustive list of possible items. A consideration of some of the principal heads of damage likely to arise in connection with the taking of a strip of land through an agricultural estate will indicate the general nature of items that may be included with this part of the claim. They may be briefly summarised as follows.

1. Disturbance on the land taken

Either an owner-occupier or a tenant is entitled to compensation for any loss suffered through having to quit a portion of farm land at short notice. Such loss may arise from the forced sale of stock which can no longer be supported on the reduced acreage of the farm, or from the forced sale of agricultural implements. They may also claim for other similar consequential losses, for example, temporary grazing and storage. Improvements to the land made at a tenant's expense may have to be taken into account.

2. Damage during construction

It is almost unavoidable that during construction a certain amount of damage will be caused to crops, etc. on land immediately adjoining the works. It is also likely that certain parts of the farm, particularly those portions that are to be severed, will be more difficult to work during this period, so that a claim for increased labour costs or total loss of rental value may be justified. If the works are being carried out close to the mansion or farmhouse, the impact on the occupier may be considerable and in extreme cases it may be necessary for the occupier to obtain temporary accommodation elsewhere.

(d) Assessment of compensation

It is common practice to assess a global figure for compensation for both the land taken and injurious affection. This is achieved by the "before and after" method. The "before" value is determined by assessing the market value of the whole interest, including the land to be taken as if no scheme of acquisition existed. The "after" value is the market value of the interest retained reflecting the full impact of the scheme.

The difference between these figures represents the value of the land taken and the injurious affection to the land retained. It is necessary to determine a value of the land taken which, when deducted from the full difference, leaves the balance as compensation for injurious affection.

However, a more technically accurate approach is to assess the value of the land taken separately, and then to assess severance and injurious affection by undertaking a before and after valuation on the retained land only. This is considered more fully later in this chapter.

4. Compensation where no part of the land is taken

(a) Injurious affection under section 10

Section 10 of the 1965 Act gives a right to compensation for injurious affection where no land is taken and as a result there is no right to a claim under s.7 of the Act. The claimant must prove four things:

1.   The works causing the injurious affection must be authorised by statute.
2.   The harm must be of a kind that would be actionable in the civil courts but for that authorisation.
3.   The loss to be compensated must be confined to depreciation in land value (e.g. not business loss or other "disturbance").
4.   The cause of the harm must be the execution of the works and not their use (e.g. the building of a new road, not the effects of the traffic on it when opened).

These four rules arise from the decision of the House of Lords in *Metropolitan Board of Works* v *McCarthy* (1874).[10] It is evident that these cases are treated very much less generously than those cases in which some land has been taken from the claimant, as discussed above. Whereas in those cases (under the 1965 Act, section 7) compensation is paid in full for "any loss or damage" resulting from injurious affection that is proved to occur, in these cases where no land is taken proof of loss is quite insufficient if the four McCarthy rules cannot be satisfied in full. Even if the rules are met, they restrict compensation to depreciation in land value due to the execution of the works. Any depreciation caused by the use of the works, and loss other than depreciation in land value, is not allowable under section 10, even where that loss is a clear and unavoidable consequence of the scheme.

Typical examples of successful claims are where development authorised by the acquiring body deprives adjoining owners of the benefit of some easement attached to their land such as a private right of light, as in *Eagle* v *Charing Cross Railway Co*, or a private right of way, or is in breach of some restrictive covenant to the benefit of which the adjoining owner is entitled, as in *Re Simeon and the Isle of Wight Rural District Council.*[11]

An important recent case in the Court of Appeal is *Moto Hospitality Ltd* v *Secretary of State for Transport.*[12] There was a complex realignment of major roads in an intersection with the M40 near Oxford, at junction 10 (the Ardley interchange). The unsuccessful claimant asserted that alterations to the road layout resulted in making the access to its motorway service area less convenient for

---

10 (1874) LR 7 HL 243.
11 [1937] 3 All ER 149.
12 [2008] 2 All ER 718.

motorists (though only by a certain amount of deviation). No land was compulsorily acquired from the motorway service area (held by the claimant on a sub-lease) but there was a CPO for other land. It was held that this could have been sufficient to support a claim under section 10 *(McCarthy* rule 1) except that *McCarthy* rule 2 was not satisfied because there would be no actionable nuisance as such to the motorway service area. Any harm was purely general.

Compensation for infringing a restrictive covenant must not exceed the diminution in value of the land to which the benefit of the covenant attaches: *Wrotham Park Settled Estates* v *Hertsmere Borough Council.*[13]

(b) Compensation for harm from "physical factors"

As the fourth of the McCarthy rules restricts compensation under s.10 to losses arising from execution of the works,[14] Parliament enacted Part I of the Land Compensation Act 1973 to give owners a right to compensation for depreciation to the land caused by the use (as distinct from the execution) of public works. The use must give rise to depreciation by reason of certain prescribed "physical factors", namely "noise, vibration, smell, fumes, smoke and artificial lighting and the discharge on to the land in respect of which the claim is made of any solid or liquid substance". In practice, it is usually noise which is the major cause of depreciation, although the other factors do contribute. In *Blower* v *Suffolk County Council*[15] the sole contributory factor was the artificial lighting, where the depreciation due to light emanating from a new road led to compensation even though the amount of light that reached the claimant's home was minimal. The "responsible authority" which has promoted or carried out the works is liable, including the Crown (but no claim may be made in respect of "any aerodrome in the occupation of a government department"). The claimant must hold an "owner's interest" in the depreciated land, namely the freehold or a leasehold with three or more years to run. If it is a dwelling it must be the owner's residence if the interest entitles the owner to occupy it. Landlords of dwellings may also claim. If it is a farm the owner must *occupy* the whole even though the *interest* need only be in part. If it is any other kind of property the interest and occupation must relate to all or a "substantial part" of it and the rateable value must not exceed the prescribed amount (£34,800 from 2010, but subject to regular review).

The period for submitting a claim begins one year after the date when the use of the works began, called the "relevant date", subject to certain exceptions, and the normal six-year limitation period under the Limitation Act 1980 then starts

---

13 [1993] 2 EGLR 15.
14 [1974] 1 All ER 201, (1974) 229 EG 1589. In *Wildtree Hotels Ltd* v *Harrow London Borough Council* [1998] 3 All ER 638, the Court of Appeal held that no compensation was payable for noise, dust and vibrations, as such (*Clift* v *Welsh Office*).
15 [1994] 2 EGLR 204.

to run. The valuation date is the relevant date. Corresponding benefit to other land of the claimant will be "set off" against the compensation (if any). There must be no duplication of any compensation otherwise payable, e.g. under section 7 of the 1965 Act, or as damages in tort for nuisance. It was held in *Vickers* v *Dover District Council*[16] that liability in nuisance is not excluded if the authority is acquiring under a power (i.e. permissively) and not a duty, so that Part I of the 1973 Act does not apply in such a case. Any soundproofing undertaken in accordance with Part II of the 1973 Act must be taken into account.

The use of the public works giving rise to compensation must be taken to include reasonably foreseeable future intensification. Beyond this, subsequent alterations to works, or changes of use apart from mere intensification and apart from aerodromes or highways, will give an additional right to compensation. "Physical factors" caused by aircraft as such give no right to compensation except in respect of new or altered runways, "taxiways" or "aprons".

Once the legal right to compensation is established, it is then a question of determining the depreciation in the market value of the land due to the injury complained of. In *Hickmott* v *Dorset County Council*[17] a claim made under Part I of the Land Compensation Act 1973 failed because the claimant could show nothing beyond an increased fear of danger to property from passing traffic in consequence of road widening.[18] It is clear from *Pepys* v *London Transport Executive*[19] that in order to establish depreciation in market value it is not sufficient to show merely that a sale at a higher figure fell through by reason of the authority's works while a sale at a lower figure went ahead subsequently. It must also be demonstrated that this was caused by physical factors arising from the use of the works.

Two methods are commonly adopted to determine the amount of injurious affection. One is to make two valuations of the property: (i) as it previously stood, and (ii) after the interference with the legal right in question. This "before and after" method, as it is called, can be used both in this type of case and also in cases where part of the land has been taken. This method is to be preferred, but in practice reliable comparable evidence is rarely available.

The other is to determine depreciation as a percentage of the value of the land retained due to the factors to be taken into account. This more subjective approach, while technically inferior, is much more commonly used.

---

16 [1993] 1 EGLR 193.
17 [1977] 2 EGLR 15.
18 See also *Streak and Streak* v *Royal County of Berkshire* (1976) 32 P&CR 435: failure to attribute depreciation to proximity of new motorway. But the claimant in *Davies* v *Mid-Glamorgan County Council* [1979] 2 EGLR 158 succeeded in a claim based on depreciation attributed to successive extensions to an airfield near his land.
19 [1975] 1 All ER 748.

5. "Before and after" method of valuation

This method is used to assess the quantum of compensation payable for severance and injurious affection where part of a claimant's land is taken, or where no land is taken, though in the latter case only injurious affection will apply, and only that injurious affection arising from the physical factors under Part I or the execution of the works under s.10. In practice it is very often difficult if not impossible to separate the value of the land taken from the injurious affection likely to be caused to the rest of the land by the construction and use of the undertaking.

The method will involve two valuations:

1. of the property in its present condition unaffected by statutory powers; and
2. of the property as it will be after the part has been taken and in contemplation of the construction and use of the undertaking in its entirety.

The amount by which valuation 1 exceeds valuation 2 will represent the financial loss suffered by the claimant for the loss of the land taken and also for injurious affection to the remainder. In addition, the owner might be entitled to claim for injury during the carrying out of the works, and it would be necessary to come to an understanding as to the works of reinstatement which the acquiring body is prepared to provide.

An example is the taking of a strip of land forming part of the garden or grounds of a house. Here it is usually very difficult to assign a value to the strip of land taken without at the same time considering all the consequences of the taking. Suppose that the strip forms part of a garden of a fair-sized house and is to be used in the construction of a new road; the following are some of the questions that will naturally suggest themselves in assessing the fair compensation to the owner:

- Will the land remaining be reasonably sufficient for a house of this size and type?
- Will the proposed road be on the level, on an embankment, or in a cutting?
- Will it be visible from the house, and if so from what parts of it?
- What volume of traffic may be expected?
- How close will the house be to the road and to what extent is it likely to suffer from loss of privacy, noise, fumes and other inconvenience both during construction and when in use?

It is obvious that the most realistic approach to the problem is that of comparing the value of the house as it now stands (i.e. with no road to be built) with its estimated value after a portion of the garden has been taken for the new road (as in the *Duke of Buccleuch*'s case, above). But in theory, at any rate, there are two statutory obstacles to the use of this method:

- the fact that the statutory assumptions as to planning permission apply to land taken, but not to land injuriously affected;

•   the fact that section 4(2) of the Land Compensation Act 1961 requires
     that the owner's statement of claim shall distinguish the amount claimed
     under separate heads and show how the amount claimed under each head
     is calculated.

Where no land is taken and a claim is made under s.10 1965 Act or Part I, 1973
Act, the compensation provisions are limited, so the full loss suffered may not
always be claimable. Even so, it is probable that the valuer's first approach to a
problem of this kind will be along the lines of a "before and after" method and that,
having in this way arrived at what he considers a reliable figure, the valuer will
then proceed to apportion it if necessary as between compensation for land taken
and compensation for severance and injurious affection to the remaining land. He
will then need to assess any element of that loss that is not allowable under the
basis of claim.

The Lands Tribunal accepted the method in *Budgen* v *Secretary of State for
Wales*[20] and in *Landlink Two Ltd* v *Sevenoaks District Council*,[21] but in *ADP&E
Farmers* v *Department of Transport*[22] preferred that there should be separate
valuations of land taken and land retained.

The main problem facing the valuer undertaking a before and after valuation
is in finding sound comparable evidence on which the valuation can be based.
This is particularly difficult where no land is taken. Take, for example, a house
in a rural location which has no land taken, but which has a new bypass road
built nearby. A claim can be made under Part I, 1973 Act for depreciation caused
by the use of the road, but no compensation is payable for the construction of
the road, and therefore for the visual intrusion. It may be difficult enough to find
comparable evidence to support the after valuation – sales of similar houses in rural
locations with busy roads outside are required. For the before valuation, however,
comparable evidence is required of similar houses in rural locations with major
roads outside, but which suffer none of the noise, vibration, fumes, etc. which will
affect the subject property. This will be impossible. In the case of claims under
Part I of the 1973 Act, the injurious affection is therefore often determined by the
alternative direct approach and expressed as a percentage of the open market value
of the house in the absence of the scheme. This practice seems to have derived from
rating practice. Prior to the 1973 Act the only remedy available to householders
affected by the use of public works was an application for a reduction in rateable
value, as residential property was still, at that time, in the rating system. Percentage
deductions in RV was the approach generally adopted by rates tribunals.

Consequently, compensation for depreciation is often expressed as a percentage
of value. The percentage is derived from the total diminution in value and depends

20 [1985] 2 EGLR 203.
21 [1985] 1 EGLR 196.
22 [1988] 1 EGLR 209.

on the impact of the use factors rather than proximity to the scheme, although in most cases, of course, the closer the property is to the scheme the more likely is the use to have a greater adverse impact. The percentages may range from a modest 1% (awarded in *Clark* v *Highways Agency LT*, ref LCA92/1999) to 30% or more. In practice, a pattern of percentages will emerge so that compensation becomes dependent on the level of comparable settlements, although the Lands Tribunal has made it clear in many cases that market evidence, where available, will always prevail against evidence based on settlements.

Worked example of claims for severance and injurious affection

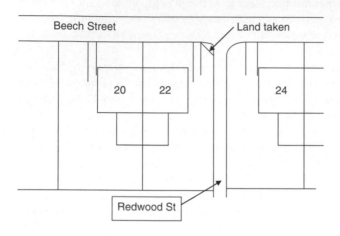

22 Beech Street

Land taken

10 m² of garden land @ £10 per m²   £100

Would have little value if offered for sale in the open market by a hypothetical willing seller

Severance and Injurious Affection under Section 7, Compulsory Purchase Act 1965

Before – open market value of retained land and building in the no-scheme world (strictly £200,000 less £100 for land taken = £199,900) but, say,	£200,000
After – open market value after construction of the motorway, and taking into account any sound insulation works undertaken under Part II of the Act	£180,000
Depreciation	£20,000

Disturbance

e.g. additional costs of cleaning carpets, windows and driveway, due to dust and dirt from the construction work, say	£500
	£20,600

Plus basic loss payment, $7\frac{1}{2}$ % of land taken	£7.50
Occupier's loss payment	£300*
Total compensation	£20,907.50

plus legal and valuation fees (rule 6, s5 Land Compensation Act 1961)

* The occupier's loss payment is the greater of $2^1/2$% or the building's value (@ $£25/m^2$ GEA) or the land value – for non-agricultural land the greater of £2,000 (but £300 if only part taken) or $£2.50/m^2$ (s33C Land Compensation Act 1973)

Part II, Land Compensation Act 1973

The local authority will carry out sound insulation works in accordance with the sound insulation regulations.

Also consider accommodation works. The local authority will make good the driveway and boundary fences, etc., or otherwise the costs will be claimable.

Consider also section 8 of the 1965 Act. Can the part of the garden be taken "without seriously affecting the amenity or convenience of the house"? If not, the highway authority can be required to purchase the whole of the premises.

20 Beech Street

Severance and Injurious Affection only

Part I, Land Compensation Act 1973

Open market value in the no-scheme world	£200,000
Open market value after construction of the motorway, and taking into account any sound insulation works offered under Part II of the Act	£180,000
Depreciation (10% of open market value)	£20,000
But the depreciation is caused partly by the execution of the works, and partly by physical factors arising from the subsequent use. Under Part I, only losses due to physical factors are compensable. Allow two-thirds, say	**£13,300**
Does not qualify under s.10 1965 Compulsory Purchase Act as there has been no interference with a private legal right.	

Part II, Land Compensation Act 1973

The local authority will carry out sound insulation works in accordance with the sound insulation regulations, produced under Part II. Any works offered will be taken into account in assessing the Part I claim, whether the offer is accepted or not.

24 Beech Street

Severance and Injurious Affection only

Part I, Land Compensation Act 1973

Open market value in the no-scheme world	£200,000
Open market value after construction of the motorway, and taking into account any sound insulation works offered under Part II of the Act	£180,000
Depreciation (10% of open market value)	£20,000

But the depreciation is caused partly by the execution of the works, and
partly by physical factors arising from the subsequent use. Under Part I,
only losses due to physical factors are compensable.
Allow two-thirds, say   £13,300

Section 10, 1965 Compulsory Purchase Act

The claimant qualifies under all four of the McCarthy rules, because, in
particular, he has suffered interference with a private right of access.
Depreciation due to the interference with the private legal right
(the access) is allowable, say   £5,000

Total compensation   **£18,300**
Part II, Land Compensation Act 1973
The local authority will carry out sound insulation works in accordance with
the sound insulation regulations produced under Part II. Any works
offered will be taken into account in assessing the Part I claim, whether
the offer is accepted or not.

Consider also accommodation works. The local authority will make good the
driveway and boundary fences, etc., or otherwise the claim for depreciation will
be higher as the after value will reflect the value of a house without a driveway.

Chapter 25

Compulsory purchase compensation III

Compensation for disturbance

1. Disturbance compensation and disturbance payments

Disturbance compensation is an element of compulsory purchase compensation; that is to say, it is a sum added to the purchase price of land compulsorily acquired. It is a separate head of compensation from land taken and severance and injurious affection. It has been evolved as a principle of case law, without express authorisation in any statute (apart from rule 6 of section 5 of the Land Compensation Act 1961, which provides that the rules governing the assessment of land value "shall not affect the assessment of compensation for disturbance ..."). Nevertheless, it has a statutory origin by implication, dating back to *Jubb* v *Hull Dock Co,*[1] in which a brewer obtained compensation for loss of business in addition to the purchase price of the compulsorily acquired brewery, the claim having been upheld on an interpretation of a private Act containing provisions similar to those of the Lands Clauses Consolidation Act 1845.

It is therefore in a sense "parasitic", because there must first be an acquisition price and, if there is, disturbance compensation can then be included in it. In *Inland Revenue Commissioners* v *Glasgow & South Western Railway Co,*[2] Lord Halsbury LC referred to acceptable claims for such well-known disturbance items as "damages for loss of business" and "compensation for the goodwill", but added: "in strictness the thing which is to be ascertained is the price to be paid for the land ...".

In *Hughes* v *Doncaster Metropolitan Borough Council,*[3] Lord Bridge said:

> It is well-settled law that whatever compensation is payable to an owner on compulsory acquisition of his land, disturbance is an element in assessing the value of the land to him, not a distinct and independent head of compensation.

---

1 (1846) 9 QB 443.
2 (1887) 12 App Cas 315.
3 [1991] 1 EGLR 31.

It may be that the acquisition price is nominal, particularly in the case of lease-hold interests, so that no actual sum is paid for the interest. This does not preclude a claim for disturbance.

The kinds of loss that are compensable as "disturbance" are not exhaustively listed, and perhaps the only discernible principle is that they can constitute anything other than land value (because that is already compensatable as market value of land taken, or depreciation of land retained). Claimants must "mitigate" their loss by taking care not to incur, and claim for, unreasonable and unnecessary expenditure. In *K&B Metals* v *Birmingham City Council*,[4] the claimant firm had bought equipment which could not be used after a draft discontinuance order was confirmed; the Lands Tribunal held that the claim for the ensuing loss was reasonable (but it would not have been if the purchase was subsequent to the confirmation of the order).

Removal costs are the most obvious example of disturbance; but it will be seen that there are many other examples of loss, particularly in regard to business premises.

The connection between this loss and the compulsory acquisition of the land in respect of which it is suffered is causal and not chronological.

Costs incurred in advance of dispossession brought about by the compulsory acquisition are thus claimable provided that they are in no rational sense premature but satisfy the essential requirements of being (i) reasonably attributable to the dispossession and (ii) not too remote. The Court of Appeal so held in *Prasad* v *Wolverhampton Borough Council*,[5] as did the Court of Session in *Aberdeen City District Council* v *Sim*.[6] These decisions, upholding claims for anticipating disturbance items, have been approved and followed by the Privy Council in an appeal from Hong Kong, *Director of Buildings and Lands* v *Shun Fung Ironworks Ltd*,[7] the relevant principles being the same in all three jurisdictions.

Disturbance under rule 5 is normally payable where a claimant has been expropriated of an interest in land, generally defined for this purpose as a freehold or a lease with over one year unexpired.

If periodic tenants, or any tenants whose term does not exceed a year, are in fact turned out (by notice of entry) before their tenancy is terminable at common law, then section 20 of the Compulsory Purchase Act 1965 requires the acquiring authority to compensate the tenant accordingly, since they are being expropriated. The market value of the tenant's interest may well be small, or nominal, but the right to compensation in principle carries with it the right to include disturbance items, which may well constitute the bulk of the claim in such a case.

---

4 [1976] 2 EGLR 180, a case concerning a discontinuance order, though the principle is equally applicable to compulsory purchase.
5 [1983] 1 EGLR 10.
6 [1982] 2 EGLR 22.
7 [1995] 1 All ER 846.

The requirement to receive compensation for land taken as a precondition of a claim for disturbance excludes many claimants who suffer genuine losses as a result of compulsory purchase. This inequity was considerably reduced by the introduction of disturbance payments in 1973. Disturbance payments, unlike disturbance compensation, are authorised expressly by statute (sections 37–38 of the Land Compensation Act 1973) in certain cases in which disturbance compensation is not payable because, although a compulsory acquisition has in fact taken place (or some comparable occurrence, see below), the claimants themselves have not had an interest compulsorily acquired from them, but instead have merely been dispossessed. These include situations in which a tenant holding under a short-term leasehold has not had that term renewed by the acquiring authority, which has taken the landlord's place on acquiring the latter's reversion.

Licensees cannot claim these payments as of right, though authorities have discretion to pay them.

Section 38(1) of the 1973 Act provides for compensation payments as follows:

(a) the reasonable expenses of the persons entitled to the payment in removing from the land from which they are displaced;
(b) if they were carrying on a trade or business on that land, the loss they will sustain by reason of the disturbance of that trade or business consequent upon their having to quit the land.

Business tenants who would also be entitled to compensation for loss of security of tenure under section 37 of the Landlord and Tenant Act 1954 can choose whichever amount is the higher, the payment under that section or the disturbance payment, but not both (1973 Act, section 37(4)). It should be observed that this provision for a disturbance payment is worded in pretty restrictive terms but over the years the Lands Tribunal (which settles disputes as to amounts, in the first instance) and the courts have interpreted it more broadly, as constituting the same right as afforded by disturbance, not a reduced or lesser right. Regard, however, must be had to the probable length of time for which the land from which claimants have been dispossessed would have been available for their use, and also to the availability of other land, when "estimating the loss" (section 38(2)).

It is important to remember that disturbance *payments* (under the 1973 Act) are claimable by tenants who have been dispossessed but not expropriated, whereas disturbance *compensation* is claimable by any tenant or freeholder who has been *both* dispossessed and expropriated because it is included in (and treated as part of) their expropriation compensation. It follows from this that it is rare for landlords to justify a claim for disturbance compensation, since they are not being dispossessed even though they may be expropriated. In *Lee* v *Minister of Transport*,[8] Davies LJ said: "Disturbance must, in my judgment, refer to the fact of having to vacate

8 [1966] 1 QB 111.

the premises". Thus, if premises are let by a freeholder to a periodic tenant, the former may be expropriated but not dispossessed while the latter (if served with a notice to quit and not notice to treat or notice of entry) is dispossessed but not expropriated. The tenant would now qualify for a disturbance payment; but neither would be entitled to claim disturbance compensation.

An exception may, however, be recognised when both landlord and tenant are corporate bodies sufficiently closely connected with one another to be treated virtually as the same person, for example if they are related companies with the same directors and their separate identity is a mere matter of form. In *DHN Food Distributors Ltd* v *London Borough of Tower Hamlets*,[9] the Court of Appeal held this to be the case for two companies with the same directors, of which one company owned the freehold of the acquired premises and the other occupied them as licensee; and the directors were awarded disturbance compensation. But in *Woolfson* v *Strathclyde Regional Council*,[10] the House of Lords insisted that normally where there are two companies they must expect to be treated separately, since they are in law two separate "persons". Their Lordships distinguished the *DHN* case narrowly upon its own special facts. In *Woolfson* there were two companies, as in *DHN*, one owning the acquired premises and the other occupying them under a licence; ownership and membership of the two companies was not, however, identical, even though there was a very considerable overlap as regards both shareholders and directors. The *DHN* type of case will be rarely met, though in practice it is not that uncommon, for example, for the senior partner or director of a firm to hold the premises as a "pension fund", and acquiring authorities tend to apply *DHN*, "lift the corporate veil" and not press the issue.

A more recent exception, of express statutory origin, is that owners not in occupation, i.e. landlords and reversionary owners in general, can include in claims for the market value of their reversionary interests when expropriated, reimbursement of costs incurred in obtaining new premises (Land Compensation Act 1961, section 10A, inserted by the Planning and Compensation Act 1991, section 70). Such a claim must be made within a year of the entry on to the land by the acquiring authority. It is analogous to disturbance, even though there is no physical dispossession but merely deprivation of an investment interest in landed property. As with physical disturbance, the claim is parasitic on the relevant purchase price of the reversion: no purchase price, no reimbursement of expenses. There may be difficulties in practice if only an advance payment has been made within the 12-month period. Although the implication is that the incidental costs and expenses incurred in acquiring the new premises are for a property of similar value to the acquired premises, the Act refers only to costs of "acquiring ... an interest in other land in the United Kingdom".

9 (1975) 30 P&CR 251.
10 [1978] 2 RGLR 19.

2. Judicial approach to quantification of disturbance compensation

Lord Nicholls, delivering the majority decision of the Privy Council in *Director of Buildings and Lands (Hong Kong)* v *Shun Fung Ironworks Ltd*,[11] included several statements of basic principle about disturbance, all on the unstated assumption that the claimant is an owner-occupier. "To qualify for compensation a loss suffered … must satisfy the three conditions of being:

a)   'causally connected' – the loss must be a natural and reasonable consequence of dispossession,
b)   'not too remote', and
c)   'not a loss which a reasonable person would have avoided'."

The third condition is usually expressed as a requirement that the claimants must "mitigate" their loss, which means that they must not unreasonably inflate the loss; it does not mean they must make excessive efforts to assist the acquiring authority.

It follows that, if reasonable, a claim to cover the cost of relocating a business can be acceptable notwithstanding that this may exceed the cost of total extinguishment of the business. "It all depends on how a reasonable businessman, using his own money, would behave in the circumstances." The same discount rates for capitalising lost profits should be used for relocation cases and extinguishment cases alike.

Relocation must not be confused with "equivalent reinstatement", which only applies when there is no general demand or market for the current use of the land.[12]

The duty to mitigate a loss arises when there is certainty that the compulsory acquisition will proceed, usually accepted as being when a notice to treat or a General Vesting Declaration is served. So, if a notice to treat and notice of entry are served some while after an order is confirmed, such that relocation of a business is impossible in the time between the service of the notices and actual entry, it is not open to an authority to argue that the claimant should have anticipated the situation and looked for other premises following confirmation of the order. Conversely, where a house subject to compulsory purchase was repossessed for default in mortgage repayments, the owner's losses did not count as "disturbance": *McTaggart* v *Bristol & West Bldg Soc and Avon County Council*.[13]

Any loss which meets the three conditions set out above can be claimed, but the following section sets out some of the items most commonly included in a disturbance claim.

11 See n.8, *supra.*
12 Land Compensation Act 1961, Section 5, rule (v). See Chapter 23.
13 (1985) 19 P&CR, 184.

3. Disturbance compensation: items which are typically claimed

(a) Residential premises

Where dwelling-houses are compulsorily acquired, occupiers who are freeholders, or lessees enjoying an appreciable profit rent, will be entitled to compensation based on the market value of their interest in the property, estimated in accordance with the statutory provisions already described. More likely, however, they will be tenants holding under a contractual tenancy and paying approximately the full rack rental of the premise, so the value of their interest in the land will be negligible and they will receive no compensation for being forced to quit before the expiration of their term. Similar principles apply when the interests are no greater than those of a yearly tenant and tenants are required to give up possession before the expiration of their interest. If the acquiring body has compulsorily purchased the reversion it will become the tenant's new landlord, and may well be content to await the termination of the tenancy contractually by effluxion of time or notice to quit, in which case there will be no purchase price in which disturbance can be included.

Regardless of the value of their interest in the land, the occupiers will be entitled to disturbance compensation, under rule 5 of the 1961 Act, section 20 of the 1965 Act or a disturbance payment under section 37 of the 1973 Act, which will include:

- loss on disposal of fixtures and other items which cannot be removed from the property being acquired other than at an unreasonable cost;
- cost of removal from the property being acquired to the new home; and
- expenses incurred in acquiring the new property and other incidental expenses caused by having to leave the premises.

Non-removable items

Occupiers may remove any fixtures and other items to which they are entitled. Items which typically come within this category are carpets, curtains, cookers and other kitchen equipment.

If the rule that claimants must take all reasonable steps to mitigate their loss is strictly applied, it would seem that a claimant is entitled to claim the lesser of either the value of these items or the costs of removing them to the new home, including any costs of adaptation that may be required to fit them into their new surroundings and including costs of plumbing them in or otherwise getting them into working order. In practice, acquiring authorities are often happy to let claimants decide whether they wish to keep these items by transferring them to the new home or to leave them, though a stricter line may be taken in respect of major items such as a modern American-style plumbed-in fridge-freezer.

Where items cannot be transferred, for example fitted carpets or curtains which will not fit into the new property, claimants are not entitled to the cost of buying new carpets or curtains since they are treated as receiving value for money for their purchases. This is the "betterment" principle – a claimant is entitled to like for like compensation, not new for old. In practice, the surveyors acting for the parties are likely to agree the value of items not transferred. A deduction should be made for the resale value of those items, though this will often be nil or nominal. The basis of valuation is the price that a purchaser would pay in a normal open market transaction. There is an exception to the betterment principle in that a claimant may be entitled to new for old where he has no alternative but to replace the old item with a new one, and he obtains no financial benefit from the additional expenditure. In *Allen* v *Doncaster* (1996) the Lands Tribunal awarded the full £300 cost of acquiring new curtains because the 10-year-old curtains they replaced were still serviceable and could not be refitted at the new home. It has to be said there does appear to be betterment in having new curtains compared to 10-year-old ones, and this may be a generous interpretation of the principle which may not be followed where larger sums are involved.

Cost of removal

This item will include not only the actual cost of moving the owner's furniture to other premises but also any other expense incidental thereto, such as temporary storage of furniture and refitting appliances. An authority normally requires three quotations to be obtained to show that the claimant is mitigating the cost. Storage will arise where there is a gap between giving possession and moving into a new home. Where claimants are unable to find another home before entry by the authority they may claim the cost of renting furnished accommodation or living in a hotel until a new home is found. A problem of renting accommodation is that landlords normally offer tenancies of a minimum of six months, which may seem excessive if a new home is found within that period.

Where claimants hire a van and carry out the move themselves or with the assistance of friends or family, they cannot claim the cost they would have incurred if they had employed a firm of furniture removers. On the other hand, the duty to mitigate the loss does not require claimants to indulge in a self-help approach.

Incidental expenses

In addition to cost of removal, there may be other incidental expenses incurred by owner-occupiers in consequence of being dispossessed of their premises. In *Harvey* v *Crawley Development Corporation*,[14] Romer LJ summarised the principle governing the inclusion of such items in the claim as follows:

> The authorities ... establish that any loss sustained by a dispossessed owner (at all events one who occupies his house) which flows from a compulsory

14 [1957] 1 QB 485.

acquisition may properly be regarded as the subject of compensation for disturbance provided, first, it is not too remote and secondly, that it is the natural and reasonable consequence of the dispossession of the owner.

In the case in question the claimant, who was required to sell and give up possession of her house, was entitled to recover legal costs, surveyor's fees and travelling expenses incurred in finding another house to live in, together with similar expenses incurred to no purpose in connection with the proposed purchase of a house on which she received an unfavourable report from her surveyor. Such fees are a genuine loss, whereas a purchase price paid for new premises, large or small, represents not a loss but "value for money".

Items within this category will depend upon the circumstances of each particular case and will include such matters as value of stationery made redundant, costs of notifying friends and organisations of the change of address, costs of installing or transferring a telephone, the cost of uniforms if children have to change schools, and so on. In the case of claimants who are disabled, the costs of adapting the new property to their needs, or to those of a member of their family, are recoverable since they are not costs of adaptation which are reflected in the value of the new house, in contrast to, say, the costs of upgrading an outdated kitchen or bathroom. Alternatively, section 45 of the Land Compensation Act 1973 allows a claim under rule 5 for equivalent reinstatement where premises have been substantially modified for a disabled person.

Legal and other professional fees incurred after notice to treat in connection with the preparation of the owner's claim are also recoverable under rule 6 of section 5 of the Land Compensation Act 1961. This practice was sanctioned by the Court of Appeal in *London County Council* v *Tobin.*[15] But the same court, in *Lee* v *Minister of Transport,*[16] held that these items, though within rule 6, are not "disturbance" but are covered by the words "any other matters not directly based on the value of the land". Thus, solicitors' and surveyors' fees incurred in obtaining new premises count as "disturbance" items; but solicitors' and surveyors' fees incurred in preparing the compulsory purchase claim itself count as "any other matter", and will be paid even where a claimant does not qualify for disturbance compensation.

(b) Business and professional premises

Like householders, occupiers of business premises will be entitled on compulsory purchase to compensation for the value of their occupier's interest in the land and, subject to the qualifying rules, to disturbance comprising any loss that is a natural and reasonable consequence of dispossession and is not too remote.

15 [1959] 1 All ER 649.
16 [1966] 1 QB 111.

In addition to losses of the type suffered by a residential claimant, a trader or businessman may have a claim on a broad range of business losses such as forced sale of stock and trade disturbance. Unlike domestic claimants, the occupier of business premises may be unable to find suitable alternative premises. In such cases, the claim for disturbance will be based on the loss suffered by the total extinguishment of the business rather than the costs and losses suffered by its removal to other premises. The claimant does not normally have a choice. If suitable alternative premises are available and the costs of relocation are less than the costs of extinguishment, as is almost always the case, then the business should be transferred to the new premises. If the claimant chooses not to move in these circumstances, the compensation will be assessed on the basis of a notional move – see, for example, *Rowley* v *Southampton Corporation*[17] and *Bailey* v *Derby Corporation*.[18] If the claimant does move to new premises but other premises were available which were suitable and would have resulted in lower compensation being payable, then compensation is assessed as if the claimant had moved to the other premises – *Appleby & Ireland Ltd* v *Hampshire County Council*[19] and *Landrebe* v *Lambeth London Borough Council*.[20]

Compensation for interest in land

As in the case of residential property, if lessees are enjoying a substantial profit rent they will be entitled to compensation for loss of their leasehold interest based on the ordinary principles of compulsory purchase valuation, as explained in Chapter 22.

The Land Compensation Act 1973, section 47, requires that both the business tenancy and the reversion must be valued on the basis that the tenancy enjoys statutory protection under the provisions of the Landlord and Tenant Act 1954. In practice this is unlikely to affect the claim for land taken, as any new tenancy will be at a market rent so there will be no profit rent after any renewal. It may, however, affect the disturbance claim, particularly in respect of any claim for goodwill. As a result of this provision it will be difficult to argue for reduced compensation for extinguishment of a business on the grounds that the lease was nearing its expiry date.

In the case of tenants from year to year, or for any lesser interest, whose right to compensation depends on section 20 of the Compulsory Purchase Act 1965 (replacing section 121 of the Lands Clauses Consolidation Act 1845), section 39 of the Landlord and Tenant Act 1954, as amended by section 47 of the Land Compensation Act 1973 (also see section 48 for agricultural holdings), expressly provides that the total compensation paid to them is not to be less than that which,

17 (1959) 10 P&CR 172.
18 [1965] 1 All ER 443.
19 [1978] 2 EGLR 180.
20 [1996] 36 RVR 112.

in certain circumstances, they might have received under section 37 of the 1954 Act if a new tenancy had been refused, i.e. an amount currently equal to the rateable value, or to twice the rateable value if they or their predecessors in business have been in occupation of the holding for the past 14 years or more.

Fixtures and fittings, plant and machinery

Subject to meeting his obligation to mitigate his loss, the occupier may require fixtures and fittings and plant and machinery to be taken with the premises, particularly where the possibility of transferring and adapting them to other premises is somewhat problematical. This choice is available on the authority of *Gibson* v *Hammersmith and City Rail Co.*[21] In *Tamplin's Brewery Ltd* v *County Borough of Brighton*[22] the acquiring authority made new premises available to the claimants, and the outstanding question of compensation related to equipment. The claimants bought new equipment and relinquished the old. They then claimed the cost of the new equipment as an item of compensation, reduced only by an amount representing a saving in operating costs. The Lands Tribunal accepted this in principle, though with a reduction in the total sum. Yet it would seem that the claimants, in obtaining new equipment, got "value for money" in so doing, and this may cast doubt on the decision.

Normally the compensation will be the value of the existing fixtures to an incoming tenant for the business or, if capable of being sold by the claimant, the loss on forced sale being the difference between the value to an incoming tenant and the price achieved on a sale.

Value to an incoming tenant is the price that a purchaser of the business would pay for these items, so that they represent the value to the business. The price is likely to be above that payable for similar items available elsewhere, if only because they are in place and immediately available for use.

The underlying principle in determining value within a disturbance claim is that it is value to the owner/claimant, not market value. In *Shevlin* v *Trafford Park Development Corporation*,[23] the authority's valuer adopted the provisions of the Royal Institution of Chartered Surveyors (RICS) *Appraisal and Valuation Manual* in valuing the plant and machinery. This provided that:

> Where a valuation of plant and machinery is intended to reflect its work to an undertaking which is expected to continue in operation for the foreseeable future, the normal basis of valuation will be "value to the business" which is defined as the "value based upon the assumption that the plant and machinery will continue in its present existing use in the business of the company".

21 (1863) 2 Dr & Sns 603.
22 (1971) P&CR 746.
23 [1998] 1 EGLR 115.

The Tribunal member took the view that "value to the business" is consistent with value to the owner and accepted the approach adopted by the authority's valuer, which was to determine the net current replacement cost at £92,965. However, it was accepted that the value to the claimant was above the replacement cost and a figure of £100,000 determined by "a robust approach" was adopted.

It is common for specialist plant and machinery valuers to be instructed to agree the values. The claimant's valuer will also be able to organise disposal of the items. Where there is a significant amount of items, an auction may even be held on the premises. The price achieved at auction, or on a sale by other means, is deducted from the value to the owner. Since the price will be depressed because of the circumstances this is referred to as a forced sale price, and thus the claim is the loss on forced sale.

Cost of removal

This will often be substantial, including in the case of industrial premises such items as the cost of moving machinery and its installation in new premises. Acquiring authorities commonly require two or three quotations to be obtained.

Incidental expenses

Other items of claim may include notification of change of address to customer and suppliers, new sign writing on vehicles, value of redundant stationery, statutory redundancy payments where staff cannot be re-employed, as well as accrued holiday pay and statutory compensation for short-term notice where applicable, travelling expenses and management time in seeking new premises, and so on. Each claim turns on its own particular facts and the acquiring authority is liable to meet all reasonable expenses incurred which are not too remote. The authority in effect takes claimants as it finds them. So, for example, where a claimant who owns an affected business lives on the other side of the world, the fares and other costs incurred in coming to the UK to deal with the removal of the business may be reasonably incurred, even though the costs are high. As in the case of private houses, there is authority for the inclusion of legal and surveyors' fees incurred in connection with the acquisition of new premises, and also legal and other professional fees such as accountants' or specialist valuers' fees incurred in connection with the preparation of the claim. See for example *LCC* v *Tobin*.[24] In the *Shevlin* case referred to above, the Tribunal accepted accountants' fees for advice on the taxation implications of the receipt of compensation.

24 [1959] 1 All ER 649.

Loss on stock

This may consist either of depreciation in the process of removal to other premises, or loss on forced sale in those cases where no other premises are available, or where the trade likely to be done in new premises will be of a somewhat different class, or where the stock will not bear removal.

Loss on depreciation may be expressed as a percentage of the value in the trader's books, but may be covered by the cost of taking out an insurance policy before the move. As purchasers are aware of the circumstances behind the sale, loss on forced sale is often high: significantly less than wholesale prices. Where the turnover of stock is high, as with, for example, retail grocers, a valuation of the stock just prior to removal may be required. This may require the payment of fees for a specialist stock valuer, which would be included in the claim.

Trade disturbance

"Trade disturbance" is the term applied to any loss of profits likely to result from traders being dispossessed of their premises. On sale in the open market, a business will usually command a value which is quite different from that of the premises. This is its "goodwill" and depends on its potential future profitability in the course of trading. There may be other, temporary, loss of profits caused by disruption of the business prior to closure or before, during and after relocation. "Goodwill" is capital, and profits are income; but since the one is derived from the other it is important always to avoid double-counting and ensure that the same loss is not compensated twice over.

Compensation paid for loss of goodwill does not mean that the acquiring authority has purchased the goodwill. Any goodwill that exists following the displacement from the premises remains with the claimant. Acquiring authorities paying compensation for the loss of goodwill may seek undertakings from the claimant not to open a new business within a certain time or a certain distance whereby the claimant can resurrect the goodwill. Where claimants procure a claim for total extinguishment of their business as of right under section 46 of the 1973 Act (see below), the section provides for the acquiring authority to seek such undertakings as a condition for exercising the claimant's right under the section.

It is also important to remember that the value of goodwill is not its open market value as such but the value to the owner.[25] Hence evidence from trade valuers as to the price for which a business would be sold is a good starting point, but is not necessarily the value to the owner and thus the compensation to which they are entitled.

---

25 See, for example, *Sceneout Ltd* v *Central Manchester Development Corporation* [1995] 2 EGLR 179.

It will be convenient to discuss the nature of the loss of goodwill in relation to retail trades, and then to see to what extent the same considerations apply to the wholesale trader, the manufacturer or the professional firm.

The "goodwill" of a business has been defined as the probability of the business being carried on at a certain level of profit. In effect, it is the value of the business itself as a capital asset, distinct from (a) the premises, (b) the stock-in-trade and other chattels, and (c) the profits and income from the business. Traders in business in certain premises for a number of years will have acquired a circle of regular customers on whose patronage they can rely, and will also be able to count on a fairly steady volume of casual custom. As a result, by striking an average over three or four normal years they can make a fairly accurate estimate of their annual profits. It is the probability of those profits being maintained, or even increased, in the future that constitutes the "goodwill" of the business. Any reasonable purchaser would pay more for this established business than he would pay for vacant premises from which to set up a business from scratch.

Although goodwill is a capital item assessable on the prospect of continuing profits in future years, the profits to be earned in the course of current trading are another matter. They are the subject of compensation in their own right (*Watson* v *Secretary of State for Air*[26]) if the compulsory acquisition prevents their being earned because it cuts short or disrupts the commercial activity which would, if continued, have produced them.

In *Bailey* v *Derby Corporation*,[27] the then Master of the Rolls, Lord Denning, in regard to the claimant's demand for compensation in respect of the goodwill of his business, said: "all that is acquired is the land. The compensation is given for the value of the land, not for the value of the business." But the reason why the claim for goodwill failed in this case was that the claimant could have disposed of it on the open market. The excuse for not trying, namely age and ill-health, was not in fact valid, because age and ill-health need not prevent such a transaction being put in hand.

The sequel to this case was the enactment of section 46 of the Land Compensation Act 1973. This allows claimants aged 60 or over (irrespective of health) to require compensation to include extinguishment of goodwill if they have not independently disposed of that asset. They must give an undertaking not to dispose of it, nor to carry on the business themselves within a reasonable area and period, to be specified, on pain of having to repay the amount in question. The premises must not have a rateable value in excess of a specified amount, currently £34,800 but subject to regular review. Special rules apply where a claim is made by a partnership or company.

26 [1954] 2 All ER 582 – in this case, farming profits for the current year.
27 [1965] 1 All ER 443.

The various factors which go to make goodwill can usually be divided into two classes:

1. those which, being of a personal nature, do not depend on the situation of the premises, e.g. the name and reputation of the firm or the personality of the proprietor; and
2. those which are dependent on location, e.g. the advantage derived from being on the main shopping street of a town or in a street or district which is the recognised centre for businesses of a particular type.

When business premises are taken under compulsory powers, the trader's goodwill is not necessarily destroyed but it may be seriously damaged even if the business is transferred to other premises. How seriously depends upon whether the goodwill is mainly personal in character or largely dependent upon the situation of the premises. The test of whether goodwill can legitimately be said to have been partially or totally extinguished, thus justifying a claim for compensation, is to consider whether or not it could reasonably be sold on the open market and whether its marketability, if any, has been reduced because of the circumstances of the compulsory purchase.

The usual method of assessing this class of loss is by multiplying the average annual net profits by a figure of Years' Purchase (YP), which varies according to the extent of the injury likely to be suffered. In most cases the figure of net profits is based on the past three years' trading as shown by the trader's books. However, if the profits have been adversely affected by the scheme for a period, then such profits would be ignored. For example, if the last year's accounts have been adversely affected, the previous three years "untainted" accounts might be adopted. The accounts are normally supplied to the valuer by the claimant's accountant. They are commonly accounts prepared for tax purposes, rather than to show the true profits. Consequently, adjustments need to be made to the accounts.

For example, the figures given will not usually have taken into account interest on the capital employed in the business, and it is argued that, as traders could earn interest upon their capital by investing it elsewhere, the annual profit due to the business is really not the figure taken from the books, but that figure less interest on capital. The tenant's capital would comprise the value of the fixtures and fittings, stock, etc., and the necessary sum to be kept in hand for working expenses. The latter item would obviously vary considerably, depending on the nature of the business.

The valuers must also ascertain what rent has been charged in the books. Usually the actual rent paid is charged in the accounts, and if valuers estimate that the property is worth more than the rent paid, they must deduct the estimated profit rent. In the case of a freehold occupier, if no rent has been charged, then notional rent, the estimated annual rental used for arriving at the value of the property, must be deducted. The reason for this is that, if the business ceased, the profit rent or rental value could continue to be obtained by letting the property; therefore, part of the profit apparently shown in the accounts is actually a return on the ownership of the property. To put it another way, the accounts of a business operating from freehold premises will show higher profits than the same business operating from leasehold premises, so this adjustment is necessary to ensure both businesses receive the same compensation.

The wages of assistants employed in a firm will naturally be deducted in finding the net profits. Where the owners work in the business but don't take a wage it is necessary, in theory at least, to deduct a notional wage for their time, again to ensure that two identical businesses, one staffed by paid employees and one staffed by the owners, receive the same compensation. However, it was held by the Lands Tribunal in *Perezic* v *Bristol Corporation*[28] that there was no evidence that it was customary to deduct a sum in respect of the remuneration of the working proprietor of a one-man business. It was apparent that to do so would result in a business that made virtually no profits at all, yet it was evident that such businesses were valued by their proprietors, and often sold for respectable prices in the open market – presumably because of the value placed on being "your own boss" rather than an employee. This approach was extended to the remuneration for a husband and wife operating a business (*Zarraga* v *Newcastle-upon-Tyne Corporation*[29]). Therefore, any deduction made in the accounts for such remuneration will be added back. This also applies to any deductions for director's wages of pension payments where it is a family business owned and run by the director or directors. This is particularly common for businesses owned and run by a husband and wife.

In *Appleby & Ireland Ltd* v *Hampshire County Council*[30] a considerable number of adjustments to the net profits shown in the accounts were made, including adjustments in respect of directors' reasonable remuneration (as against actual salaries paid), bad debts, and exceptional expenditure on research. Indeed, the accounts need to be examined and adjustments made to remove any factors that distort the true profit levels. Such of these deductions and adjustments as are applicable having been made, and true net profits having been found, the next step is to fix the multiplier that should be applied to this figure in order to compensate traders fairly for injury to their goodwill.

Before considering the level of multipliers that may be appropriate in any particular case, a general comment is appropriate. In practice, the multipliers range from 0.5–5. These are reflected in the many Lands Tribunal decisions on the value of goodwill but, as the Tribunal frequently stresses, each case turns on its facts and the multiplier adopted in one case cannot be used as a "comparable" for another.

28 (1955) 5 P&CR 237.
29 (1968) 205 EG 1163.
30 [1978] 2 EGLR 180.

The problem in practice is that there are no true comparables. It may be possible to discover a multiplier from an actual market sale of a business, but this is evidence of market value whereas what is sought is value to the owner. It is agreed that value to the owner may be equal to market value, but it is often higher, and cannot be lower – if the owner valued his business at less than the market value, then he would have sold it at the market price. It is common practice for a pattern of settlements to emerge which become comparables within the scheme. Awards in past Lands Tribunal cases can also be helpful in assessing the appropriate years' purchase.

Where the taking of the premises will mean total extinction of trade, the full value of the goodwill will be allowed as compensation. Where the trade will not be totally lost but may be seriously injured by removal to other premises, a fair compensation will be a proportion of that which would be paid for total extinction. A claimant who is a cash tobacconist adjoining the entrance of a busy railway station may lose the whole trade if the premises upon which the trade is carried on are no longer available. The claimant may have to relocate in a quite different location and will have to start up a new business. In such a case, up to the full value of the business may be the appropriate compensation – possibly three to four years' purchase of the net profits.

A milkman, on the contrary, knowing all his customers by reason of delivery of milk daily at the houses, may face removal with little risk of loss of that portion of his trade, and a bespoke tailor may be much in the same category, although he may suffer some loss of casual custom and also in respect of some items sold over the counter. In such cases as little as half a year's profits might be proper compensation.

It is obvious that, since the variety of trades is so great and the circumstances of each case may vary so widely, every claim of this sort must be judged on its merits. But the following are among the most important considerations likely to affect the figure of compensation:

1. *The nature of the trade and the extent to which it depends upon the position of the premises*

For instance, grocers or confectioners or drapers whose business is done entirely over the counter will suffer considerable loss if they have to move to premises in an inferior trading location. Probably two and a half to three years' purchase of the net profits would not be excessive compensation. On the other hand, a bespoke tailor would be far less likely to be injured by removal since the customers might be expected to follow, assuming that premises convenient to them can be found. A business that has poor long-term prospects – perhaps a video rental store in these days of easily available downloads – can expect a relatively low YP, whereas a multiplier at the higher end of the range may apply to a business, such as a scrapyard, which has a degree of monopoly due to the difficulty in obtaining planning consent to set up a new business.

2. *The terms on which the premises are held*

The compensation payable to traders holding on a lease with a number of years to run will probably be substantially the same as that of claimants in similar circumstances owning the freehold of their premises – particularly in view of the additional security of tenure enjoyed under the Landlord and Tenant Act 1954. But a shopkeeper who holds on a yearly tenancy only can expect as high a level of compensation if he has a strong prospect of the tenancy being renewed almost indefinitely. Where a lease contains a redevelopment clause or other power for repossession by the landlord, the effect on the years' purchase will depend on the likelihood of the power being exercised.

3. *The pattern of profits*

The value of goodwill depends on the prospects for making profits in the future. The prospects can be extrapolated from recent evidence – normally, as previously related, the last three years' accounts. If these show strong and sharply rising profits, then the prospects are bright, supporting a high YP, whereas low and/or falling profits have the opposite effect.

Temporary loss of profits

In addition to compensation for the value of the goodwill or the decrease in its value, a trader may have lost profits in the period running up to closure or during transfer of the business. For example, a retailer selling women's clothes would aim to dispose of as much stock as possible if the business is to close. Profits would be below normal levels in the period as a result of both price cutting and also not having new season designs in stock. Thus a claim for temporary loss of profits would be made in addition to the value of goodwill. This is also the sort of case where the final year's profits would be ignored (or adjusted) in determining the average net profits.

The difficulty of determining temporary loss of profits lies in establishing what the profits would otherwise have been. One approach is to extrapolate the profits from previous years, if there is an identifiable trend. Another is to establish the pattern of profits from other premises to produce an index which can be applied to the subject premises. For example, if the retailer selling ladies' clothes is a multiple retailer, the profit patterns from similar outlets can assist in assessing the likely profits that would have been earned in the subject premises. The difference between these and the actual profits equals the temporary loss.

The question of compensation payable to wholesale traders and manufacturers involves similar considerations to those already discussed. It will usually be found, however, that the probable injury to trade is much less than in the case of retail business, since profits are not so dependent upon position. Provided other suitable premises are available, the claim is usually one for so many months' temporary loss of profits during the time it will take to move and establish the business elsewhere, rather than any permanent injury to goodwill (e.g. see *Appleby & Ireland Ltd* v *Hampshire County Council*[31]). In practice, alternative premises will need to be taken before the loss of the existing property. This will result in a period when the trader is incurring two payments of rent, perhaps rates, insurance and similar outgoings. The additional payments will be compensated as part of the temporary loss of profits, as they reduce profits. They are referred to as double overheads. This is an area where the risk of "double counting" is very strong. Where a claim is submitted for temporary loss of profits, this loss will include any double overheads. A separate claim for an extra set of bills for rates, rent, etc. may be considered as an alternative, but should not be allowed in addition.

In this type of case the sum claimed for temporary loss of income has been regarded as not subject to income tax because compulsory purchase compensation, regardless of its particular components, is supposed to be "one sum" in the nature of a "price to be paid for the land" (*Inland Revenue Commissioners* v *Glasgow & South Western Railway Co*[32] – a decision of the House of Lords). It was therefore held in *West Suffolk Council* v *Rought Ltd*[33] that the Lands Tribunal, in assessing the compensation payable for temporary loss of profits, should have deducted their estimate of the additional taxation which the claimant company would have had to bear if it had actually earned the amount which the interruption to its business prevented it from earning, because, if earned in fact, that amount would have been liable to tax as income. On this basis, acquiring authorities for some time paid compensation to companies net of tax. However, changes in the tax status of compensation payments flowed from provisions in the Finance Acts 1965 and 1969 and the Income and Corporation Taxes Act 1970, and the current position, exemplified in the decision of the Court of Appeal in *Stoke-on-Trent City Council* v *Wood Mitchell & Co Ltd*,[34] is that temporary loss of profits, loss on stock and such revenue items as removal expenses and interest are paid gross by the acquiring authority and then taxed as receipts in the hands of the claimant company.

In the case of professional men and women, such as solicitors or surveyors, the goodwill of the business is mainly personal and is not very likely to be injured by removal. But the special circumstances of the case must be considered, particularly the question of alternative accommodation. For instance, a firm of solicitors in a big city usually requires premises in the recognised legal quarter and in close touch with the courts.

A claimant forced to vacate his factory, shop, etc. may be obliged to move to other premises which are the only ones available, and will base a claim for trade

31 [1978] 2 EGLR 180.
32 (1887) 12 App Cas 315 (the words of Lord Halsbury LC).
33 [1957] AC 403 – a decision of the House of Lords following the House's decision in *British Transport Commission* v *Gourley* [1956] AC 185, which applied this principle to payments of damages in tort.
34 [1978] 2 EGLR 21.

disturbance on the cost of adapting the new premises plus the reduced value of the goodwill in those premises.

The latter figure may be due to reduction in profits on account of increased running costs in the new premises. For instance, heating and lighting may cost more, transport costs may be higher, or it may be necessary to employ more labour owing to the more difficult layout of a production line. Or, again, the new position may not be as convenient for some customers as was the old. For instance, in *LCC* v *Tobin*[35] the Court of Appeal approved a figure for trade disturbance although it appeared that the profits in the new premises were likely to be higher than in the old. The compensation was based on the difference between the value of the profits in the old premises multiplied by three years' purchase and those in the new premises multiplied by one and a half years' purchase, the lower figure being justified by changes in the general circumstances of the business in the new premises.

Notwithstanding the emphasis given to the generally accepted approach of applying a multiplier to average profits, there will be cases where the profits are small or where the business has been in recession and shows small profits over recent years, or even losses, following earlier years of prosperity. A strict application of the accepted approach may produce a small amount for goodwill, or even nil. Depending on the facts of the case, this may be recognised as an inadequate figure – for example, if the years before acquisition were a bad period but the prospects for recovery are good. At its extreme, a loss-making business may nonetheless have a valuable goodwill. As a departure from the norm, the valuers may agree that it is more appropriate to adopt a different approach – such as discounting projected future profits[36] – or simply plumping for a spot figure or a high multiplier with no evidence to support it, termed as "adopting a robust approach" by the Lands Tribunal.[37] The aim is to arrive at what is seen to be and accepted as a fair figure of compensation.

It is common for the disturbance claim to be prepared some time after the claimant has moved from the acquired premises. If so, information can be obtained from the trading figures in the alternative premises which provide strong evidence as to the loss of profits suffered by the claimant in the period following the move. Regard will be had to such evidence, applying the *Bwllfa* principle of the weight to be given to evidence of events after the valuation date.[38]

To justify a claim based on removal of the business to other premises, claimants must show that they have acted reasonably and taken only such premises as a prudent person would take in order to safeguard their interests and mitigate loss. In *Appleby & Ireland Ltd* v *Hampshire County Council*[39] the Lands Tribunal held that the claimants should have moved to premises in the town where they were

35 [1959] 1 All ER 649.
36 *Reed Employment Ltd* v *London Transport Executive* [1978] 1 EGLR 166.
37 *W. Clibbert Ltd* v *Avon County Council* [1976] 1 EGLR 171.
38 *Bwllfa & Merthyr Dare Steam Collieries Ltd* v *Pontypridd Waterworks Co* [1903] AC 426.
39 [1978] 2 EGLR 180.

when acquired, and compensation was assessed on the notional move within the town rather than costs and losses incurred in the actual move to another town.

An acquiring body may sometimes offer alternative accommodation with a view to reducing a claim for trade disturbance.

Thus, if retail traders claim that their business will suffer complete loss by the taking of the premises in which it is carried on, it is open to the acquiring authority to indicate premises that can be acquired for the reinstatement of the trade, or even to offer to build premises on land available for the purpose, and to transfer the premises to the claimant, in reduction of their claim. Arrangements of this kind are a matter of agreement between the parties. No claimant can be forced to accept alternative accommodation, although it may be difficult to substantiate a heavy claim for trade disturbance in cases where it is offered.

In *Lindon Print Ltd* v *West Midlands County Council*,[40] the Lands Tribunal accepted that the onus of proof is on the acquiring authority to show that claimants have not mitigated their loss.

Worked example of a claim for disturbance on total extinguishment of a business

1. Goodwill

Market Traders Ltd is a company owned by Mr and Mrs Davies. The company is run by Mr Davies while Mrs Davies acts as company secretary.

The company holds a leasehold interest in a warehouse. The lease is for 80 years from 1945 at a fixed rent of £200 p.a. The current rental value is £10,000 p.a. The lease is being acquired compulsorily and, as no suitable alternative premises are available, the business will be extinguished. The profit and loss account for the past year is set out below. Plant and machinery, fixtures and fittings and stock have been valued by a trade valuer at £180,000.

Profit and loss account for past year

Turnover			1,189,284
Cost of sales			
Opening stock		139,869	
Purchases		944,632	
Closing stock		115,440	
			969,061
Gross profit			220,223
Less Overheads			
Salaries and wages	59,197		
Telephone charges	704		
Printing, postage and stationery	2,056		
Staff refreshments	558		
Directors' bonus	27,000		

40 [1987] 2 EGLR 200.

Directors' remuneration	54,670		
Directors' national insurance	8,330		
Directors' pension costs	1,355		
Heating and lighting	4,837		
Repairs and renewals	15,771		
Insurances	4,397		
Rent and rates	11,581		
Sundry expenses	221		
Cleaning	4,448		
Bank charges	2,560		
Bad debt provision	170		
Auditors' remuneration	3,750		
Accountancy fees	962		
Depreciation	20,545		223,112
Net trading loss for the year			2,889

Assess the adjusted net profit

Annual profit (loss) from accounts		2,889	
Add directors' remuneration		91,355	
			88,466

Deduct

Profit rent 10,000 – 200		9,800	
Reasonable remuneration or work done by			
directors (this is not a small business), say		36,000	
Interest on capital			
Current value of vehicles, P&M, F&F	180,000		
Cash in hand say, 2 weeks' purchases, say	40,000		
	220,000		
Interest at 4%	0.04	8,800	54,600
Adjusted net profit			33866

The net profits need to be assessed for each of the previous three years. Assume the net profits are:

Year 1: Net profit	28,576	
Year 2: Net profit	33,866	
Year 3: Net profit	36,811	
Average net profits		33,084
Profits have risen by 50% over 3 years, so strong business		
performance, say Years' Purchase (YP)		4
		132,366

2. Loss on forced sale

Value to incoming tenant of vehicles, F&F, P&M	180,000	
Price realised at auction (assume)	73,900	
Loss on forced sale	106,100	
Stock (assume 40% of £115,440)	46,176	£152,276

3. Redundancy payments

Payments to staff calculated on statutory basis (assume)		£37,000
Total claim up to this point		321,642

4. Miscellaneous

Depends on circumstances of each case but will include a
   wide range of items such as temporary loss of profits,
   disconnecting services and equipment, notifying
   customers of closure, etc.

5. Fees

(a) Reasonable legal fees on conveyance of lease
(b) Reasonable surveyors' fees

4. Disturbance in relation to development value

In *Horn* v *Sunderland Corporation*,[41] Lord Greene MR said that rule 6:

> does not confer a right to claim compensation for disturbance. It merely leaves
> unaffected the right which the owner would, before the Act of 1919, have had
> in a proper case to claim that the compensation to be paid for the land should
> be increased on the ground that he has been disturbed.

In the same case the right was described as "the right to receive a money payment
not less than the loss imposed on him in the public interest, but on the other hand
no greater".

It is true that, in practice, the value of the land itself will be assessed in accordance
with rule 2 and another figure for disturbance may be arrived at under rule 6. But
these two figures are, in fact, merely the elements which go to build up the single
total figure of price or compensation to which the owner is fairly entitled in all the
circumstances of the case, and which should represent the loss the owner suffers
in consequence of the land being taken.

It follows that an owner cannot claim a figure for disturbance which is inconsistent with the basis adopted for the assessment of the value of the land under
rule 2.

For instance, in *Mizzen Bros* v *Mitcham UDC* (1929)[42] it was held that claimants
were not entitled to combine in the same claim a valuation of the land on the
basis of an immediate sale for building purposes and a claim for disturbance and
consequential damage upon the footing of interference with a continuing market
garden business, since they could not realise the building value of the land in the
open market unless they were themselves prepared to abandon their market garden
business. The existing-use value of the land as a market garden was about £12,000
and the disturbance items totalled £4,640. These items were aggregated but still
fell short of the development value of the land for building, which was £17,280.
The claimants could not justify any claim by way of disturbance, even though they

---

41 [1941] 2 KB 26; 1 All ER 480.
42 *Estates Gazette Digest*, 1929, p. 258; reprinted in *Estates Gazette*, 25 January 1941, p. 102.

were in fact being disturbed, because these losses would be incurred in any event if the land were developed. On the other hand, where a valuation on the basis of the present use of the land, plus compensation for disturbance, may exceed a valuation of the land based on a new and more profitable user, the owner has been held entitled to claim the former figure, though not both.

The facts in *Horn v Sunderland Corporation* were that land having prospective building value and containing deposits of sand, limestone and gravel was compulsorily acquired for housing purposes. The owner occupied the land as farm land, chiefly for the rearing of pedigree horses.

The owner claimed the market value of the land as a building estate ripe for immediate development and also a substantial sum in respect of the disturbance of his farming business.

The official arbitrator awarded a sum of £22,700 in respect of the value of the land as building land, but disallowed any compensation in respect of disturbance of the claimant's business on the grounds that the sum assessed could not be realised in the open market unless vacant possession were given to the purchaser for the purpose of building development.

It was held by a majority of the Court of Appeal that the arbitrator's award was right in law provided that the sum of £22,700 equalled or exceeded:

(i) the value of the land as farm land; plus
(ii) whatever value should be attributable to the minerals if the land were treated as farm land; plus
(iii) the loss by disturbance of the farming business.

If, however, the aggregate of items (i), (ii) and (iii) exceeded the figure of £22,700, the claimant was entitled to be paid the excess as part of the compensation for the loss of his land.

Claims for disturbance must relate to losses which are the direct result of the compulsory taking of the land and which are not remote or purely speculative in character. Loss of profits in connection with a business carried on the premises and which will be directly injured by the dispossession of the owner is a permissible subject of a claim. But where a speculator claimed, in addition to the market value of building land, the profits that the speculator hoped to make from the erection of houses on the land, the latter item was disallowed.[43] It was quite distinct from the development value of the land, which was a legitimate item of claim on the basis just described. In any event this would be "double counting" – the developer could either sell the land at development value or carry out the development and make a profit: he could not do both. The point at issue was succinctly put by Lord Moulton in a decision of the Privy Council when he said that "no man would pay

---

43 *Collins v Feltham UDC* [1937] 4 All ER 189; and see *D McEwing & Sons Ltd v Renfrew County Council* (1959) 11 P&CR 306.

for the land, in addition to its market value, the capitalised value of the savings and additional profits which he would hope to make by the use of it."[44] Again, where an owner of the business premises compulsorily acquired was also the principal shareholder in the company that occupied the premises on a short-term tenancy, a claim in respect of the depreciation in the value of the owner's shares which might result if the acquiring body were to give the company notice to quit was held to be "too remote" for compensation.[45]

44 *Pastoral Finance Association Ltd* v *The Minister (New South Wales)* [1914] AC 1083.
45 *Roberts* v *Coventry Corporation* [1947] 1 All ER 308 (the words of Lord Goddard CJ).

Chapter 26

Blight notices

1. Compulsory purchase instigated by land owners: blight notices

Planning *proposals* which will eventually involve the compulsory acquisition of land may very well depreciate the value of a property or even make it virtually unsaleable. Such depreciation will be ignored, under section 9 of the Land Compensation Act 1961, in assessing the compensation when the property is actually acquired. But this will not help the owner to sell the property in the open market in the meantime. For instance, the owner-occupier of a house affected by such proposals may, for personal reasons, be obliged to move elsewhere and may suffer considerable hardship in having to sell at a very much reduced price.

In partial mitigation of such cases, a strictly limited class of owner-occupiers is given power to compel the acquisition of their interests in land, provided that the claimants in each case can prove that they have made reasonable efforts to sell their property but have been unable to do so except at a price substantially lower than they might reasonably have expected to obtain but for the threat of compulsory acquisition *inherent in the proposals*. The power is derived from sections 149–171 of the Town and Country Planning Act 1990.[1] It is a common misconception that a blight notice can be used to force an acquiring authority to compulsorily purchase land which the owner considers will be adversely affected by a scheme due to its proximity. In fact, a blight notice can only be used where there are plans to compulsorily purchase the land in the future, and the effect of a successful notice is merely to bring forward the time of that compulsory purchase.

(a) Blighted land

The first requirement that the claimant must meet (1990 Act, section 149) is to show that the land in question comes within one of the categories set out in Schedule 13 to the 1990 Act and is therefore "blighted land". Schedule 13 lists 23 kinds

---

1 Part VI, Chapter II; also the Planning Act 2008, section 175 (see Chapter 14).

of situation in which official proposals have reached a stage that points to the eventual acquisition of the claimant's land by some public authority, referred to as the "appropriate authority".[2] It is important to be aware that even before a scheme reaches the stage of falling within one of these categories, the saleability of a property required for the scheme may be severely affected. Also, it is quite possible for a property to be thoroughly blighted in the real world, and yet not fall within any of the legal definitions of "blight". For example, In *Halliday* v *Secretary of State for Transport*[3] it was held that a consultative document reviewing policy options is not a "plan".

An "agricultural unit" may be only partially "blighted". The part blighted is the "affected area" and the remainder is the "unaffected area". In cases of this sort, owner-occupiers can serve a blight notice which extends to the unaffected area as well as to the affected area if they can show that the unaffected area as such is not capable of being reasonably farmed even with any other available land which the claimant can add to it (1990 Act, section 158).

(b) Qualifying interests

The second requirement that claimants must meet, under section 149(2) of the 1990 Act, is to show that they have a "qualifying interest" in the land. They must be either the owner-occupier or the mortgagee[4] (with a power of sale currently exercisable) or the personal representatives[5] of the owner-occupier, who is:

- an individual "resident owner-occupier" of the whole or part of a dwelling-house, or
- the "owner-occupier" (person or partnership) of the whole or part of any other hereditament of which the net annual value for rating does not exceed a certain prescribed limit (at present £34,800 but subject to regular review)[6], or
- the "owner-occupier" (person or partnership) of the whole or part of an agricultural unit.

The term "owner-occupier" is defined in some detail in section 168 of the 1990 Act. The claimant must have an "owner's interest" in the land, i.e. the freehold or a lease with at least three years to run, and must have been in physical occupation of the property for at least six months up to the date of service of the blight notice, or up to a date not more than 12 months before such service, provided that (in such

---

2 Town and Country Planning Act 1990, section 169.
3 (2003) RVR 12.
4 1990 Act, section 162. The mortgagee cannot serve a blight notice if the mortgagor has already served one which is still under consideration (and the converse is also true).
5 1990 Act, section 161.
6 Town and Country Planning (Blight Provisions) Order 2010/498 (England) and Town and Country Planning (Blight Provisions) (Wales) Order 2011.

a case) the property (unless it is a farm) was unoccupied in the intervening period. For an owner-occupier's mortgagee, who is entitled to serve a blight notice, the relevant period is extended by six months for the mortgagee's particular benefit.

2. Blight notice procedure

The notice to purchase the owner-occupier's qualifying interest must, under section 150 of the 1990 Act, relate to the whole of that interest in the land, be in the prescribed form,[7] and be served on the "appropriate authority" (that is, the authority which, it appears from the proposals, will be acquiring the land).

Within two months of the receipt of the claimant's notice, the appropriate authority may, under section 151, serve a counternotice of objection, in the prescribed form,[8] on any of the following grounds:

(a) that no part of the hereditament or agricultural unit is blighted land;
(b) that they do not intend to acquire any part of the hereditament, or of the affected area of the agricultural unit, affected by the proposals;
(c) that they intend to acquire a part of the hereditament or affected area but do not intend to acquire any other part;
(d) that, in the case of blighted land within paragraphs (1), (3) or (13) of Schedule 13, but not paragraphs (14), (15) or (16), they do not intend to acquire any part of the hereditament or the affected area within 15 years[9] from the date of the counternotice;
(e) that the claimant is not currently entitled to an interest in any part of the hereditament or agricultural unit;
(f) that the claimant's interest is not a "qualifying interest"; and
(g) that the claimant has not made "reasonable endeavours" to sell (except when the land is included in a CPO and so falls within paragraphs (21) or (22) of Schedule 13); or that the price at which it could be sold is *not* "substantially lower" than that which the claimant might reasonably have expected to get but for the planning proposals.[10]

Within two months of the counternotice, the claimant may, under section 153 of the 1990 Act, require the authority's objection to be referred to the Lands Tribunal, who may uphold the objection or declare that the claimant's original notice is a valid one. If, in case (c), the authority's objection is upheld, or is accepted by the claimant, the authority will, of course, have to purchase the part referred to in the counternotice as land it intends to acquire.

---

7 Town and Country Planning General Regulations 1992 (SI 1992/1492).
8 1990 Act, section 151(1).
9 They can specify a longer period.
10 1990 Act, section 150(1).

If the authority objects on grounds (a), (e), (f) or (g) above, the claimant must satisfy the Lands Tribunal that its objection is "not well-founded". In other words, the burden of proof lies on claimants to show that the land is "blighted land",[11] that their interest is a "qualifying interest", and that they have made reasonable but unsuccessful attempts to sell at a fair market price.

However, if the authority objects on grounds (b), (c) or (d), it is for the authority to satisfy the Lands Tribunal that its objection is "well-founded". In other words, the burden of proof lies on the authority to show that it has no intention to acquire the land (or part of the land) in question, within 15 years or at all. In *Duke of Wellington Social Club and Institute Ltd* v *Blyth Borough Council*,[12] the Lands Tribunal held that an objection that the authority does not intend to acquire can be held to be "not well-founded" if there would be hardship on the claimant as a result; but in *Mancini* v *Coventry City Council*[13] the Court of Appeal disapproved the assertion that hardship could be relevant to an authority's intention not to acquire.

Section 159 of the 1990 Act empowers the authority serving a counternotice in response to a blight notice relating to a farm which includes an unaffected area (i.e. when part of the farm is not blighted land) to challenge the assertion that the unaffected area is incapable of being reasonably farmed on its own (or with any other available land added to it). Indeed, the authority must challenge that assertion if it is claiming an intention to take part only of the affected area (i.e. the blighted part of the farm). If in either circumstance the Lands Tribunal upholds the counternotice it must make clear to which (if any) land the blight notice will apply, and specify a date for the deemed notice to treat.

Where no counternotice is served, or where the claimant's notice is upheld by the Lands Tribunal, the appropriate authority will be deemed under section 154 of the 1990 Act to be authorised to acquire the claimant's interest in the hereditament, or (in the case of an agricultural unit) in that part of the unit to which the proposals that gave rise to the claim relate. But if the authority succeeds in an objection that it should take only part of a claimant's property, section 166 of the Act preserves the right to compel it to take all or none in accordance with section 8 of the Compulsory Purchase Act 1965, on proof that otherwise the property, if it is a "house, building or manufactory", will suffer "material detriment", or that, if it is "a park or garden belonging to a house", the "amenity or convenience" of the house will be seriously affected.

Notice to treat will be deemed to have been served at the expiration of two months from the service of the claimant's notice or, where an objection has been made, on a date fixed by the Lands Tribunal (section 154 (2), (3)).

---

11 In *Bolton Corporation* v *Owen* [1962] 1 QB 470, the Court of Appeal held that statements in a development plan which "zone" the claimant's land in general terms for residential use are not sufficient to justify a finding that it is required for the "functions of a local authority", even if the likelihood is that it may be redeveloped for social housing and not private housing.
12 (1964) 15 P&CR 212.
13 (1982) 44 P&CR 114.

The general basis of compensation, where an authority is obliged to purchase an owner-occupier's interest under the 1990 Act, is that prescribed by the Land Compensation Act 1961. Thus, in principle, the market value figure that claimants have tried in vain to secure by their own efforts in a private sale has now become payable to them by the "appropriate authority", together with, as appropriate, a Home Loss, Basic Loss and Occupier's Loss payment, severance and injurious affection and disturbance.

Except where the authority has already entered and taken possession under the deemed notice to treat, a party which has served a blight notice may under section 156 withdraw it at any time before compensation has been determined, or within six weeks of its determination. In this case any notice to treat deemed to have been served will be deemed to have been withdrawn.

Chapter 27

Purchase notices

1. Compulsory purchase instigated by landowners

The provisions of Part VI of the Town and Country Planning Act 1990 provide a remedy for landowners who suffer from adverse planning decisions or proposals. This involves procedures under which local authorities and other public bodies are, in certain circumstances, compelled by a purchase notice to buy land affected by such decisions or proposals.

Where a purchase notice takes effect, the land is transferred and not retained, and the recipient of the adverse planning decision is compensated, not directly in the form of planning compensation, but indirectly in the form of compensation for compulsory purchase. This does not in fact prevent claimants from receiving lost development value in terms of money, just as they do in planning compensation for revocation, modification and discontinuance orders, but limits that compensation to the maximum the principles of compulsory purchase compensation allow.

2. Purchase notices

(Sections 137–48 of the Town and Country Planning Act 1990, i.e. Part VI, Chapter I of the Act).

A purchase notice applies where planning permission is refused, either by the local planning authority or by the Secretary of State, or is granted subject to conditions, and the owners of the land can show that it has become incapable of reasonably beneficial use:

- in its existing state; or
- even if developed in accordance with such conditions (if any); or
- even if developed in any other way for which permission has been, or is deemed to be, granted under the Act, or for which the Secretary of State or the local planning authority has undertaken to grant permission.

"Owner" means the person entitled to receive the rack rent of the land, or who would be entitled to receive it if the land were so let – section 336(1) (see *London Corporation* v *Cusack-Smith* [1955] AC 337).

The owner may, under the Town and Country Planning General Regulations, within 12 months of the planning decision in question, serve a purchase notice on the district council or (as the case may be) the London Borough Council or the Common Council of the City of London, depending on where the land is situated, requiring them to purchase their interest in the land.

The council on whom a purchase notice is served shall, within three months of such service, serve a counternotice on the owner to the effect that:

(a) it is willing to comply with the purchase notice; or
(b) another specified local authority, or statutory undertaker, has agreed to comply with it in the council's place; or
(c) for reasons specified in the notice, neither the council nor any other local authority or statutory undertaker is willing to comply with the purchase notice, and that a copy of the purchase notice has been sent to the Secretary of State on a date specified in the counternotice, together with a statement of the reasons for their refusal to purchase.

In cases (a) and (b) above, the council on whom the purchase notice was served, or the specified local authority or statutory undertaker, as the case may be, will be deemed to be authorised to acquire the owner's interest in the land and to have served a notice to treat on the date of the service of the counternotice, as if the land were being acquired under a compulsory purchase order.

In case (c), where the purchase notice is forwarded to the Secretary of State, the following courses of action are open to him:

• to confirm the notice, if satisfied that the land is in fact incapable of reasonably beneficial use in the circumstances specified in section 137; and
• to confirm the notice but to substitute another local authority or statutory undertaker.

Even in these cases, the Secretary of State is not obliged to confirm the notice in certain special circumstances relating to land forming part of a larger area which has a restricted amenity use by virtue of a previous planning permission (Town and Country Planning Act 1990, section 142). Thus, if planning permission is given to build houses on a field, subject to a condition that for amenity reasons trees are to be planted along the road frontage, the developers cannot, after building and selling the houses, claim that the road frontage land which they retain is "incapable of reasonably beneficial use" so as to justify serving a purchase notice.

Other options open to the Secretary of State are:

• not to confirm the notice but to grant permission for the required development, or to revoke or amend any conditions imposed;

- not to confirm the notice but to direct that permission shall be given for some other form of development; and
- to refuse to confirm the notice, or to take any action, if satisfied that the land has not in fact been rendered incapable of reasonably beneficial use.

"Reasonably beneficial use" is to be judged in relation to the present use of the land, but also having regard to any alternative use to which it can reasonably be put without planning permission. In *Hudscott Estates (East) Ltd v SoS* [2000] 2 PLR 11 it was decided that the Secretary of State was entitled to conclude that land was capable of reasonably beneficial use in its existing state as it had a potential for use for grazing with other land, following the carrying out of works at reasonable expense.

In considering whether or not the use of any land is "reasonably beneficial", the Secretary of State can take account of a use not related to development.[1] In *Colley v Secretary of State for the Environment, Transport and the Regions*,[2] the Court of Appeal accepted a submission by the local planning authority and the Secretary of State that a woodland site subject to a tree preservation order had a "reasonably beneficial" use *as a public amenity*. In principle, this may well be valid; yet on the facts the small area of woodland was an untended tangle of briars, brushwood and nondescript trees and shrubs. It needs to be added that there were special circumstances in the Colley case which would have entitled the claimant to unreasonably high compensation if the notice had been accepted. This precedent may not, therefore, be followed in future cases.

"Incapable of reasonably beneficial use" signifies that the land in question must be virtually useless and not merely less useful, because practically any planning restriction will result in land being less useful (*R v Minister of Housing and Local Government, ex parte Chichester Rural District Council*).[3]

Before taking any of the steps enumerated above, including refusal to confirm, the Secretary of State must give notice of his proposed action to the person who served the purchase notice, the local authority on whom it was served, the local planning authority and any other local authority or statutory undertaker who may have been substituted for the authority on whom the notice was served. He must also give such person or bodies an opportunity of a hearing if they so desire. If after such hearing it appears to the Secretary of State to be expedient to take some action under the section other than that specified in his notice, he may do so without any further hearing.

If the Secretary of State confirms the purchase notice, or has not taken any action within nine months from the date when it was served on the local authority,

---

1 As amended by the Planning and Compensation Act 1991, section 31 and Schedule 6.
2 (1999) 77 P&CR 190.
3 [1960] 2 All ER 407.

or within six months of its transmission to him by the local authority (whichever period is the shorter), the local authority on whom it was served will be deemed to be authorised to acquire the land compulsorily. The authority is then deemed to have served a notice to treat on such date as the Secretary of State may direct, or at the end of the period of six months, as the case may be. The notice to treat is deemed to be authorised as if the acquiring authority required the land "for planning purposes" under Part IX of the 1990 Act.

Any party aggrieved by the decision of the Secretary of State on the purchase notice may, within six weeks, make an application to the High Court to quash the decision on the grounds that either (i) the decision is not within the powers of the Town and Country Planning Act 1990, or (ii) the interests of the applicant have been substantially prejudiced by a failure to comply with the relevant requirements (1990 Act, Part XII, section 288), but cannot otherwise challenge the validity of his decision (section 284).

If the Secretary of State's decision is quashed, the purchase notice is treated as cancelled; but the owner may serve a further purchase notice in its place as if the date of the High Court's decision to quash were the date of the adverse decision (section 143).

(a) Compensation as in compulsory purchase cases

Where a purchase notice takes effect, i.e. the authority is deemed to have served a notice to treat, the compensation payable to the owner will be assessed, as in any case of compulsory acquisition, under the provisions of the Land Compensation Act 1961 (as described in Chapters 23–25).

Where the Secretary of State does not confirm the purchase notice but directs instead that permission should be given for some alternative development, then if the value of the land with permission for such alternative development is less than its value with permission for development within Schedule 3 to the 1990 Act, the owner is entitled to compensation equal to the difference between the two values (section 144). This is subject to any direction by the Secretary of State excluding compensation in respect of any conditions as to design, external appearance, size or height of buildings, or number of buildings to the acre.

The obvious question is: what is the benefit of having a purchase notice accepted and the land compulsorily purchased, given the prerequisite that the subject land must be incapable of reasonable beneficial use? The answer lies in the assumptions which apply once the purchase notice becomes a CPO. If, for example, the reason planning permission is refused is that it is affected by road building proposals, then the road scheme will be left out of account in assessing the compensation payable. This is a useful option for a landowner whose property is blighted by such a road scheme but who does not qualify to serve a blight notice because he does not meet the qualifying rules, e.g. he is not in occupation of the land.

(b) Revocation, modification and discontinuance orders and other special cases

Purchase notices may also be served when land has been rendered "incapable of reasonably beneficial use" in consequence of a revocation or modification or discontinuance order (section 144 of the 1990 Act), or a refusal (or restrictive grant) or revocation or modification of a "listed building" consent (sections 32–36 of the Planning (Listed Buildings and Conservation Areas) Act 1990). In theory they may be made available under advertisement regulations or tree preservation orders, but only if appropriate provisions were to be included in the regulations issued, which is extremely unlikely.

Chapter 28

Compensation under the Town and Country Planning Acts

Revocation, modification and discontinuance orders, etc.

Until 1991, the law on this subject was in its essentials a direct survival from the immediate post-war period when the Town and Country Planning Act 1947 was passed. The Planning and Compensation Act 1991 swept away most of the post-war system of compensation for planning restrictions – one of the two pillars of the "compensation/betterment" question – except for cases of revocation and discontinuance (which occur but rarely in practice). There is now no trace remaining of the concept which emerged with the Uthwatt Report during the Second World War, namely that some kind of national balance should be achieved between (i) the proportion of prospective development value which should be transferred to the state and not left to landowners in the event of a grant of planning permission, and (ii) the proportion of that value which should be conceded to landowners as compensation in the event of a refusal of planning permission.

1. Revocation, modification and discontinuance cases

The survival of planning compensation in revocation, modification and discontinuance cases represents something quite different, which is the loss not of *prospective* development value but of *authorised* development value. In revocation cases (though not discontinuance cases) this distinction is probably more apparent than real; but it can be traced back to the Town and Country Planning Act 1954 which gave effect to the then government's belief that some planning compensation should be available within modest limits.

The real practical distinction is that local planning authorities cannot avoid making decisions on a vast number of planning applications each year, many of which must inevitably be refused (or granted subject to burdensome conditions), but they can certainly avoid making more than a very few decisions to revoke planning permissions already granted (though the Secretary of State can do it for them and saddle them with the consequences). The initiative lies with them in revocation cases, but most certainly not in cases of refusal of a planning application. As a revocation involves removal of a landowner's legal rights, this continues to involve a heavy burden of compensation.

2. Revocation and modification orders

Under sections 97–100 of the Town and Country Planning Act 1990 a local planning authority may (by means of an order which must be confirmed by the Secretary
of State if it is opposed by the owners or occupiers of the land in question or anyone
else with any property interest in the land, but does not require such confirmation in
other cases) revoke or modify a planning permission already given. A revocation
or modification order can only be made (a) before building or other operations
authorised by the permission have been completed, in which case work already
done will not be affected by the order, or (b) before any change of use authorised
by the permission has taken place.

If the proposed order is unopposed, the local planning authority need not submit
it to the Secretary of State for confirmation. Alternatively, the Secretary of State
may himself make a revocation order, provided that he first consults the local
planning authority.

Not less than 28 days' notice of the proposed order must be given to the owner
and occupier of the land affected and to any other person who, in the opinion of
the local planning authority, will be affected by the order, to enable them to apply
to be heard by a person appointed by the Secretary of State.

When the order takes effect, section 107 of the 1990 Act provides for the payment
by the local planning authority of compensation under the following heads:

- Expenditure on work which is rendered abortive by the revocation or modification, including the cost of preparing plans in connection with such work or
  other similar matters preparatory thereto. But no compensation will be paid in
  respect of work done before the grant of the relevant planning permission, e.g.
  in anticipation of such permission being granted. There is a clear contradiction
  here in that work done before the grant of planning permission is excluded,
  but the cost of preparing plans, which will be needed for the application, is
  specifically allowed. This can be interpreted as meaning that no costs incurred
  prior to the grant of planning permission can be claimed *except* for the cost
  of preparing plans and related matters, such as the planning application fee.
- Any other loss or damage directly attributable to the revocation or modification, which includes depreciation of the land value in consequence of the total
  or partial revocation of planning permission.

Claims for compensation must be served (in writing) on the local planning
authority within 12 months of the revocation or modification order, or of the
adverse decision under section 108 (below).[1] In accordance with the provisions
of section 117 of the 1990 Act, compensation for depreciation in the value of
an interest in land will be calculated on a market value basis as prescribed by

---

1 Town and Country Planning General Regulations 1992 (SI 1992/1492), Regulation 12.

section 5 of the Land Compensation Act 1961. The measure of compensation will be the amount by which the value of the claimant's interest with the benefit of the planning permission exceeded the value of that interest with the planning permission revoked or modified.

In calculating such depreciation, it is to be assumed that planning permission would be granted for development within paragraphs 1 and 2 of Schedule 3 of the 1990 Act. These comprise development that would involve:

- *rebuilding* any building built on or before 1 July 1948 which was demolished after 7 January 1937, or any more recently built building "in existence at a material date", provided that the cubic content "is not substantially exceeded"; and
- *conversion* of a single dwelling into two or more dwellings.

The gross floor space to be used for a given purpose must not be increased by more than 10% in the rebuilt building, if the previous building was original to the site or built before 1 July 1948, and must not be increased at all in other cases (1990 Act, Schedule 10).

Thus, if planning permission for development is revoked or modified, compensation must be assessed on the assumption that development value within Schedule 3 is enjoyed by the claimant, which is most unlikely to be the case in the real world. In the absence of compulsory purchase (by purchase notice or otherwise), there can be no redress. The purpose of Schedule 3 is of historical interest only and, until its well overdue repeal, it will continue to frustrate the goal of fair compensation.

The problems of Schedule 3 are illustrated by the case of *Canterbury City Council* v *Colley*,[2] in which the House of Lords dealt with the question of compensation for revocation. The facts were that in 1961 planning permission had been granted to build a house on a plot of land near Whitstable in Kent. The foundations were laid, so the planning permission was part implemented and did not lapse through passage of time, but the house was never built. The planning permission was revoked in 1987, depriving the owners of the valuable right to build a house on the site. They therefore claimed compensation under what is now section 107 of the 1990 Act. The existing-use value of the vacant site was then £8,250. Full development value with the benefit of the 1961 planning permission (had it not been revoked) was £115,000. The difference caused by revocation was the measure of loss of land value, i.e. £106,750. However, on the artificial Schedule 3 assumption that the site must be treated as having the benefit of planning permission to rebuild "any building" which had stood on the site in 1947, the value of the site was £70,000, reducing the depreciation from £106,750 to £45,000. This result is clearly unfair, as no such permission had been granted, and in reality all development potential had been lost. However, the statutory obligation to assume

Schedule 3 development is clear. The House of Lords held that on the plain meaning of the statutory wording the compensation must be restricted to £45,000, pointing out that however absurd the result, there is absolutely nothing to suggest that the application of the Schedule 3 planning assumption is discretionary. The Localism Act 2011 removed Schedule 3 as an assumption to be made when assessing compensation for compulsory purchase. It's removal as an assumption in planning compensation, or its complete repeal, may not be long in coming.

Section 108 of the 1990 Act applies the above provisions as to compensation where planning permission given by a development order is withdrawn, whether by revoking or amending the order or by making a direction under it.[3] Compensation becomes payable when, on an express planning application being made, permission is refused or is granted subject to conditions other than those imposed by the development order. In cases where the development order is itself revoked or modified, the express planning application must be made within a year thereafter.[4] However, this time limit was avoided in *Green* v *City of Durham* 2007 (Lands Tribunal ref. LCA/170/2006) by the simple expedient of making a second identical planning application and appealing against the inevitable second refusal.

Disputes as to the compensation payable under sections 107 and 108 are referable to the Lands Tribunal (section 118). Special provisions are laid down for compensation in cases of revocation or modification of planning permission for mineral development, whether under section 97 of the 1990 Act or by way of various mineral orders under Schedule 9 of the Act (section 116).

Where revocation or modification of planning permission renders land "incapable of reasonable beneficial use", it may entitle the owner to serve a purchase notice under Part VI of the Act.[5]

Provision is made[6] for the apportionment of compensation for depreciation in value exceeding £20 between different parts of the land, where practicable, and for the registration of the "compensation notice" as a land charge. Compensation so registered will be repayable in whole or in part if permission is subsequently given for development that consists (wholly or largely) of the construction of residential, commercial or industrial buildings or any combination thereof, or of any other activity to which the Secretary of State believes these provisions should apply by reason of the likely development value. This applies whether the development occurs on the whole or only part of the land affected by the order, but not where the registered compensation was paid in respect of a local planning authority's order modifying planning permission previously given, and the subsequent development is in accordance with that permission as modified.[7]

---

3 For example, a direction made by the Secretary of State or the local planning authority under article 4 of the General Permitted Development Order 1995.
4 Town and Country Planning Act 1990, section 108(2).
5 Section 137(1)(b). See Chapter 27.
6 Sections 109–110.
7 Sections 111–112.

Worked example of claim for compensation for revocation of planning permission

Developers have acquired and demolished an office building with a prominent main road position. The offices had stood on the site since 1935 and the site extends to around 2,500 m². The developers have been granted planning consent to develop the site as a petrol filling station, subject to a condition requiring improvement of the access to the site. Terms have been provisionally agreed for the letting of the completed development to a major oil company at a rental of £85,000 p.a. Similar petrol station investments currently sell at prices which reflect a yield of 7%.

Prior to the grant of planning consent the developers incurred the following costs:

a)	planning application fees	£15,000
b)	architect's fees for preparing plans	£3,000
c)	demolition	£11,000

Since obtaining the planning consent they have already spent £18,000 on constructing foundations.

Due to a change in policy, the local planning authority has recently revoked the planning permission, and has indicated that it would allow industrial development of the site.

Following service of the revocation order the developers lodged an appeal and completed the steel framework for the building at a cost of £30,000. However, their appeal has been unsuccessful.

The cost to complete construction of the building is estimated to be around £260,000.

Industrial development land sells in the area for around £400,000 per ha. Office development land sells for around £1,200,000 per ha.

Making references to the appropriate legal provisions, advise the developers as to their likely entitlement to compensation.

Under s.107 of the Town and Country Planning Act 1990, compensation is payable for losses arising from a revocation order made under s.97 of the Act.

Compensation will include depreciation in the value of the land, together with abortive costs incurred subsequent to the grant of consent and prior to the revocation. The exception is the costs of preparatory plans, including the application fee, which are payable regardless of when they are incurred. Costs incurred after the revocation order cannot be claimed as any such works are unauthorised.

Depreciation can be calculated on a before-and-after basis. In this example a much simplified version of a residual valuation is used for the "before" valuation – in reality a much more detailed residual valuation would be required.

Before value (this is a much simplified version of the residual valuation which would be required in practice)

Value of completed development			
Rental value	£85,000		
YP in perp. @ 7%		14.29	
		£1,214,650	
Less			
Costs to complete	£260,000		
Plus other costs – finance, fees, etc.	£100,000		
Developers profit @ 15% of gross development value	£182,200		
Total costs		£542,200	
Site value			£672,450

After value

Take higher of industrial value or office value reflecting the planning assumptions, including Schedule 3 development (though only if the assumed planning permission would have a real market value, bearing in mind that the building would have to have a similar footprint to the original).

Industrial – 0.25 ha @ £400,000	£100,000	
Office – 0.25 ha @ £1,200,000	£300,000	
Take after value of		£300,000
Loss incurred in the form of depreciation		£372,450

Abortive costs

Allow only costs incurred following planning approval and prior to revocation, plus planning fee and costs of preparatory plans.

Planning fee	£15,000	
Preparatory plans	£3,000	
Foundations	£18,000	
Total abortive costs		£36,000
Total compensation claim		

Disallow any of the costs which are of value to the alternative development, as such costs are not abortive.

However, note that in fact none of the abortive costs should be allowed. Because of the way the depreciation has been calculated, to do so would be double counting. If this money had not been spent, then the cost of completing the development would be higher, and therefore the residual site value would be lower.

Therefore total claim   £372,450

3. Discontinuance orders

Under sections 102–4 of the Act of 1990, a local planning authority may, by means of an order confirmed by the Secretary of State, at any time require the discontinuance or modification of an *authorised* use of land (as in *Blow* v *Norfolk County Council*,[8] in which the authorised use of land as a caravan site was required to be discontinued) or the alteration or removal of authorised buildings or works.[9]

Section 115 provides that any landowner who suffers loss in consequence of a discontinuance order, either through depreciation in the value of the land, or by disturbance, or by expense incurred in complying with the order, is entitled to compensation provided the claim is made within 12 months.[10] Interest at a prescribed statutory rate is payable on the compensation from the date that the discontinuance takes effect.[11] As an alternative to claiming compensation for depreciation in the value of the land, if the discontinuance order renders the land "incapable of reasonably beneficial use", the owner of the interest may be able to serve a purchase notice under Part VI of the Act.[12] Special provisions apply to certain discontinuance orders relating to mineral development (see section 116).

4. Other cases of planning compensation

Compensation on account of planning requirements may similarly be payable in connection with the following matters under the Town and Country Planning Act 1990.

* Refusals, or restrictive grants of consents; or directions for replanting trees *without financial assistance from the Forestry Commission* under tree preservation orders (sections 203–4).
* Cost of removal of advertisements in certain special and limited circumstances (section 223).
* Stop notices (section 186).
* Pedestrian precinct orders (section 249 in Part X).

Compensation is payable in these cases, as in those mentioned earlier in this chapter, by the local planning authority to whom application has to be made within

---

8 [1966] 3 All ER 579.
9 Where, on the other hand, uses or works are "unauthorised", i.e. have been begun or carried out without a grant of planning permission or contrary to conditions attached to such permission, the local planning authority will proceed by means of an "enforcement notice" under Part VII of the Town and Country Planning Act 1990, so that no compensation is payable. On this, see Chapter 14.
10 Town and Country Planning General Regulations 1992 (SI 1992/1492), Regulation 12.
11 Planning and Compensation Act 1991, Schedule 18, Part I.
12 Section 137(1)(c).

six months of the official decision giving rise to the claim (12 months in the case of stop notices and tree preservation orders).[13]

The making of pedestrian precinct orders (in so far as they cause properties to lose vehicular access to highways) entitles owners to compensation for depreciation in value of the properties affected. Stop notices are used by planning authorities to put an immediate stop to uses or development they consider to be in breach of planning permission. They can only be used in conjunction with an enforcement notice which otherwise allows the breach to continue while any appeal is considered. Should the enforcement notice be withdrawn or quashed on appeal, persons affected by the stop notice are entitled to compensation for resultant "loss or damage". Compensation is also payable for restrictions imposed on applications for consents under tree preservation orders, except where the refusal, or restrictive grant, of a consent under a tree preservation order is accompanied by a certificate (not quashed on appeal) that specific trees are to be preserved "in the interests of good forestry" or by reason of their "outstanding or special amenity value". In *Bell* v *Canterbury City Council* (1988),[14] the Court of Appeal held that "loss or damage" in relation to the refusal of consent to remove protected trees included not only the value of the timber but also the reduction in the existing-use value of the land on which the trees stood. Compensation in respect of advertisements is payable only "in respect of any expenses reasonably incurred" in carrying out works (i) to remove advertisements, or (ii) to discontinue use of any site for advertising, dating back (in either case) to 1 August 1948.

5. Listed buildings and ancient monuments

In the case of listed buildings of special architectural or historic interest, section 28 of the Planning (Listed Buildings and Conservation Areas) Act 1990 provides for payment of compensation for orders revoking or modifying listed building consents (except for unopposed orders) on the same basis as revocation or modification order compensation under the Town and Country Planning Act 1990, section 107, described above.

Section 29 authorises payment of compensation for loss caused by the issue of a building preservation notice, on a similar basis to compensation for stop notices (above). Claims must be made within six months of the making of the order.[15]

Part I of the Ancient Monuments and Archaeological Areas Act 1979 empowers the Secretary of State to make a "schedule of monuments" (rather on the lines of

---

13 Town and Country Planning General Regulations 1992 (SI 1992/1492), Regulations 12 and 14; Town and Country Planning (Control of Advertisements) Regulations 1992 (SI 1992/666), Regulation 20; Town and Country Planning (Tree Preservation Orders) Regulations 1969 (SI 1969/17), Model Order.
14 [1988] 1 EGLR 205.
15 Planning (Listed Buildings and Conservation Areas) Regulations 1990 (SI 1990/1519), Regulation 9.

listed buildings) in order to protect ancient monuments. Consents are then required for carrying out works affecting ancient monuments, and the Secretary of State must pay compensation to anyone who "incurs expenditure or otherwise sustains any loss or damage" in consequence of a refusal, or conditional or short-term grant, of consent by him (sections 7–9).

Index

Landlord and Tenant Act (1927) 296–7
Landlord and Tenant Act (1954) 297–305
landlords' rights 278–9
lease terms: and rent 53–5; for retail
premises 324
leasehold interests: capital gains tax on sale
of 398; with income below market rate
119; valuation of 109–20, 168–80
Leasehold Reform Act (1967) 269–70,
294–5
Leasehold Reform, Housing and Urban
Development Act (1993) 270, 276,
279–80
leaseholds 18, 20–2; *see also* building
leases; occupation leases
leases: extension of 277–8; for life 24;
stamp duty payable on 417; surrenders
and renewals of 128–33; *see also*
hypothetical leases
lettings of non-residential property, VAT
related to 410–11
licensed premises, rating assessment
for 382
licenses to enter upon land 24–5
listed buildings 225–6, 508–9
local development orders 210
Local Government Act (1989),
Schedule 10 268, 277
location of property 36, 321–2, 333
loss of capital, provision for 111–12

management charges 65
management of investments 197–201
management powers, retention of 279
market approach to valuation 12, 15,
36–40; for development land 39–40;
for residential property 38–9
market instability, valuation in periods of
180–3
market rent (MR) 41–56, 95; effect of
capital improvements on 55–6;
estimation of 49–50; freehold lets *above,
at* or *below* 96–109; hierarchy of
evidence on 56; for industrial premises
329–31; for office premises 336; for
retail premises 324–7
market transactions used in valuation
274–6
market value: for commercial properties
generally 317; and compulsory purchase
compensation 428–35; of ground rents
261–2; for industrial premises 329–31;

for office premises 337–40; for retail
premises 327–8
market value rule on disposal of assets 398
'marriage value' 121–2, 276–7
mineral heriditaments, rating assessment
for 383
modification orders 502–4
mortgage finance 8, 348–54; mortgagee's
security 350–1; second and subsequent
mortgages 354; valuations for purposes
of 351–4

net present value (NPV) 136–40; compared
with internal rate of return 142–3
non-domestic rates 61–2, 358–83;
exemptions and reliefs from 361; on
unoccupied properties 367; *see also*
rating valuations

occasions of charge for capital gains
purposes 387
occupation: *actual, exclusive* and *beneficial*
362–3; transience of 363
occupation leases 23–4
occupiers' loss, compensation for 448–9
office premises 335–40; market value of
337–40; rating assessment for 375;
tenancy terms for 336–7; yields from 74
"options" contracts 27
outgoings 57–66, 200; on blocks of flats
290–2; on office premises 337; in
relation to rent 66; on residential
property 282–3
outline planning permission 211–12
owner-occupiers: assessment of worth for
114; exemption from capital gains tax
392–3

Parry's Valuation and Investment Tables
79–81, 86–90, 94
part disposals of assets 396–7
"peppercorn" rent 23
physical state of property 36–7
places of worship 294–5
planning appeals 217–19; against
enforcement notices 223–4; *overlapping*
or *repeated* 218–19
planning applications 211–15
planning authorities 203
planning conditions and planning
obligations 216–17
planning contravention notices 221